Nickel and Chromium Plating

J. K. DENNIS, B.Sc., Ph.D., A.I.M.
*Department of Metallurgy,
University of Aston in Birmingham*

T. E. SUCH, B.Sc., F.R.I.C., F.I.M.F.
*Director of Research (Processes),
W. Canning and Co. Ltd.*

LONDON
NEWNES-BUTTERWORTHS

THE BUTTERWORTH GROUP

ENGLAND
Butterworth & Co (Publishers) Ltd
London: 88 Kingsway, WC2B 6AB

AUSTRALIA
Butterworth & Co (Australia) Ltd
Sydney: 586 Pacific Highway, Chatswood, NSW 2067
Melbourne: 343 Little Collins Street, 3000
Brisbane: 240 Queen Street, 4000

CANADA
Butterworth & Co (Canada) Ltd
Toronto: 14 Curity Avenue, 374

NEW ZEALAND
Butterworth & Co (New Zealand) Ltd
Wellington: 26–28 Waring Taylor Street, 1

SOUTH AFRICA
Butterworth & Co (South Africa) (Pty) Ltd
Durban: 152–154 Gale Street

First published in 1972 by
Newnes-Butterworths, an imprint
of the Butterworth Group

© Butterworth & Co. (Publishers) Ltd., 1972

ISBN 0 408 00086 4

Filmset and printed in England by Page Bros (Norwich) Ltd, Norwich

Preface

This book is written for scientists, technologists and students who may have diverse backgrounds but who wish to acquire a knowledge of the fundamentals on which the important industrial processes of nickel and chromium plating are based. Its scope includes descriptions of the methods of deposition and the properties of the coatings of these metals obtained from aqueous solutions (with or without electric current) and either applied individually or superimposed. (The latter is the case for most of the 'chrome' plating that is carried out today.) The emphasis is placed on the modern techniques employed for the deposition of these metals, whether it is for decoration, corrosion protection or engineering applications. However, obsolete processes are briefly described in the first chapter, which it is felt makes an essential and interesting introduction to current practice.

We hope that scientists who are already working in this field will find our review of this modern technology to be of assistance to them, whether they are in fundamental or applied research, or employed in the industry. It must be pointed out that the book is intended to provide information of a standard that is also suitable for undergraduates reading for degrees in industrial metallurgy or chemistry and graduates taking specialised M.Sc. courses in corrosion and protection; it is not meant to serve as a practical handbook to give shop-floor guidance on how to prepare and operate production plating baths. It will also be of assistance to students studying for qualifications awarded by bodies such as the Institute of Metal Finishing, Institution of Corrosion Technology, Royal Institute of Chemistry and Institution of Metallurgists. We have therefore included chemical, electrochemical and metallurgical theory where such knowledge is essential to the full understanding of the processes described. The most significant features of the organic compounds added to produce bright nickel plate are summarised in Appendix 1, so that the effect of these addition agents can be understood more easily by the reader, who otherwise could be confused by the proprietary nature of these processes and the large number of patents in existence. Those involved in production plating should derive considerable help from the detailed descriptions and critical comparisons of alternative plating processes now available and of the control techniques that can be used for the solutions and the coatings they produce. The analytical appendix should assist those plating in either 1 litre beakers or 1 000 gal vats. In view of the arrival of S.I. units, these have been used wherever possible, but have been slightly adapted where it seemed desirable.

Our main reason for writing this book was to fill a gap that we felt existed in metal-finishing literature. We considered that there was no up-to-date English text dealing in depth with both nickel and chromium plating, except from the operational approach. This might well encourage newcomers to the field to consider it to be more an art than a science. The subject matter of this book has been included in some previous publications, but it has usually been treated in a limited manner based mainly on the particular scientific discipline in which the author was trained. Metal deposition does not fit neatly into the sphere of any pure science, for it is concerned both with the properties of the electrolyte solution (chemistry) and of the metal deposit (metallurgy) and, of course, the process of obtaining one from the other (electrochemistry). We considered that our experience could be combined to enable us to write a text suitable not only for those first encountering this technology but also for those experienced in the field, by reviewing, as far as possible, all of the most important information published in widely scattered papers, including very recent ones. Some unpublished work and many new optical and electron photomicrographs are also included.

We hope that we have achieved our aim of providing our readers with the opportunity of obtaining in one book detailed scientific information on nickel and chromium plating, previously available only in scattered and unconnected texts.

We are most grateful to those colleagues who aided us by supplying information, assisting with photographic work or taking part in helpful discussions. We also wish to thank all those organisations who supplied or allowed us to reproduce illustrations and graphs. Reference to their original source is given in the captions or accompanying text. However, we are particularly grateful to the following for their permission to include a number of figures that were originally published by them:

> American Electroplaters' Society
> *Electroplating and Metal Finishing*
> Institute of Metal Finishing
> International Nickel Co. Ltd.
> *Metalloberfläche*
> W. Canning and Co. Ltd.

Particular thanks are due to Dr. L. L. Shreir who first realised the need for this book and who subsequently carried out a critical perusal of the manuscript.

J.K.D.
T.E.S.

Contents

1	Introduction and Historical Review	1
2	Electrochemical Aspects of Electrodeposition	10
3	Metallurgical Aspects of Electrodeposition	33
4	Plating Baths and Anodes Used for Industrial Nickel Deposition	57
5	Engineering Applications	75
6	Bright Nickel Plating	92
7	Control and Purification of Nickel Plating Solutions	122
8	Physical and Mechanical Properties of Electrodeposits and Methods of Determination	147
9	Chromium Plating	184
10	Decorative Nickel Plus Chromium Coating Combinations	208
11	Corrosion Resistance and Testing of Nickel Plus Chromium Coatings	235
12	Recent Developments	266
Appendix 1:	Combinations of Organic Compounds that Produce Semi-bright or Bright Nickel Plate	303
2:	Analysis of Deposition Solutions—Selected Methods	306
3:	Properties of Chromium, Nickel and Copper	315
Index		317

Chapter 1

Introduction and Historical Review

Electrodeposited metals can often be employed as an ideal means of providing a thin surface coating which has some property (or properties) superior to that of the substrate. It may, for example, be possible to employ a cheaper or stronger substrate than could otherwise be used and yet achieve good corrosion resistance by applying a suitable electrodeposited coating. Electrodeposited nickel is typical of metals which can be included in the above category. It is often applied for decorative and protective purposes to cheap mild-steel pressings and to die-cast zinc or aluminium alloy components. Die-casting is an economical means of mass-producing exact dimensional replicates of the original, but the alloys used are not suitable for service in a corrosive atmosphere without some form of protective coating. About 98% of the estimated 59000 t of nickel consumed during 1971 in the electroplating industry of the non-communist countries was used in the form of thin, corrosion resistant and often also decorative coatings on cheaply produced or strong substrates. Most nickel coatings of this type are subsequently chromium plated to form the familiar composite nickel plus chromium system. While nickel coatings may be applied solely for corrosion resistance where their inherent dullness is of no importance, the majority have to provide both decorative and protective functions. If a final bright appearance is required, dull deposits have to be polished to a high lustre before chromium plating and, since this polishing operation is so expensive, the major part of nickel plated for decorative applications is now deposited in a fully-bright condition. Such coatings are obtained from solutions which contain organic chemicals in addition to the inorganic constituents. Not only can these additions modify the structure of the nickel deposit so that polishing is unnecessary, but many also have scratch-filling (the so-called 'levelling') properties which also eliminate or reduce the amount of polishing of the basis metal which is required. Variants of these solutions give bright, levelling deposits and form the vast majority of nickel baths now in industrial use.

However, it is not intended to minimise the importance of the remaining uses of nickel electrodeposition. Indeed, they are of great economic importance. Often a comparatively thin nickel coating, of which the weight and cost expressed as anode metal are quite small, may be used to repair a most expensive component which would otherwise have to be scrapped.

Large engineering components which have involved much costly machining and heat-treatment, and have been damaged, worn or over-machined, perhaps only on small portions of their surface, can be salvaged by building up these specific portions with nickel to restore their original dimensions. If a large nickel thickness is necessary, this reclamation involves the deposition of more nickel than required, followed by machining to size. Obviously, the economics of these operations must be compared with those of fabricating an entirely new component.

The benefits conferred on the surface by thick or 'heavy' nickel plate are not only better corrosion protection, but also the greater abrasion resistance obtainable from certain types of nickel coatings. These advantages are now utilised on many new as well as reclaimed parts; this will be discussed in Chapter 5. The superior wear-resistant properties of these nickel electrodeposits are often further enhanced by the deposition of fairly thick chromium coatings. This is the 'hard' chromium plate of the engineer.

Electroforming is the fabrication of articles entirely by electrodeposition. Nickel is a popular metal for this purpose since it can be plated in a ductile and low stressed form which has moderate hardness. Sometimes the thick skin of the electroform is backed by even thicker copper plate, which although softer than nickel, can be deposited at a faster rate than can most types of nickel. Often the 'working' surface of the electroform is chromium plated after removal from the mandrel.

It will be noted that for most decorative/corrosion protective purposes, engineering uses and even for electroforming, electrodeposited nickel is given a top coat of chromium. For this reason, it is impossible to dissociate chromium electrodeposition from that of nickel. Therefore, although the main purpose of this text is to discuss the deposition of nickel, the technology of chromium plating will be discussed wherever it is relevant.

The chief emphasis will be on electrodeposition of nickel, but mention will be made of electroless plating of chemically reduced nickel, as this is a valuable technique for applying uniform coatings to articles of complex shape, where the inherent limitations of the electrolytic process form an impossible barrier to obtaining a uniform coating of nickel over the whole surface.

Although the science of electroplating is a comparatively young one, being only 170 years old, nickel was first deposited 130 years ago and chromium some 20 years later. However, the great advances in its technology have been achieved only in the last 60 years. Both these and the early pioneering work form a fascinating historical introduction to the present processes used in industry.

HISTORY OF NICKEL PLATING

The electrodeposition of nickel was first described in 1837. G. Bird[1] electrolysed solutions of nickel chloride or sulphate for some hours and so obtained a crust of metallic nickel on a platinum electrode. In 1840, the first patent for commercial nickel plating was granted to J. Shore[2] of England who specified a solution of nickel nitrate. Soon afterwards a number of investigators published the results of their experiments. A. Smee of England (1841), Ruolz of France (1843) and Bottger of Germany (1843) were the first of these. Ruolz used nickel chloride or nitrate but Bottger's was the first publication to mention an electrolyte solution based on 'acid ammonium sulphate'; this bath, with variations, was the one mainly used in commerce for the next 70 years. However, G. Gore (1855) seems to have been the first to publish[3] details of the neutral nickel ammonium sulphate bath (nickel ammonium sulphate is colloquially known as *double nickel salts*). Becquerel also published this process in 1862, using a concentration of 70–80 g/l, although it is possible he used an excess of ammonia. Therefore, undue credit has perhaps been given to Dr. I. Adams of the U.S.A. for developing this process, although he did apparently use it in the laboratory at Harvard University in 1858–1860[4]. However, he certainly appears to have been the first to commercialise it by plating gas burner tips in 1866[5].

In 1868, W. H. Remington of Boston also commenced the deposition of nickel on a commercial scale using a nickel ammonium chloride solution but ran into difficulties, probably due to the use of an excess of ammonia in the bath. He was the first to describe[6] the use of electrolytic nickel anodes and the use of insoluble baskets (in this case made of platinum) to contain nickel cubes, although these were impure, containing 5% copper and nearly 1% iron.

Meanwhile, Dr. Adams was endeavouring to perfect his own technique which resulted in his master patent for nickel ammonium sulphate baths being published in 1869[7]. This, together with his business ability, gave his company a virtual monopoly of commercial nickel plating for the next 17 years.

This patent's principal claim was that 'the nickel solution should be free from the presence of potash, soda, alumina, lime or nitric acid, or from any acid or alkaline reaction'. Methods of preparing this solution were detailed. His competitors claimed that in essence this only implied that pure nickel ammonium sulphate should be used, but Adams' reference to the necessity for neutrality is a vital and possibly a novel feature. This type of bath has certainly always been associated with Adams and it is from the time of its introduction by him, that nickel plating first became a commercially feasible operation. Because of this and Adams' energetic publicity, it rapidly became utilised throughout the then industrial world.

In the autumn of 1869, Adams visited Europe to sell his process there. A small experimental bath was soon set up in Liverpool and one of about 2000 litre was set up a little later (December) in Paris by A. Gaiffe. A bath of the same size was installed in Birmingham by the spring of 1870. This may have been at the works of Bouse and Muncher, for in 1873 they were reported to be using the Adams solution imported from the U.S.A., with cast nickel anodes containing 6–10% iron. Up to 15 hours' plating time was necessary to obtain a good coating. In 1873 Adams patented a plating process based on nickel sulphate[8]. In 1878 Weston patented[9] the addition of boric acid, while in 1879 Powell patented the addition of citric or benzoic acids as additives.

About 1880, nickel plating salts were being made in Vienna by Pfanhauser and in Birmingham by Canning. In the latter's first handbook, published in 1889, the preparation of a neutral plating bath from 100 g/l double nickel salts is described. In their catalogue, published concurrently with the handbook, single nickel salts (nickel sulphate) were also listed.

In their 1891 catalogue, Canning were able to announce that their salts and anodes were free from copper and iron. During the 1890s the use of boric acid as a buffer and chlorides as anode corrodants became more popular, but far from universal. Although in 1900 Canning listed double and single nickel salts, boric acid, citric acid and sodium chloride as being available for nickel plating, they still preferred the neutral nickel ammonium sulphate solution, this to be used at room temperature and at low current densities, a plating time of 4–5 h being required. In contemporary publications Langbein and Pfanhauser mentioned the use of boric acid and chloride only in some of their formulations. However, in 1906 Bancroft spoke out strongly in favour of the necessity of chloride ions in the nickel bath to ensure satisfactory anode corrosion. Foerster in 1897 and 1905 had already described the use of nickel sulphate or chloride in hot solutions at high current densities.

In 1910, Canning offered a proprietary mixture which contained nickel sulphate, sodium chloride and boric acid and was to be maintained 'slightly acid'. This was to be used at a concentration of 237 g/l at a temperature of not less than 16°C and could be worked at a current density twice that of a double nickel salts solution, but a voltmeter and ammeter were said to be essential and so were anodes of 99% purity. This bath, although its composition was not revealed, was in all essentials similar to the Watts bath, formulated in 1916 by Professor O. P. Watts of the University of Wisconsin. He published[10] the formula for a bath which has stood the test of time and even now is used with little modification for a large percentage of commercial electroplating operations. He recommended the following solution to be used hot at much higher current densities than then employed:

$$\begin{array}{ll}\text{Nickel sulphate, } NiSO_4 \cdot 7H_2O & 240 \text{ g/l} \\ \text{Nickel chloride, } NiCl_2 \cdot 6H_2O & 20 \text{ g/l} \\ \text{Boric acid, } H_3BO_3 & 20 \text{ g/l}\end{array}$$

This solution and its modifications have been endowed with the name of the man who first thoroughly described its benefits over the double nickel salt process. While Professor Watts strongly recommended that this bath be used hot, he did not mention agitation. Therefore, although the Watts bath with variations, which usually contain increased concentrations of one, two or all three of its constituents, was gradually adopted over the course of years as the almost universal basis of industrial plating processes—whether dull or bright, there have been, until recently, differences in the U.S. and U.K. practice of agitation. By the end of the 1920s British platers were starting to adopt air agitation, which did not achieve popularity in the U.S.A. until thirty years later.

In 1931, the similarities and differences between the best U.S. and U.K. practice can be seen by comparing two contemporary papers, in which the English author[11] praised the use of air agitation, with the concomitant necessity for filtration, while the American[12] feared the troubles that agitation might cause. However, in 1931 a large proportion of the nickel baths in Europe and U.S.A. were still operated at room temperature with no agitation.

It will be noticed that values for acidity on the pH scale are given in

Table 1.1 which gives typical operating conditions. The use of pH measurements for controlling nickel baths was first suggested in 1921 and by the end of that decade was common industrial practice.

Table 1.1 OPERATING CONDITIONS FORMERLY USED FOR NICKEL PLATING

Country	Temperature	Current density	pH
U.K.	32–35°C	2–2·5 A/dm^2	5·6–5·8
U.S.A.	50–55°C	3–4 A/dm^2	5·2–5·5

Both the U.K. and U.S.A. were using the Watts nickel bath and it is from this type of bath that bright nickel deposits are now obtained. However, it appears that the first commercial bright nickel plate was obtained from a double nickel salt solution, probably containing small amounts of cadmium as a brightener. Certainly in their 1910 catalogue, Canning described a nickel plating process named *Velete*, which was said to give a 'brilliant bright deposit in 5 to 10 min without further polishing and for any period up to 30 min gives the brightest and whitest deposit obtainable'. This may well have been the same type of solution described[13] as being used at Elkingtons in Birmingham in 1912. It was soon found that small additions of zinc salts had the same brightening effect. Other investigators showed that glucose. glycerine, gum tragacanth or gum arabic also helped to produce a bright nickel deposit. These deposits were only fully bright when very thin and their wider application was limited by their great brittleness. Aromatic sulphonates were to prove far superior brighteners in both these aspects. Lutz and Westbrook were the first to take out a patent[14] for these compounds. However, not until Schlötter[15] marketed his process in 1934 did bright nickel become a commercial reality for mass production of all classes of plated goods. His bath and variants were quickly adopted in the industrial world. Another bright nickel process was put forward[16] in 1936 and this was based on the deposition of a cobalt-nickel alloy. This proved a serious competitor to the organic type process although its greater cost has gradually resulted in its almost total elimination; one point in its favour is that it has reasonably good ductility.

The ousting of the cobalt-nickel alloy bath has resulted in the present domination of the market by organic brighteners. These have been greatly improved by investigations which have resulted in the achievement of many of the properties of the ideal bright nickel process as first postulated by Eckelmann[17] in 1934. These developments are discussed in Chapter 6. Suffice it here to say that the first truly levelling solution was introduced in 1945, but this gave only a semi-bright plate which needed polishing. A little later, levelling and fully bright processes were developed and these are now used for most commercial nickel plating. Semi-bright levelling deposits still have an important use in that they constitute the major part of double-layer nickel coatings and so confer improved corrosion resistance onto these. Bright levelling processes may have almost reached their limit as regards brightening and levelling properties and the tendency now is for investigation of new processes to give these desired properties without the deleterious effects of brittleness, stress and darkness, the latter being particularly troublesome in low current density areas. All these faults are accentuated if the organic compounds decompose during electrolysis, as will be discussed later.

Many other electrolyte solutions have been mentioned in the literature and some of these, in particular pyrophosphate, are still being pursued in the laboratory. However, apart from baths based on nickel sulphamate and fluoborate they appear to offer no advantage for the deposition of nickel as distinct from nickel alloys, and hence do not appear likely to replace the Watts solution even after 50 years of use.

This statement also applies to baths used for thick coatings of nickel deposited for engineering purposes. The first work on this application was probably carried out using the double nickel salt solution. During World War I heavy coatings of various metals—iron, nickel, cobalt and copper—were applied for salvage and repair purposes by workshops of the British Army and Air Force. Fletcher, Havelock and McLare[18] were important workers in this field and with others were responsible for developing much improved methods of cleaning and deoxidising steel so as to obtain good adherence of the thick nickel deposits, the former[19] in particular developing the process for anodically etching steel in sulphuric acid solution. After World War I, these men used their acquired knowledge to exploit these new techniques for the benefit of industry.

Another outcome of the success of this technique was the setting up by the War Office of a Research Laboratory at Woolwich to investigate the physical and mechanical properties of electrodeposited metals and the way in which these were affected by composition of the solutions and operating conditions. Much of this work was on nickel and the 'Woolwich School' was the first to systematically research into these relationships. The results are published in the papers of MacNaughton, Hothersall, Hammond and others of their teams. Their findings have been summarised in the first Hothersall Memorial Lecture[20]. They have permeated and influenced nickel plating practice in every satisfactory plating shop.

NICKEL ANODES

The development of all aspects of nickel deposition was assisted by the introduction of better anodes. The importance of their high purity became increasingly recognised in the 1920s. However, the purer anodes became, the more difficult they were to dissolve. In 1929 the depolarised anode was patented by Harshaw[21] and this was a great advance since it dissolved smoothly under almost all conditions. 1931 is often considered the year in which cast carbon-containing anodes were introduced to the industry, yet their merits were known in 1904[22]. Cast or rolled anodes containing carbon were found to be eminently suitable for bright nickel plating baths in which depolarised anodes often do not dissolve uniformly. Now the wheel has turned full cycle with the introduction of titanium baskets to hold anode pieces. At first these were employed to use up anode scrap, but in 1959 anode slugs were produced, and so-called 'primary' nickel has now become very popular for economic reasons. This nickel is usually in the form of rectangular pieces cut from electrodeposited sheet or as small discs electrodeposited in that shape. Thus a century has elapsed and a technique is being used which was originally proposed in 1868 by Remington, who in his patent[6] described the use of a basket, woven from platinum wire, or any other electrical conductor not materially affected by electric current or the solution employed, to hold particles of nickel.

HISTORY OF CHROMIUM PLATING

Bunsen in 1854 was the first to mention[23] the electrodeposition of chromium, but more credit is due to Dr. Geuther[24] who published the first detailed account of the electroplating of this metal at Göttingen, Germany in 1856. It is most interesting to note that he used a chromic acid solution, which presumably contained some residual sulphuric acid. Professor H. Buff of Giessen tried to repeat this work[25]. As he was unable to do so, he cast doubt on Geuther's results.

This may have lead to the concentration on the electrolysis of trivalent chromium salts during the next 40 years. Many workers devoted much fruitless time to investigations of these types of solutions. Amongst these were Placet[26] and Bonnet in France. However, they also used chromate solutions as revealed in their various patents published in 1891 and it may have been from these that they obtained the kilogramme of metallic chromium which they exhibited. Another Frenchman, M. LeBlanc[27] was most sceptical about their claims, since he found he could not electroplate any chromium from solutions of chromium sulphate; accordingly some controversy raged about the feasibility of depositing even a little chromium from any bath, until in 1905 Carveth and Curry[28] published their findings.

These investigators worked under Professor W. D. Bancroft at Cornell University and concluded that not only were some of the findings of Placet and Bonnet correct but they also readily produced plate from chromic acid baths. This lead to Professor Bancroft stating in 1906[29] that 'the real solution from which to deposit chromium is not chrome alum nor sulphate, it is chromic acid'.

In Budapest, Dr. F. Salzer worked on solutions of chromic acid and reported[30] the benefits of adding chromium sulphate to obtain electrodeposits of chromium.

From 1912 to 1914, Dr. Sargent worked at Cornell, also under Bancroft, carrying out a systematic investigation into the electrolytic behaviour of various mixtures of chromic acid and chromium sulphate but did not report his results until 1920[31]. In the early 1920s, Dr. Liebreich studied the same topic in Berlin. This led to his taking out a number of patents, those of 1924[32] being recognised as most suitable for ready electrodeposition of chromium. Thus the commercial electroplating of this metal was being brought ever nearer, and further impetus was given by the work of Professor C. Fink and Dr. W. Pfanhauser.

Fink[33] with his co-workers, Schwartz[34], Eldridge and Dubpernell, did most valuable work at Columbia University. This and Liebreich's work resulted in the first commercial electrodeposition of chromium in 1924, almost simultaneously in the U.S.A. and Germany. In 1923 Fletcher[35] was working in England on the deposition of chromium and both Ollard[36] and Mac-

Naughton in 1925 reported the use of Liebreich's bath, the former at Metropolitan Vickers Electrical Co. Ltd. and the latter at Woolwich Research Department. Nevertheless, it was not until 1928 that chromium plating processes were marketed commercially in the U.K. At first, these chromium deposits, which were plated from a cold solution, were often rather thick but dull, and had to be polished to obtain a high lustre. It was soon found that chromium could not be used as a total replacement for nickel plating but as a thin bright top layer over the nickel to preserve the reflectivity of the plated part.

Since 1924, many advances have been made in chromium plating for corrosion prevention, as will be detailed in later chapters, but essentially the solutions are much the same, since they are still based on chromic acid, trivalent chromium solutions being confined at present to electrowinning. One of the biggest advances was the use of silicofluoride ions as catalysts in addition to sulphate ions[37, 38]. This type of solution has about 18 to 20% cathodic efficiency compared with the 10 to 12% of the chromic acid bath containing only sulphate as catalyst. This higher efficiency is still a long way from those obtainable with most other electroplating processes and it is the aim of investigators to increase it.

REFERENCES

1. BIRD, G., *Phil. Trans.*, **127**, 37 (1837)
2. SHORE, J., U.K. Pat. 8 407 (1840)
3. GORE, G.. *The Art of Electro-Metallurgy. Including All Known Processes of Electrodeposition.* 236 (1855)
4. ADAMS, I., *Trans. Am. Electrochem. Soc.* **9**, 211 (1906)
5. ADAMS, I., U.S. Pat. 52 271 (1866)
6. REMINGTON, W. H., U.S. Pat. 82 877 (1868)
7. ADAMS, I., U.S. Pat. 93 157 (1869)
8. ADAMS, I., U.S. Pat. 136 634 (1874)
9. WESTON, E., U.S. Pat 211 071 (1878)
10. WATTS, O. P., *Trans. Am. Electrochem. Soc.* **29**, 395 (1916)
11. CANNING, E. R., *Trans. Am. Electrochem. Soc.*, **59**, 371 (1931)
12. WATTS, O. P., *Trans. Am. Electrochem. Soc.*, **59**, 379 (1931)
13. PROCTOR, C. H., *The Metal Industry*, **7** No. 4, 124 (1915)
14. LUTZ, G. and WESTBROOK, R. L.. U.S. Pat. 1818 229 (1928)
15. SCHLÖTTER, M., U.S. Pat. 1972 693 (1932)
16. WEISBERG, L. and STODDARD, W. B., U.S. Pat. 2026 718 (1936)
17. ECKELMANN, L. E., *Mon. Rev. Am. Electroplaters' Soc.*, **21**, 18 (1934)
18. MCLARE, J. P., *Trans. Faraday Soc.*, **23**, 87 (1924–1925)
19. FLETCHER, R. J., U.K. Pat. 162 391 (1920)
20. GARDAM, G. E., *Trans. Inst. Metal Finishing*, **29**, 78 (1952–1953)
21. HARSHAW, W. J., SAVAGE, P. and BEZZEMBERGER, K., U.S. Pat. No. 1751 630 (1929)
22. HAWKINS, H. J., 'The Polishing & Plating of Metals', Hazlitt and Walker, Chicago, 274 (1904)
23. BUNSEN, R., *Ann. Phys.*, **91**, 619 (1854)
24. GEUTHER, A., *Annalen*, **99**, 314 (1856)
25. BUFF, H., *Annalen*, **109**, 129 (1859)
26. PLACET, E., *Comptes Rendus*, **115**, 945 (1892)
27. LEBLANC, M., *Trans. Am. Electrochem. Soc.*, **9**, 315 (1906)
28. CARVETH, H. R. and CURRY, B. E., *J. Phys. Chem.*, **9**, 353 (1905)
29. BANCROFT, W. D., Discussion on LeBlanc's paper (Reference 27)
30. SALZER, F., German Pat. 225 769 (1909)
31. SARGENT, G. J., *Trans. Am. Electrochem. Soc.*, **37**, 479 (1920)
32. LIEBREICH, E., German Pat. 448 526 (1924) and U.K. Pat. 243 046 (1924)

33. FINK, C. G., U.S. Pat. Nos. 1 581 188 (1926) and 1 802 463 (1931)
34. SCHWARTZ, K. W., *Trans. Am. Electrochem. Soc.*, **44**, 451 (1923)
35. Private Communication from National Physical Laboratory
36. OLLARD, E. A., *The Metal Industry*, **27**, 235 (1925)
37. FINK, C. G. and MCLEESE, U.S. Pat. 1 844 751 (1932)
38. STARECK, J. E., PASSAL, F. and MAHLSTEDT, H., *Proc. Am. Electroplaters' Soc.*, **37**, 31 (1950)

BIBLIOGRAPHY

DUBPERNELL, G., 'The Story of Nickel Plating', *Plating*, **46** No. 6, 599 (1959)
DUBPERNELL, G., 'The Development of Chromium Plating', *Plating*, **47** No. 1, 35 (1960)
JOHNSON, L. W., 'Bright Nickel Plating. A Resumé of the Technical Literature on the Subject'. *J. Electrodep. Tech. Soc.*, **12**, 93 (1937)
MCKAY, R. J., 'The History of Nickel Plating Developments in the U.S.A.', *Plating*, **38** No. 41, 147 (1951)
PAVLOVA, O. I., 'Istoriya Tekhniki Elektroosazhdeniya Metallov', Moscow (1963). (English translation into 'Electrodeposition of Metals: A Historical Survey' by Israel Programme for Scientific Translations)

Chapter 2

Electrochemical Aspects of Electrodeposition

Electrochemistry is adequately covered in numerous textbooks including one by Davies[1]. However, it is not possible to discuss the electrodeposition of nickel and chromium without including at least one chapter on this topic, and, as far as possible, this will be related specifically to the two metals of immediate interest.

POTENTIALS OF METALS IN AQUEOUS SOLUTIONS

Before considering reactions that take place during electrodeposition, it is necessary to consider the electrode equilibria which are possible when a metal is immersed in an aqueous solution of its ions. The electrode equilibria most relevant to electroplating are:

$$M^{z+} + ze \rightleftharpoons M \qquad \ldots(2.1)$$

$$H^+ + e^- \rightleftharpoons \tfrac{1}{2}H_2 \qquad \ldots(2.2)$$

$$O_2 + H_2O + 2e \rightleftharpoons 2OH^- \qquad \ldots(2.3)$$

In the case of equation (2.1), e.g. $Cu^{2+} + 2e \rightleftharpoons Cu$, or $Zn^{2+} + 2e \rightleftharpoons Zn$, the metal, when immersed in a solution of its ions, attains a specific electrical potential which is characteristic of that particular metal and the concentration of metal ions. These systems are reversible and can be treated thermodynamically, which means that they conform with the Nernst equation that has been derived from thermodynamic considerations:

$$E = E^0_{M^{z+}/M} + \frac{RT}{zF} \ln a_{M^{z+}}$$

$$E = E^0_{M^{z+}/M} + \frac{0.059}{z} \log_{10} a_{M^{z+}} \qquad \ldots(2.4)$$

where
E = Electrode potential at 25°C (V)
$R = 8\cdot315$ J degC^{-1} mol^{-1}

$$F = 96486 \text{ C and}$$

$$\frac{RT}{F} \ln a = 0.059 \log_{10} a \text{ at } 25°C$$

Thus the electrode potential of copper in copper sulphate solution varies logarithmically with the activity of cupric ions as predicted by the Nernst equation. The values for electrode potentials normally quoted in reference tables, the so called e.m.f. series, are the standard reversible electrode potentials E^0 for $M^{z+} + ze \rightarrow M$, which is defined as the potential of a metal in equilibrium with a solution of its ions at unit molar activity, at a temperature of 25°C. The molar activity, which corresponds with the molar concentration only in very dilute solutions, is the product of the molar concentration of the ions and their activity coefficient.

It should be noted that potentials of half-cells, such as M^{z+}/M, cannot be measured unless they are coupled to another half-cell. For this reason the hydrogen equilibrium (equation 2.2) is taken as the standard, and when $a_{H^+} = P_{H_2} = 1$, the standard electrode potential $E^0_{H^+/\frac{1}{2}H_2}$ of the hydrogen electrode is given an arbitrary value of 0.00 V.

In the case of reversible systems, the standard electrode potential E^0 can be evaluated by an e.m.f. method, and for these systems the potential of the metal in a solution of its ions will conform to the Nernst equation (2.4). However, metals such as nickel, iron, chromium, aluminium, etc., when immersed in solutions of their ions do not give reversible systems and the standard electrode potentials must be evaluated from the standard free energy ΔG^0, which in turn is evaluated from calorimetric data.

Since $\qquad \Delta G^0 = \Delta H^0 - T \Delta S \qquad \ldots (2.5)$

and $\qquad -\Delta G^0 = zFE^0 \qquad \ldots (2.6)$

it follows that E^0 can be evaluated from the heat of reactior ΔH^0 and the entropy change ΔS.

These considerations show that the standard electrode potential is a thermodynamic quantity that has little relevance in considering the cathodic deposition and anodic dissolution of irreversible metals such as nickel and chromium.

It must be emphasised therefore that the reversible potential E, as defined by the Nernst equation, is a thermodynamic quantity that defines the activity of the metal ions in a solution when the system is at equilibrium. When $E = E^0$ then $a_{M^{z+}} = 1$ and when $E \lessgtr E^0$ then $a_{M^{z+}} \lessgtr 1$. It follows that when E is made more negative, i.e. cathodically polarised to E_P, the polarised potential, then the $a_{M^{z+}}$ in solution will decrease by deposition of metal until it corresponds with E_P; conversely, making the potential more positive will result in dissolution of the metal to provide an increased concentration of M^{z+}.

These considerations provide an explanation of the cathodic and anodic processes that occur in electroplating from the viewpoint of the final position of equilibria. If the depletion of metal ions resulting from cathodic deposition is not replaced by the replenishment of ions by a corresponding anodic process, then the former process will cease when $a_{M^{z+}}$ attains a value corresponding with E_P, which is overcome in practice by the use of a soluble anode or by additions of metal salts to the bath.

It is also apparent from equation 2.4 that the more negative E_P, the greater is the tendency for the reaction

$$M^{z+} + ze \rightarrow M$$

to proceed in the direction of metal deposition. This aspect of electro-deposition will be considered subsequently.

SIGN CONVENTION

The sign of the standard electrode potential E^0 is arbitrary, and is dependent on the convention adopted. Thus for a number of years textbooks published in the U.S.A. have written the equilibrium with the electrons on the right-hand side:

$$M \rightleftharpoons M^{z+} + ze,$$

thus adopting the same convention as that used by W. M. Latimer in his reference book entitled *Oxidation Potentials*[2]. On the other hand, the converse convention has been adopted by European authors who have written the equilibrium as a reduction: $M^{z+} + ze \rightarrow M$, a convention that has now been adopted internationally* and which will be used throughout this book. Nevertheless, a number of books will still contain standard electrode potentials based on the U.S. convention, which gives rise to some confusion. The position can be summarised as follows:

Internationally accepted convention		*Original U.S. convention*	
Equilibria	$E^0(V)$	Equilibria	$E^0(V)$
$Cu^{2+} + 2e = Cu$	$+0.34$	$Cu = Cu^{2+} + 2e$	-0.34
$H^+ + e = \frac{1}{2}H_2$	0.00	$\frac{1}{2}H_2 = H^+ + e$	0.00
$Ni^{2+} + 2e = Ni$	-0.25	$Ni = Ni^{2+} + 2e$	$+0.25$
$Cr^{3+} + 3e = Cr$	-0.74	$Cr = Cr^{3+} + 3e$	$+0.74$
$Zn^{2+} + 2e = Zn$	-0.76	$Zn = Zn^{2+} + 2e$	$+0.76$

MEASUREMENT OF ELECTRODE POTENTIAL

In most instances it is impracticable to use a hydrogen electrode as a reference electrode to evaluate electrode potentials, and there are a variety of reversible systems available that can be used to give reproduceable reference electrodes. Thus the equilibrium $Cu^{2+} + 2e \rightleftharpoons Cu$ can be used to provide a $Cu/CuSO_4$ reference electrode, whose potential depends on the activity of Cu^{2+} (see equation 2.4). A number of reference electrodes which are referred to as *electrodes of the second kind*, are based upon a metal which is in contact with a sparingly soluble salt of the metal, the activity of metal

---

* International Union of Pure and Applied Chemistry, Stockholm Convention (1953)

ions being controlled by the activity of the anion of the salt. Thus in the case of the Ag/AgCl/HCl electrode:

$$a_{Ag^+} \times a_{Cl^-} = K_S = 1{\cdot}76 \times 10^{-10} \text{ mol}^2 \text{ l}^{-2} \qquad \ldots(2.7)$$

and $\qquad a_{Ag^+} = K_S \times a_{Cl^-}^{-1} = 1{\cdot}70 \times 10^{-10} \times a_{Cl^-}^{-1} \text{ mol l}^{-1} \qquad \ldots(2.8)$

so that the potential will vary with the concentration of chloride ions. Similar considerations apply to the calomel electrode $Hg/Hg_2Cl_2/KCl$, which consists of a mixture of mercury and mercurous chloride in potassium chloride. Commercial electrodes are available which can be immersed into the solution under study and electrical contact with the solution effected by means of a porous plug or ground glass collar. It is essential to ensure that solution does not diffuse into the electrode causing contamination or poisoning. Electrodes of this type can be used in conjunction with a Luggin capillary when it is necessary to determine the potential of a polarised electrode, the capillary being placed close to the surface of the electrode to minimise the *IR* drop. The potential of electrodes of the second kind thus depend upon the activity of the anion and upon temperature. In the case of the calomel electrode, it is advantageous to use an 0·1N potassium chloride solution for accurate work. A saturated solution of potassium chloride is easier to make up and use, as a few crystals can be left in the electrode to ensure saturation, but the temperature coefficient of this electrode is higher than that of the one containing a dilute solution. The potential of the saturated calomel electrode (S.C.E.), $Hg/Hg_2Cl_2/KCl$ (sat.) at 25°C is +0·242 V with respect to the standard hydrogen electrode (S.H.E.), and the potential becomes less positive as the temperature rises. Other secondary standards can be used as reference electrodes and these are described in the appropriate literature. The voltmeter used in the potential measuring circuit should be such that it does not draw current from the circuit, e.g. a valve voltmeter or Vibron electrometer.

HELMHOLTZ DOUBLE LAYER

The reactions occurring at the metal/solution interface are of importance when discussing mechanisms of deposition. Ionisation of the metal results in an excess of electrons in the metal and an increase in the number of cations in the layer of solution adjacent to the metal. Various hypotheses have been put forward to explain the processes occurring in this region, which is known as the *electrical double layer*. The first investigation was carried out by Quincke[3] in 1861 and later Helmholtz[4], after whom the region was named, continued the investigation. The early and simplest interpretations of the phenomenon were based on the idea that positive and negative charges were arranged opposite each other in parallel layers, as in a condenser. Gouy[5] and Chapman[6] regarded this explanation as being too simple, since it did not take into account the thermal motion of ions. They suggested that a diffuse array of ions would be more likely to occur and provided mathematical evidence to support their theory. Stern[7] and Graham[8] have suggested amendments to the theory to take into account the finite dimensions of ions and the occurrence of solvated and desolvated ions which make possible the existence of two layers. These two layers are known as the inner and outer Helmholtz layers, and play an important part in electrochemical reactions taking place in the vicinity of the interface. A diagrammatical representation of the film is shown in Fig. 2.1. Adsorption of foreign ions or neutral molecules can disturb the equilibrium in the

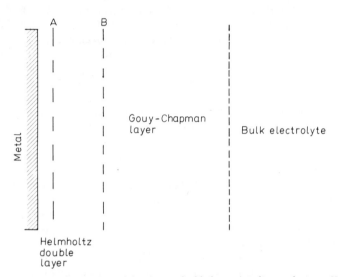

Figure 2.1. Schematic representation of the electric double layer. A *indicates the inner Helmholtz plane and* B *the outer Helmholtz plane*

double layer, and since organic additions are likely to produce these types of effects, the mechanisms of deposition and brightener incorporation are closely associated with the phenomenon of the electrical double layer. Transfer of hydrated metal ions through the double layer is another important aspect which influences the deposition process.

APPLICATION OF EXTERNAL E.M.F.

The simplest cell to consider is that in which two identical reversible metal electrodes are placed in a solution containing their ions and are connected by an external circuit to a potential source. Application of a very small e.m.f. displaces the electrode potentials from the equilibrium value, current flows and metal is either deposited on the cathode, or, an alternative cathode reaction occurs. Usually the alternative reaction is the libration of hydrogen. Concurrently, at the anode, metal dissolves or oxygen is liberated.

Gas liberation takes place at electrodes when the conditions favour this reaction in preference to metal deposition or dissolution, e.g. in decinormal acid solution the a_{H^+} is approximately 1 g ion/l and if hydrogen liberation can occur at a less negative potential than that necessary for metal deposition, then the former process takes place preferentially. In neutral or alkaline solutions a_{H^+} is very small (10^{-7} g ion/l for the former and between that value and 10^{-14} g ion/l for alkaline solutions of Normal and lower strength) but hydrogen can then be discharged directly from the water molecule:

$$H_2O + e \rightarrow \tfrac{1}{2}H_2 + OH^-$$

If the two potentials of metal discharge and hydrogen evolution are fairly close together, the two reactions occur simultaneously resulting in a low metal deposition efficiency, i.e. $<100\%$. If hydrogen is liberated at the cathode, the pH of the electrolyte solution will rise due to removal of hydrogen ions from solution. Initially, this will be confined to the film of solution adjacent to the cathode, but the pH of the bulk solution will eventually increase as diffusion takes place. Alternatively, if the anode efficiency is low and oxygen is liberated at the anode, the hydroxyl ion concentration is reduced and the pH decreases. In commercial nickel plating processes the pH usually rises, since the cathode efficiency is rarely 100%.

In a simple system consisting of copper electrodes in copper sulphate, deposition occurs when the potential is made slightly less negative than the equilibrium value and a considerably more negative potential would be required to liberate hydrogen. This latter potential is much more negative than is apparent from the standard potential value since copper discharge occurs almost reversibly (low overpotential) whereas hydrogen discharge on copper is irreversible. The existence of an appreciable overvoltage for the discharge of hydrogen on a nickel electrode enables nickel to be deposited from nickel sulphate solution at high efficiencies, even though the reduction of nickel ions to metal is also an irreversible process.

This phenomenon is the reason for the success of the Watts nickel solution, whose electrochemistry is discussed in Chapter 4. The standard electrode potentials of nickel and hydrogen (-0.25 V and 0.00 V respectively) indicate that energetically, hydrogen discharge should occur in preference to nickel ion reduction. Nickel can in fact be deposited from sulphate solutions at a cathode efficiency approaching 100%, which is feasible theoretically when other factors are taken into consideration, i.e. the activities of the hydrogen and nickel ions and the concentration and activation overpotentials for the two electrode processes involved.

POLARISATION

When an electrode process occurs at an electrode, its potential departs from its equilibrium or steady state value, and the electrode is said to be *polarised*. The terminology is confusing and, for example, the change in potential resulting from a change of concentration has been referred to as *concentration polarisation*, *concentration overpotential* and *concentration overvoltage*. For convenience, we shall use the term *overpotential* for all effects that result in the departure of the potential of the electrode from its unpolarised value.

When an external e.m.f. is applied and a current flows at a finite rate, the potentials of the two electrodes change from their unpolarised values. The potential of the cathode becomes more negative and that of the anode becomes more positive, and the electrodes are said to be *polarised*. The polarised potential of the electrode is due to the overpotential associated with one or more electrode processes that are proceeding at a finite rate at the electrode under consideration. Thus a nickel electrode whose potential is made more negative in sulphuric acid is said to be polarised, the polarisation being due to the overpotential of the reaction

$$H^+ + e \rightarrow \tfrac{1}{2}H_2$$

On the other hand, if the nickel electrode is made cathodic in a Watts bath the overpotential will be due to the reaction

$$Ni^{2+} + 2e \rightarrow Ni$$

Polarisation is thus the departure of the potential of the electrode from its equilibrium or steady state value when a reaction proceeds at a finite rate. The various types of overpotential that result in polarisation of the electrode can be classified as follows:

(a) Concentration overpotential.
(b) Activation overpotential.
(c) Ohmic overpotential.
(d) Resistance overpotential.

Concentration overpotential is due to the change in concentration of metal ions around the electrodes and it accounts for a considerable part of

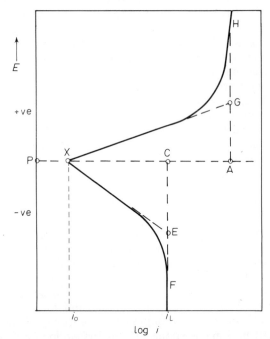

Figure 2.2. Relationship between E and i for a cathodic and anodic electrode process. Point X gives the reversible potential P of the electrode and the corresponding exchange current density i_0. Cathodic polarisation gives the linear Tafel line XE which is then affected by diffusion phenomena, so that the curve departs from linearity and attains the limiting current density i_L at F. Similar considerations apply to the anodic curve XG and its anodic limiting current density H.

the total polarisation at the current densities used in industrial plating. The solution adjacent to the cathode becomes depleted of metal ions as metal is deposited, and further deposition becomes more difficult; by reference to the Nernst equation (equation 2.4) it can be seen that the potential becomes more negative. At equilibrium on open circuit the reaction at an electrode is reversible, the rate of dissolution of metal into solution being equal to the rate of deposition. As the solution in the vicinity of the cathode becomes denuded of metal ions, the balance in these rates of reaction is disturbed and the potential becomes more negative. The greater the current density, the lower the concentration of metal ions near the cathode and so the greater the polarisation. Infinite concentration overpotential is approached when the concentration of metal ions approaches zero; the current density which results in this is known as the *limiting current density* (Fig. 2.2). This situation rarely occurs in practice, since before zero concentration of metal ions is attained an alternative electrode process takes place. Similar reactions occur at soluble anodes but in this case the metal ion concentration increases. These regions of localised variation of concentration at the anode and cathode are known respectively as the *anode* and *cathode films* or as the *diffusion films*. A concentration gradient exists across the films; they are not of uniform concentration and they extend beyond the limits of the Helmholtz double layer.

INVESTIGATION OF CATHODE FILMS

Three methods (drainage, pinhole and freezing) have been suggested for the investigation of cathode films[9]; all three methods appear to have limitations or disadvantages but nevertheless have provided a certain amount of information regarding composition and thickness of cathode films.

Haring used the drainage method to determine the composition of the cathode film formed during nickel deposition. The essential feature of this technique is the fairly rapid withdrawal of a vertical cathode from the plating bath. The adhering film has a thickness of approximately 25 μm; this liquid is collected and analysed to obtain an average composition for a film of that thickness. Graham suggested the pinhole method in which a sample of solution is withdrawn by a capillary, through a hole in the cathode, during electrolysis. The drainage method suffers from the disadvantage that the thickness of the actual cathode film cannot be measured and only the average composition of the adhering film can be determined. In the pinhole method, the exact position in the film at which solution is withdrawn is difficult to define precisely and liquid is also extracted from the bulk solution. Unless the capillary tip is moved with respect to the cathode surface, the extraction of solution will take place at one particular distance from the surface. This would not indicate the average concentration or the thickness of the film. The thickness of the cathode film could probably be estimated if the capillary tip were to be moved in small increments so that a series of concentration values could be obtained at various distances from the cathode surface. Considerable practical difficulties would be encountered in a technique such as this and disturbances of the film could also introduce errors.

Brenner introduced the freezing method which made possible the estimation of the film thickness and the concentration gradient across it. This technique involves the use of a cylindrical cathode into which a freezing mixture is introduced during electrolysis in order to freeze a layer of solution onto the cylinder. If this is maintained below the freezing point of the solution, successive layers can be turned off using a lathe and afterwards analysed. It has been shown that the cathode film, formed during electrodeposition of copper from an acid copper sulphate solution, varies in thickness between 200 μm and 250 μm. These methods can also be used to demonstrate that the pH of the cathode film is higher than that of the main bulk of the electrolyte solution.

ACTIVATION OVERPOTENTIAL

Activation overpotential is logarithmically related to the current density by the equation that was first evaluated by Tafel:

$$\eta = a \pm b \log_{10} i \qquad \ldots (2.9)$$

where the positive sign relates to the anodic process and the negative sign to the cathodic process. Since

$$a = \frac{2 \cdot 3RT \log_{10} i_0}{(1 - \beta)zF} = \frac{0 \cdot 059 \log_{10} i_0}{(1 - \beta)z}$$

and

$$b = \frac{2 \cdot 3RT}{(1 - \beta)zF} = \frac{0 \cdot 059}{(1 - \beta)z}$$

where β is the symmetry factor that frequently equals $0 \cdot 5$, the Tafel equation for a cathodic process becomes

$$\eta = \frac{0 \cdot 118 \log_{10} i_0}{z} - \frac{0 \cdot 118 \log_{10} i}{z}$$

so that η is related to i_0 the exchange current density, and the smaller i_0 the more negative the overpotential.

Studies of the deposition of nickel indicate that i_0 is in the range of $1 \cdot 7$ to $6 \cdot 9 \times 10^{-7}$ A/dm^2 and that the Tafel b is in the range of 0·070–0·090 V. However, it must be emphasised that these parameters show considerable variation, and other workers have obtained values outside the ranges given above. Nevertheless, the small value of i_0 for nickel ion discharge (i_0 for $Ag^+ + e \rightarrow Ag$ is ≈ 100 A/dm^2, $Cu^{2+} + 2e \rightarrow Cu$ is $\approx 10^{-2}$ A/dm^2) shows that the process is highly irreversible; and similar considerations apply to the discharge of hydrogen on a nickel surface.

Activation overpotential is associated with the processes that occur at the electrodes and these processes require an activation energy in order to make them proceed. Typical processes involved include hydration or dehydration of ions, discharge of ions at electrodes and the formation of crystals or molecular gases from adsorbed atoms. One of the main factors involved is associated with the passage of the potential-determining ions through the Helmholtz double layer. These mechanisms are difficult to study and at the present time far more is known about the deposited metal and the way in which nucleation and growth occurs than about the deposition process. The deposit is a much easier subject to study.

Conway and Bockris[10] have carried out fundamental research on the mechanism of electrodeposition of metals and they attempted to relate this to electroplating conditions. Suggestions were put forward to account for the manner in which the hydration sheath is stripped from the metal ion and the ion incorporated in the lattice. They have discussed the structure of the electric double layer at the metal/solution interface, the constitution of the ions in the solution, which are the source of the electrodeposited entity, and the deposition mechanism. The importance of the hydration sheath around an ion was stressed, particularly its shape and size and the effect on it of impurity or deliberate organic additions. The reaction for copper deposition ($Cu^{2+} + 2e \rightarrow Cu$) is cited as an example to illustrate the mechanisms which may be involved in a deposition process. Possible reaction paths were suggested and one of these is included here as an illustration:

1. Cu^{2+} (hydrated in solution). It diffuses to the electrode.
2. Cu^{2+} (hydrated, at electrode). It is transferred to the electrode surface.
3. Cu^{2+} (partially hydrated, attached to the electrode surface as an '*adion*'). It diffuses across the electrode surface to a crystal building site.
4. Cu^{2+} (adion at crystal building site). It becomes part of the lattice.
5. $Cu^{2+} + 2e \rightarrow Cu$. The copper becomes incorporated in the lattice.

Conway and Bockris used the term '*adion*' to describe the entity resulting from the transfer from the solution side of the double layer to the electrode. The ion retains part of its charge and is therefore an adsorbed ion.

RATE DETERMINING REACTION

One mechanism in a sequence such as that shown above will be the slowest, and hence the rate determining process. As a result of their experimental work, Conway and Bockris found that step 3, i.e. transfer of ions across the double layer, is the rate determining step. The controlling slow step is not always the same but varies with the metal, current density and environment. For example, in the case of hydrogen liberation at a platinum cathode, the slowest process is the formation of the molecular gas from adsorbed atomic hydrogen and this is the rate determining mechanism which governs the value of hydrogen overpotential. Activation overpotential is not only confined to cathodic reactions but is also involved in anodic processes (e.g. oxygen overvoltage). Activation overpotential associated with the deposition of transition metals such as iron, cobalt, nickel and chromium, is large but is much smaller for non-transition metals such as silver, copper and zinc. The value of the activation overpotential associated with the deposition of metals is influenced by the anion present in the solution, but this does not have such a great effect on the activation overpotential associated with the evolution of hydrogen.

CONCENTRATION OVERPOTENTIAL

Fig. 2.2 shows that as the potential is made more negative, the current density increases until it reaches a limiting value, viz. the limiting current density, which cannot be exceeded unless an alternative process occurs. Since water is always present in aqueous solutions, hydrogen evolution is the usual alternative process that accompanies metal discharge.

Fig. 2.3 shows the potential/$\log_{10} i$ relationship for cathodic discharge of a metal and hydrogen. The region AB represents the discharge of metal ions

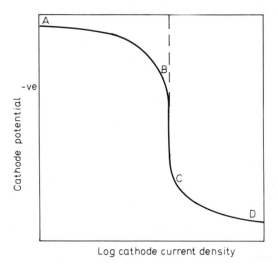

Figure 2.3. Relationship between cathode current density and cathode potential. The diagram indicates the limiting current density for metal discharge BC and the onset of a second process such as hydrogen evolution at C

at the cathode; as the potential is made more negative the rate of discharge of ions and hence the current density i is increased. However, a stage is reached (BC) at which metal ions are discharged as soon as they arrive at the cathode, and a further increase in potential cannot result in any further increase in current density. At this point the process is diffusion controlled and the current involved is the limiting current density. CD represents the occurrence of the alternative process when the potential has reached a sufficiently negative value for this to occur. The limiting current density i_L depends on the temperature of the solution, the degree of agitation and the composition of the solution. These factors are utilised in commercial plating practice to ensure that the limiting current density is as high as possible.

The transfer of metal ions in solution is governed by three mechanisms, i.e. diffusion, migration and convection. If a process is diffusion controlled the following relationships are applicable[11].

$$\eta_C = \frac{2 \cdot 3RT}{zF} \log_{10} \frac{i_L}{i_L - i} = \frac{0 \cdot 059}{z} \log_{10} \frac{i_L}{i_L - i} \qquad \ldots (2.10)$$

which shows that when $i \to i_L$, the limiting current density $\eta \to \infty$, i.e. the concentration overpotential becomes infinitely negative.

It can be shown that

$$i_L = \frac{D_{M^{z+}} zF a_{M^{z+}}}{(1-t)\delta} \times 10^{-1} \qquad \ldots (2.11)$$

where i_L is in A/dm^2, D is the diffusion coefficient of M^{z+} in cm^2/s and $a_{M^{z+}}$ its bulk activity in g ions/l, δ is the thickness of the diffusion layer in cm and t is the transport number of M^{z+}.

It can be observed from equation 2.11 that the limiting current density can be increased by increasing the diffusion coefficient D and the bulk activity of the ions $a_{M^{z+}}$ and by decreasing the thickness of the cathode film δ; the thickness of the cathode film in an unstirred solution is ≈ 0.05 cm and this can be decreased to ≈ 0.001 cm by vigorous agitation. If the limiting current density is increased then the magnitude of η_C is decreased at any current density (equation 2.10) and the onset of the conditions mentioned above which lead to an increase in limiting current density can be achieved by increasing the temperature of the solution, increasing the bulk concentration of the solution and increasing the rate of movement of the solution over the cathode surface. An increase in temperature leads to an increase in the diffusion coefficient and today most plating solutions are operated at elevated temperatures unless some overriding factor mitigates against this. The use of high temperatures will result in greater operating costs, particularly if the organic additions are volatile, and may be impracticable if any constituent of the solution decomposes at elevated temperature. The diffusion coefficient can also be increased by using an alternative anion, e.g. nickel ions diffuse faster in the presence of chloride ions than in the presence of sulphate ions. The activity of the metal ion in solution can be raised by increasing the bulk concentration of the metal salt in the solution. Concentrated solutions are unsatisfactory either if the solution is viscous causing expensive salts to be lost as 'drag-out' or if salts crystallise out due to a slight fall in temperature. Moving the electrolyte solution rapidly over the cathode, either by movement of the cathode or the solution or both, is the usual method of decreasing the thickness of the cathode film. Air agitation is probably the most common method used for stirring nickel solutions but it is unsuitable if any of the organic components of the solution are readily oxidised or if a high-foaming wetting agent is present.

Confusion over the term *limiting current density* may occur since the fundamental limiting current density described here may be confused by electroplaters with the maximum current density at which a sound deposit can be obtained. A sound deposit can be defined as one which is free from burning, treeing or powdery deposition. The limits of the sound deposit are usually determined by practical tests. Since current density varies greatly over a complex-shaped component and even over an apparently simple rectangular panel, unless certain precautions are taken, some regions may be plated satisfactorily, while burning may take place at others which receive a higher current density. This would not, of course, be acceptable in commercial practice. Burned or powdery nickel deposits are produced after hydrogen evolution has occurred causing a rise in pH at the cathode due to the local removal from solution of hydrogen ions. Metal hydroxides or basic salts are precipitated and become incorporated in the deposit; buffers help to counteract this by stabilising the pH.

It is unlikely that deposition will be controlled by one process and migration and convection may also influence the limiting current. Ions migrate towards the appropriate electrode under the influence of an electric field. Convection currents arise from two sources. The dilute cathode film is less dense than the bulk solution and upward flow results, but at the anode the concentration is higher than in the bulk solution and so downward flow occurs. Cooling takes place at various regions in the bath, particularly at the surface, and thermal currents result.

OHMIC OVERPOTENTIAL

The presence of ionically-conducting films on the surface of electrodes leads to additional polarisation effects, although they do not cause changes in reversible equilibrium potentials. In nickel solutions these may result from reduction or oxidation of organic addition agents. Sometimes this type of polarisation is classified as activation overpotential since the films inhibit the passage of the potential-determining ions through the phase boundary. It is termed *ohmic overpotential* to distinguish it from the resistance overpotential due to the IR drop across the main bulk of the solution.

The above considerations show that the potential measured by means of a Luggin-capillary, which minimises, but does not completely eliminate the IR drop in the solution, is due to a number of irreversible effects, i.e.

$$E_P = E_S + \eta_{conc.} + \eta_{act.} + \eta_{ohmic} + \eta_{resist.}$$

where E_P is the polarised potential and E_S the steady-state unpolarised potential. It follows that

$$\eta_{total} = E_P - E_S$$

which will be negative for a cathodic process and positive for an anodic process.

DECOMPOSITION VOLTAGE AND DISCHARGE POTENTIAL

The minimum applied e.m.f. necessary to produce a finite current I in a cell is given by the expression:

$$V = E_{cell} + \text{total cathode polarisation} + \text{total anode polarisation} + IR$$

E_{cell} is the e.m.f. necessary to cause the reaction to take place reversibly. Since both the anode and cathode polarisation oppose the applied voltage, they are both added to E_{cell} irrespective of sign. The voltage/current relationship is shown in Fig. 2.4. At the instant the current starts to flow the decomposition voltage should be equal to E_{cell}, but when measured experimentally it is usually somewhat higher than this due to polarisation.

The discharge potential is the potential of the cathode at which the electrode process can occur at a finite rate. In the case of reversible electrodes this corresponds with the potential given by the Nernst equation, but with irreversible electrodes it will be related to the steady-state potential and to the overpotential.

It should be noted that neither the decomposition voltage nor the discharge

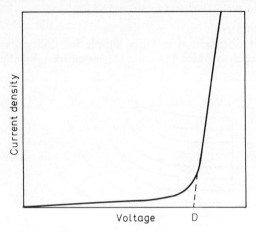

Figure 2.4. Decomposition voltage (indicated by point D)

potential are precise values, since the precise evaluation of a 'finite current' is difficult experimentally.

CURRENT DISTRIBUTION

An applied potential on a particular article of specific shape results in a primary current distribution which is dependent on the position of the article with respect to the anodes, tank walls, other cathodes or any obstructions in the solution which could influence the current flow pattern. Polarisation changes the primary current distribution which is governed only by the size, shape and position of the electrodes. Primary current distribution can be computed for simple shapes but not for complex ones. Kasper studied this problem in detail and published several papers on the subject in the early 1940s[12].

For linear conductors, equipotential surfaces are parallel to the electrodes and the current density over each electrode is uniform. In a Hull cell, which is a device used to show the variation in appearance of the electrodeposit plated from the particular electrolyte solution at varying current densities, the anode and cathode are not parallel but are positioned as shown in Fig. 2.5. The equipotential lines in this case are not parallel and the current density is higher at C than at D. The current distribution has been calculated for this arrangement and standard charts are available so that the cathode panel can be compared with a chart in order to relate the appearance to a particular current density. The current density at any point along the cathode in this cell can be calculated from the expression proposed by R. O. Hull, its inventor:

$$i = I\,(5{\cdot}10 - 5{\cdot}24 \log_{10} x)$$

where i is the current density in A/dm^2 at a point x cm from the high current density end and I is the cell current. Rousselot[13] later investigated the primary current distribution in the Hull cell and stated that the equation

$$i = I\,(3\cdot96 - 3\cdot88\,\log_{10} x)$$

was more accurate.

The current distribution is fairly simple for concentric cylinders. It is uniform over each cylinder, but since the inner one is smaller than the outer,

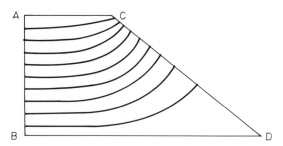

Figure 2.5. *Diagrammatic representation of current flow lines in Hull cell* (AB *is the anode and* CD *is the cathode*)

the current density on it is higher. The equipotential surfaces are closer together near the inner cylinder. A somewhat similar type of distribution holds good for concentric spheres, in that the current density on the inner sphere is higher because the areas of the two spheres vary as the squares of their radii.

Recently Rousselot[14] has described equipment which can be used to determine current distribution over three-dimensional electrodes. Earlier work by the same author dealt only with two-dimensional problems[13]. Analogical methods were used to determine the distribution over three-dimensional electrodes and conducting paper was used for two-dimensional work. Mosaic electrodes were used to facilitate the determination of the current distribution over three-dimensional electrodes and the equipment was designed so that the current received by each element could be measured. Although normally only the primary current distribution can be determined by this technique, the information is still of value. However, primary current distribution is applicable only if the current efficiency is 100% at all current densities and if polarisation does not occur. If the current efficiency changes with current density, then the secondary current distribution is more important.

Equipment of the type developed by Rousselot could be used to measure the effect of jigging arrangements, plastic shields and metal robbers without carrying out numerous thickness checks on plated articles. This type of information is clearly of value to the commercial electroplater who has to comply with minimum thickness standards on the surfaces of the parts being plated, whilst avoiding deposition of excessively thick coatings on any regions.

COVERING POWER, THROWING POWER AND LEVELLING

Confusion frequently arises over the exact meaning of the terms *covering power*, *throwing power* and *levelling*. All three phenomena are important since they are concerned with the quality and production cost of electrodeposited coatings. Each will be defined and its significance to the electroplater explained.

COVERING POWER

Covering power has not been defined precisely; it is only a qualitative measure of the ability of a plating solution to deposit metal over the whole surface of a complex shaped cathode, i.e. even where the current density approaches zero. It is solely concerned with the presence of electrodeposited metal on the surface, not with the thickness distribution of the coating. Covering power depends on the nature, pretreatment and surface condition of the substrate, in addition to the conditions of deposition. Chromium has notoriously poor covering and throwing power compared with most electrodeposits, and its ability to cover a nickel undercoat is particularly influenced by the type of deposit, the initial current density and the post-plating treatment of the nickel. Particular problems arise in the case of some bright nickels when certain organic compounds are adsorbed on the surface

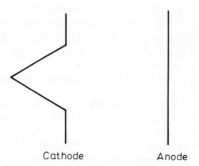

Figure 2.6. Diagrammatic representation of bent cathode used to determine covering power

of the nickel from the plating solution. These can prevent or deter chromium deposition.

Covering is usually assessed on the results of practical tests, but these are of value only when carried out under controlled specified conditions. One technique is to plate onto a cathode bent to form an "L" or a deep recess as shown in Fig. 2.6. An alternative procedure is to use a Hull cell and to note the current density at which deposition ceases. Both tests are useful for chromium, particularly the latter which is frequently termed a *chromability* test.

Good covering power would have little significance for many purposes if the distribution of metal was poor. For corrosion resistance, an adequate thickness of metal must be deposited; the mere presence of some metal would not be a guarantee of satisfactory performance in service. Covering power results should therefore be treated with discretion. Usually if covering power is poor. macro throwing power is also poor.

THROWING POWER

This must be considered as two distinct phenomena, macro and micro throwing power, since different electrochemical characteristics are necessary to achieve optimum conditions for the two phenomena.

Macro throwing power

Macro throwing power can be defined as the ability of a plating solution to produce deposits of more or less uniform thickness on irregularly shaped cathodes. From an economic viewpoint, it is essential to be able to deposit a coating as uniform as possible, even on irregularly shaped articles, so that excessive metal is not wasted on protuberances in order to obtain a certain thickness in recessed areas.

The Haring-Blum throwing-power box is the most popular technique used to obtain a quantitative evaluation, but the Hull cell can also be used and advantages for the latter technique have been claimed by Watson[15]. The Haring-Blum cell consists of two plane cathodes situated at either end of a rectangular container with a gauze anode between them so that, for example, the distances L_1 and L_2 of the cathodes from the anode are 5:1. Throwing-power determinations are only of significance if all the experimental conditions (size of cell, size of electrodes, depth of solution and solution temperature) are rigidly specified. Several expressions have been suggested for calculating throwing power. That due to Haring and Blum states that:

$$T = \frac{(P - M)}{P} \times 100 \qquad \ldots (2.12)$$

where T is macro throwing power, P is the primary current distribution (this is equal to the ratio of the distances between the anode and the two cathodes

Table 2.1 THROWING-POWER VALUES DERIVED FROM VARIOUS FORMULAE

Values of M	$T = \dfrac{100(P - M)}{P}$	$T = \dfrac{100(P - M)}{(P + M - 2)}$	$T = \dfrac{100(P - M)}{(P - 1)}$
1	80	100	100
2	60	60	75
3	40	33	50
4	20	14	25
5	0	0	0
10	−100	−39	−125
50	−900	−85	−1125

Note. $P = 5$ in all instances.

L_1/L_2) and M is the metal distribution ratio ($M = W_1/W_2$, W_1 and W_2 being the weights of metal deposited on the two cathodes respectively).

This formula expresses throwing power as the percentage improvement of the metal ratio M over the primary ratio P. The disadvantage of this interpretation of throwing power is that an erratic scale of values is obtained and the significance of the results is not immediately obvious. In Table 2.1 a comparison is made between the results obtained using this formula and two other formulae described below.

Field's[16] modification of the expression is written in the form:

$$T = \frac{100(P - M)}{(P + M - 2)} \qquad \ldots (2.13)$$

where the symbols have the same meaning as in equation 2.12. It can be seen from Table 2.1 that the values obtained in this instance range from $T = 100\%$ for perfect throwing (uniform thickness all over an irregular shaped cathode) to $T = -100\%$ when the metal distribution ratio is very great (extremely poor throwing power). This series of values is easier to interpret and understand, its main disadvantage being that a negative sign is used to represent a quantity that has no negative interpretation.

A further modification of the formula, i.e.

$$T = \frac{100(P - M)}{(P - 1)} \qquad \ldots (2.14)$$

has been suggested by which *throwing efficiency*, as it is then usually termed, can be estimated. In this case T is also equal to 100% when the metal distribution is uniform. In two recent papers the deficiencies of expressions employed to calculate throwing power have been recognised and new formulae suggested[17, 18].

The values given in the top row of Table 2.1 always represent perfect macro throwing power while those in the bottom one represent extremely poor macro throwing power; they vary only because of the different expressions used for the calculation.

Factors which influence macro throwing power

Current distribution. The primary current distribution is influenced only by the geometrical characteristics of the system, such as the shape of the anode, cathode and cell and the position of the electrodes with respect to the cell walls. If conditions of primary current distribution prevail, the current density is much higher at projections than recesses. As mentioned earlier, current distribution is extremely difficult to calculate except for a few simple shapes (e.g. a Hull cell). It is fortunate that in practice the primary distribution is rapidly transformed, due to the electrochemical factors involved in metal discharge, into the secondary distribution. Polarisation is greater at higher current densities and therefore polarisation is greater at projections than at recesses. This has an equalising effect on the current distribution and hence on the uniformity of metal distribution. It has been shown experimentally that high concentration overpotential results in good macro throwing power.

Plating conditions. The effect of the plating conditions (current density, temperature, degree of agitation and composition of electrolyte solution) depends on their effect on polarisation. Any change which leads to an increase in polarisation, such as an increase in current density, a decrease in temperature, a reduction in the amount of agitation and a low metal-ion concentration in solution, results in an improvement in macro throwing power.

Current efficiency. Macro throwing power is influenced by the relationship between current density and current efficiency if the latter is less than 100%. An efficiency of less than 100% can have an equalising effect on current distribution and can consequently lead to an improvement in throwing power. This is of no interest to the electroplater if improved throwing power is achieved only as a result of an excessive loss of current efficiency which would necessitate prolonged plating times. In the case of nickel plating solutions, current efficiency usually increases with increasing current density as shown diagrammatically in Fig. 2.7. Therefore, at low mean current density the

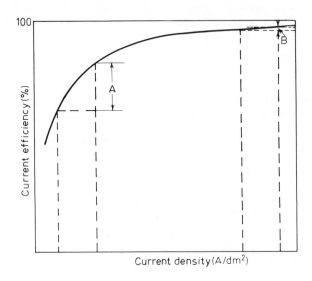

Figure 2.7. Variation of current efficiency with current density

variation in current efficiency A is greater than at high mean current density B and consequently the macro throwing power is lower at low mean-current density than at high. However, the current efficiencies of nickel solutions are high and so this aspect is not of great significance in nickel plating. It can be important for some solutions such as certain alkaline baths in which the current efficiency decreases as the current density is increased. This gives an improvement in throwing power as the current density is increased.

Conductivity. Good throwing power is favoured by high conductivity.

Throwing power of nickel solutions

Watson[15] has made a detailed study of the macro throwing power of nickel solutions and attempted to achieve improvement by making various additions to the electrolyte solution. Of the simple baths, the all chloride, fluoborate and citrate were found to have superior throwing power to the Watts. Modification of solutions to increase conductivity or to change the current efficiency/current density relationship appear unlikely to be of value in improving the throwing power of commercial solutions. A procedure more likely to succeed is to attempt to improve the polarisation characteristics. Four methods have been used in order to achieve this result:

1. Addition of salts of alkali or alkaline-earth metals, to modify the ionisation of nickel salts. (Such additions also increase conductivity with a consequent improvement in throwing power.)
2. Addition of amines or fluoride ions to complex the nickel.
3. Addition of organic substances known to affect cathode potential.
4. Additions of anions other than chloride or sulphate.

Methods 1, 2 and 4 proved rather unrewarding, the effects being small in many cases; the effect of an addition was unpredictable and low current efficiencies were recorded at high current density in some instances. The effect of organic compounds on throwing power can be related to their effects

Table 2.2 EFFECT OF SACCHARIN, COUMARIN AND THIOUREA ON THE THROWING POWER OF THE WATTS NICKEL BATH

Solution	Average current density (A/dm^2)	Throwing power (%) (Primary current ratio 1:5)
Watts	2	15
Watts	4	8
Watts + 0·005M saccharin	4	8
Watts + 0·0002M coumarin	2	−5
Watts + 0·00224M coumarin	2	8
Watts + 0·0112M coumarin	2	11
Watts + 0·000585M coumarin	4	−6
Watts + 0·00004M thiourea	2	18
Watts + 0·00045M thiourea	2	13
Watts + 0·00224M thiourea	2	29
Watts + 0·000035M thiourea	4	19

Notes
1. The results in this Table are taken from results published by Watson[15].
2. Throwing power was calculated using Field's formula (equation 2.14).

on cathode potential, which for certain compounds is discussed in Chapter 6. Saccharin has almost no effect on cathode potential and it also has no effect on throwing power. However, compounds such as coumarin and thiourea can have a considerable effect on throwing power as shown in Table 2.2.

Micro throwing power

Good micro throwing power is the ability to deposit a uniform thickness of coating at all points on the surface profile. As the name implies, it is associated with small scale surface irregularities but these can be of greater dimensions than those normally termed *micro*. Various surface profiles have been used to investigate this property; these include grooves of V-section, wire wound on to rod and electroformed gramophone record stampers.

The electrochemical characteristics necessary to achieve good micro throwing power are the opposite of those necessary for good macro throwing power. Therefore, irregularities within a certain size range must be employed to investigate micro throwing power. For example, if V-shaped grooves of increasing size are used, a point will be reached at which the conditions change from those pertinent to micro throwing power to those associated with macro throwing power.

Diffusion of metal ions is the most important process concerned with micro throwing power since this is associated with surface irregularities of such size that effective differences in the thickness of the cathode layer may be established at various parts of the profile. To obtain good micro throwing power it is essential to have an electrolyte solution with a high concentration of dischargeable metal ions so that the regions where the cathode film is thickest, i.e. the recesses, are not depleted of metal ions. If follows that there must be a high concentration of dischargeable metal ions in the bulk solution so that diffusion can readily occur to all parts of the profile. The conditions specified are those which result in low concentration overpotential (N.B. a high concentration overpotential is required for good macro throwing power). If the electrolyte solution has a low concentration of dischargeable ions (dilute solution or complex formation) concentration differences can occur within the cathode film. High concentration overpotential will occur in the recesses, the current density will be reduced and so less metal will be deposited. All factors which reduce concentration overpotential (increase in metal ion concentration, increase in agitation and increase in temperature) improve micro throwing power but reduce macro throwing power.

Solutions must have good micro throwing power if the substrate contains pits and pores, otherwise these can eventually become bridged over and cavities remain below the coating. This can obviously result in early failure of the coating in service.

LEVELLING

Levelling can be defined as the ability of an electrodeposit to produce a thicker coating in recesses than on the flat part of a surface. Providing that the micro throwing power is adequate, surface defects of simple geometric shape (such as V-notches) are preferentially filled to some extent even if the plating solution has no real levelling characteristics. This is due to the geometry of the system and is known as *geometrical* levelling: the mechanism is illustrated in Fig. 2.8. 'True' levelling (see Figs. 6.7 and 6.9 in Chapter 6) occurs when the rate of deposition is greater in the small recesses than on the flat surface and this is due to polarisation effects which usually result from the incorporation of organic compounds in the plating solution. This phenomenon is discussed in detail in Chapter 6.

Several expressions have been suggested as a means of evaluating the extent of levelling. Some of these depend on measurements of the depth of

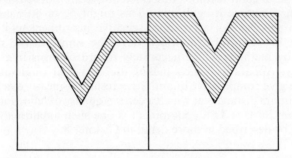

Figure 2.8. Geometrical levelling. The diagram illustrates that on a V-notch the geometrical levelling increases as the coating thickness increases

notch and the thickness of the coating, while others rely solely on measurements of surface roughness. Since levelling is dependent on the size of notch, the former procedure is more satisfactory but it is destructive and involves the preparation of a polished cross-section.

(a) Percentage levelling $= \dfrac{T_N - T_S}{D_N} \times 100$

where T_N is the deposit thickness in the notch, T_S is the deposit thickness on the surface and D_N is the depth of the notch. Levelling is therefore shown as being 100% when the profile is completely levelled and 0% when the original profile is retained (perfect micro throwing power). If the levelling value is negative the micro throwing power is obviously poor since the surface roughness has increased.

(b) Levelling $= \dfrac{T_N}{D_N + T_S}$

This expression shows the levelling to have a value of 1 when the profile is completely levelled. It also takes notch depth into account.

Evaluation of levelling by measurement of surface profile

The surface roughness before and after plating is measured using a surface analyser, e.g. by means of a 'Talysurf'. The surface quality is recorded in terms of centre line average (C.L.A.).

Percentage levelling $= \dfrac{\text{Initial C.L.A.} - \text{Final C.L.A.}}{\text{Initial C.L.A.}}$

C.L.A. is the average value of the departure of a profile from its centre line. The cross-sectional areas lying on either side of the centre line are equal, by definition[19].

While these techniques allow some quantitative expressions to be presented for the levelling power of electrodeposition processes, they do not cope with their effects on very tiny irregularities in the substrate surfaces. It requires excellent micro throwing or levelling power to fill submicroscopic blemishes, and the experimental methods described above are not sensitive enough to detect minute surface irregularities. This might be of little consequence if the human eye was not a far more sensitive instrument and thus able to distinguish between surfaces which have the same surface roughness as measured by methods which accurately record changes in coarse surface topography. Apparatus which makes use of visual observations can be adapted to give comparative quantitative results, if measurements are made of the maximum distances at which a certain quality of reflection is obtained. The Gardam Surface Truth equipment is one such apparatus and this and others will be described in more detail in Chapter 8.

REFERENCES

1. DAVIES, C. W., *Electrochemistry*, Butterworths, London (1967)
2. LATIMER. W. M., *Oxidation Potentials*, Prentice-Hall, New York (1956)
3. QUINCKE. G., *Ann. Phys. Lpz.*, **113**, 513 (1861)
4. VON HELMHOLTZ, H., *Ann. Phys. Lpz.*, **7**, 339 (1879)
5. GOUY, G., *J. Physique*, **9**, 457 (1910)
6. CHAPMAN, D. L., *Phil. Mag.*, **25**, 475 (1913)
7. STERN, O., *Z. Electrochem.*, **30**, 568 (1924)
8. GRAHAM, D. C., *Chem. Rev.*, **41**, 441 (1947)
9. GRAY, A. G., *Modern Electroplating*, John Wiley, New York (1953)
10. CONWAY, B. E. and BOCKRIS, J. O'M., *Plating*, **46**, 371 (1959)
11. POTTER. E. C., *Electrochemistry—Principles and Applications*, Cleaver-Hulme Press, London (1961)
12. KASPER, C., *Trans. Electrochem. Soc.*, **77**, 353 and 356 (1940), **78**, 131 and 147 (1940) and **82**, 153 (1942)
13. ROUSSELOT, R. H., *Metal Finishing*, **57**, 56 (Oct.) (1959)
14. ROUSSELOT, R. H., *Trans. Inst. Metal Finishing*, **42**, 100 (1964)
15. WATSON, S. A., *Trans. Inst. Metal Finishing*, **37**, 28 (1960)
16. FIELD, S. and WEILL, A. D., *Electroplating*, Pitman, London, 6th edn (1964)
17. ESIH, I., *Metalloberflache*, **23**, 161 (1969)
18. SUBRAMANIAN. R., *Electroplating & Metal Finishing*, **22** No. 10, 29 (1969)
19. *Centre-line-average Height Method for the Assessment of Surface Texture*, BS 1134: 1961

BIBLIOGRAPHY

For further information on the electrolytic growth of crystals the reader is referred to *Fundamental Aspects of Electrocrystallisation* by J. O'M. Bockris and G. A. Razumney, Plenum Press. New York (1967)

For more general information the reader is referred to the following:
EPELBOIN, I and WIART, R., 'Electrocrystallisation of Nickel and Cobalt in Acidic Solution', *J. Electrochem. Soc.*, **118**, 1577 (1971)
SCHAUS, O. O., GALE, R. J. and GAUVIN, W. H., *Plating*, **58**, 801 and 901 (1971)
SHREIR. L. L., *Corrosion*, Vols. I and II. Butterworths, 1st. edn. (1963) and 2nd. edn. (1973)
WEST. J. M., *Electrodeposition and Corrosion Processes*, Van Nostrand, London (1965)

Chapter 3

Metallurgical Aspects of Electrodeposition

The character of electrodeposits is influenced by plating conditions, nature of the substrate, composition of the solution and its purity. Extreme care is necessary in the design and operation of the experimental procedures, if consistent and reproducible results are to be obtained. Electrodeposition is beset by numerous variables, some of which are very difficult to control or standardise. Fundamental research usually involves the use of single crystal substrates, purified plating solutions and rigid control over plating conditions and solution composition. Similar restrictions are necessary when carrying out investigations at plating conditions more or less equivalent to those used for industrial plating, but polycrystalline substrates are used and the current densities employed are usually greater than those for fundamental investigations.

METHODS OF EXAMINATION OF STRUCTURE AND SURFACE TOPOGRAPHY

Optical microscopy, electron microscopy, electron probe microanalysis and X-ray techniques are the tools available, and by way of introduction it is intended to outline the scope and limitiation of these methods for the examination of electrodeposits.

OPTICAL MICROSCOPY

This is suitable for examining surfaces or cross-sections of electrodeposits (etched if necessary), but is restricted to a maximum magnification of approximately 3000 times and the depth of focus is quite limited.

Surface examination

Low-power metallurgical microscopes, particularly the binocular type, are of value for examining relatively large surface features such as pits and nodules, but are far inferior to the scanning electron microscope that provides a depth of focus 300 times greater than that of the optical microscope, thus providing a three-dimensional effect[1]. At the macro scale (rather than the micro scale) the optical microscope can provide information on nucleation and growth of deposits. A time-lapse photographic technique has proved particularly useful for this purpose[2]. The plating cell is incorporated on the stage of a projection microscope so that the developing cathode can be continuously observed and progressive growth recorded by the time-lapse ciné technique. This procedure has been successful for observing the nucleation and growth of electrodeposited copper and nickel and for observing the development of cracks in chromium[3].

Examination of structure

This is normally carried out on cross-sections of deposits mounted in Bakelite or Perspex. The choice of mounting material depends on the etch used (e.g. Perspex is attacked by a mixture of nitric acid and acetic acid, which is a popular etch for nickel). If the coating is thin, precautions must be taken to protect the outer surface of the deposit during preparation, this is particularly important if it is intended to use the microscope to measure the thickness of the coating in addition to examining its structure.

In the case of Watts nickel deposits, the choice of etching technique can influence the structure revealed as illustrated in Figs. 3.1 and 3.2. Optical

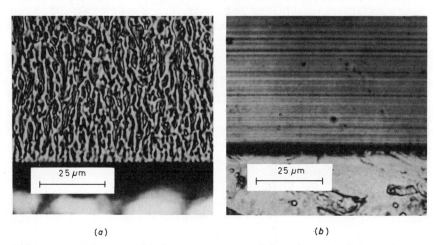

(a)   (b)

Figure 3.1. Optical micrographs of transverse sections of electrodeposits etched in a 50/50 v/v mixture of nitric and acetic acids. (a) Watts nickel and (b) bright nickel

microscopy has severe limitations as far as the examination of many electrodeposits is concerned, since they often have a very small grain size. A certain amount of information can be obtained from the examination of coarse-grained columnar deposits such as those obtained from acid copper or

Watts nickel, but bright deposits usually have an extremely fine grain size. After etching, bright nickels show characteristic laminations as illustrated in Figs. 3.1(b) and 3.3(b), but this is not associated with the grain size. The

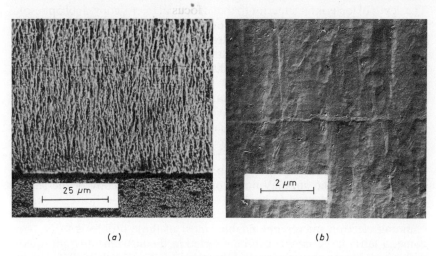

Figure 3.2. Optical and electron micrographs of transverse section of Watts nickel etched electrolytically in a solution containing 200 g/l of ammonium nitrate. With this etch, some striations are revealed in addition to the columnar grains. (a) Optical micrograph and (b) electron micrograph (replica). (After Dennis and Fuggle)

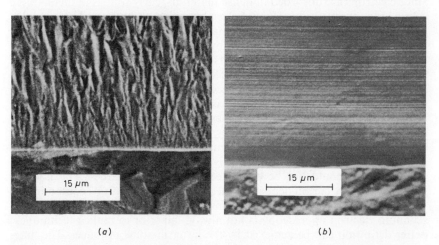

Figure 3.3. Scanning electron micrographs of transverse sections of deposits etched in a 50/50 v/v mixture of nitric and acetic acids. (a) Watts nickel and (b) bright nickel

mechanism of the formation of these striations is not completely understood, but has been attributed to the periodic incorporation of organic additives present in the plating solution.

ELECTRON MICROSCOPY

Electron microscopy is ideal for examining surface characteristics and structure[4] since it enables high magnifications to be utilised (up to × 100000). The depth of focus is far superior to that of optical microscopy, and the resolution possible is of the order of 5–10 Å for the transmission electron microscope and 150–250 Å for the scanning electron microscope.

The conventional transmission instrument can be used either to examine the surface (replica technique: × 2000 to × 10000) or the structure (thin foil technique: × 5000 to × 100000) whereas the scanning electron microscope is generally of use only for examining the surface; however, for this purpose it has many advantages over other methods. The transmission microscope has been extremely valuable in fundamental research but has proved disappointing for the examination of industrial metal-finishing problems. This is due mainly to the fact that as the surface cannot be examined directly it is necessary to prepare a replica, which is a time-consuming operation, and even so, may not provide an accurate copy of the surface features. An experienced technician is required for this delicate work, and since the specimen is so small, only selected areas can be examined. The scanning electron microscope enables most of these problems to be overcome; a fairly large sample can be inserted in the instrument (1 cm square) without elaborate preparation and the surface is examined directly.

Surface examination

Since electrons are absorbed in dense materials, early studies of metals were confined to investigations of the surface by replica techniques. These involve making a copy of the surface in a suitable low-density material; contrast of the electron image can be improved by evaporating a gold/palladium alloy onto the plastic replica[5]. If the surface has certain features, e.g. delicate spikes, complex intricate growths, deep holes or facets having re-entrant angles, it is impossible to prepare an accurate copy of the surface by stripping off a plastic replica. This procedure is therefore limited in its application, but replica techniques are still used extensively to assess qualitatively and quantitatively many surface features. For many purposes, the scanning electron microscope is more convenient. By making use of reflected primary electrons and secondary electrons it is possible to obtain information from regions which could not be examined by other techniques. The nodular growths which occur on deposits plated from nickel solutions contaminated with copper indicate the scope of the instrument (Figs. 7.3 and 7.4). However, scanning electron microscopy is not always superior to replica techniques; better definition can be achieved by the latter in some instances. The magnetic properties of the material under examination may result in low quality images when using the scanning electron microscope, but this instrument has a much wider useful range of magnification (× 20 to × 50000) than the transmission microscope.

It is essential to select the best technique for a particular purpose for it is not always an advantage to use electron microscopy in preference to optical microscopy. The two techniques are complementary to each other; useful

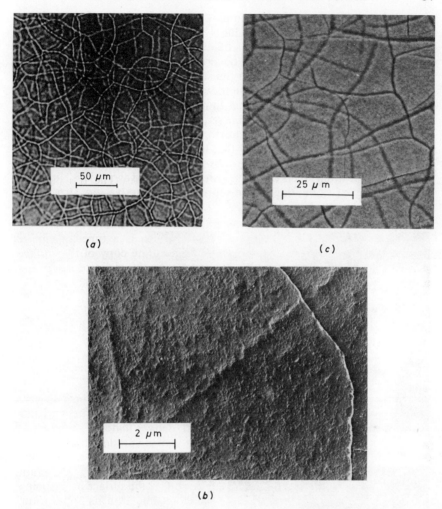

Figure 3.4. Comparison between optical and electron micrographs illustrating microcracks in chromium deposited on bright nickel undercoats under similar conditions (two sets of cracks are visible in all three photographs). (a) Optical micrograph, (b) electron micrograph (replica) and (c) scanning electron micrograph [Fig. 3.4(b) after Dennis and Fuggle]

information can be obtained by each method as illustrated by the photographs of microcracked chromium shown in Fig. 3.4.

Examination of structure

To examine the structure of a metal by transmission electron microscopy, it is essential to prepare a thin foil of the metal which is transparent to the electron beam. In the case of nickel it must be of the order of 1 000–2 000 Å thick. It is a difficult and time-consuming operation to prepare a foil of these dimensions from bulk metal, since this is likely to involve cutting, machining, spark erosion and electrolytic or chemical dissolution. However, a fairly thin sample (12·5 μm thick) can be prepared directly by the deposition process. This can be stripped from the substrate by selective dissolution or peeled off if plated onto a substrate to which it does not adhere (e.g. nickel deposited onto stainless steel). Final thinning is carried out electrolytically in a suitable solution, and so the time required to prepare a thin foil from an electrodeposited foil can be quite short[6], i.e. less than 30 min. Experimental work has been carried out to show that the structure of a particular deposit is independent of the nature of the substrate provided that when preparing the thin foil the initial layers of the deposit are dissolved away. This eliminates the possibility of examining a non-representative sample, the structure of

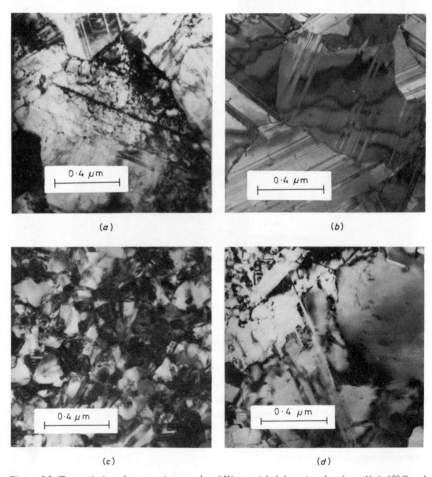

Figure 3.5. Transmission electron micrographs of Watts nickel deposits plated at pH 4, 60°C and various current densities, i.e. (a) 0·5 A/dm², (b) 4 A/dm², (c) 10 A/dm² (uniform grain size) and (d) 10 A/dm² (large grain surrounded by small grains). (After Dennis and Fuggle; see Reference 25 later)

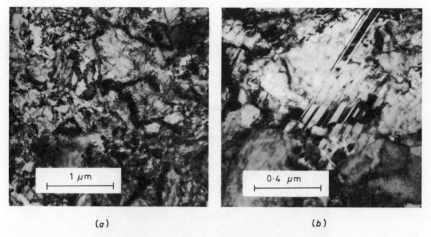

Figure 3.6. Transmission electron micrographs of semi-bright nickel deposited at 4 A/dm², pH 4 and 60°C [Fig. 3.6(b) after Dennis and Fuggle⁶]

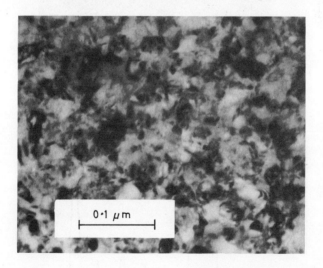

Figure 3.7. Transmission electron micrograph of bright nickel deposited at 4 A/dm², pH 4 and 60°C (after Dennis and Fuggle⁶)

which may have been influenced by the substrate. Since high magnifications are possible, the structures of Watts and semi-bright nickel deposits can be examined easily (Figs. 3.5 and 3.6), but even at the highest magnification the structure of bright nickels cannot be resolved (Fig. 3.7). Electron diffraction can also be carried out in the microscope so that certain information can be obtained even in the case of bright nickels.

ELECTRON PROBE MICROANALYSIS

The scanning electron probe microanalyser is similar in principle to the scanning electron microscope, but is designed to provide quantitative microanalysis, whereas the latter is designed to provide high resolution topographical information. The instrument has certain uses associated with the examination of electrodeposits, but these are rather more specialised than those already outlined. Its main use is to qualitatively or quantitatively

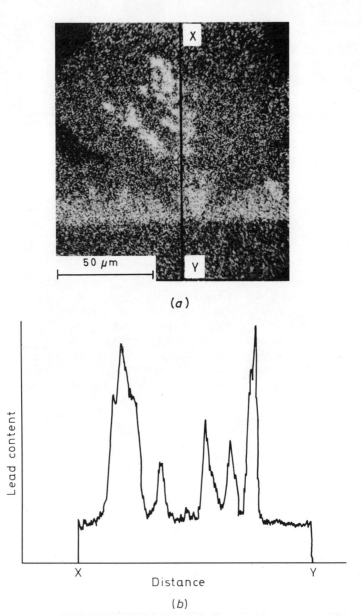

Figure 3.8. (a) *X-ray image, obtained using the electron probe microanalyser, showing the distribution of lead in a nodule formed in electrodeposited zinc/lead alloy, and* (b) *variation in the lead concentration along line* XY *in* (a)

indicate the distribution of a particular element in the sample under investigation. This is important in alloy deposits or in deposits plated from solutions containing metallic contamination. It is possible to show whether features such as nodules have a high concentration of a particular element. For example, in an electrodeposited zinc/lead alloy, nodular growths which formed at certain plating conditions have been shown to have a high lead content (Fig. 3.8). This had been suspected after examination by optical microscopy but could be confirmed only by electron probe microanalysis. Similarly, the presence of copper contamination in a nickel deposit can be detected.

The electron probe microanalyser can be used to detect discontinuities in coatings, either those that are intrinsic, i.e. present in the coating as deposited, or those formed by corrosive attack. Since electrons can penetrate only a short distance into metals, the underlying metal will be detected only at areas where the coating is absent.

Cleghorn and West[7] have used this instrument in a quantitative manner to study the early stages of deposition of chromium on nickel. They were able to determine the thickness of chromium and to show that the cathode current efficiency was constant during the period of any plating test at particular conditions.

X-RAY TECHNIQUES

X-ray techniques have been used to determine lattice parameters and the orientation of electrodeposits. Of these, the radial scan method, which is essentially a line scan through the centre of a pole figure, is particularly useful for determining the fibre texture of electrodeposits. Many electrodeposits have been found to have a certain amount of preferred orientation, although some bright deposits have random orientation. The Laué back-reflection method is suitable for determining lattice constants and is more accurate for this purpose than electron diffraction. Finch and Layton[8] carried out a fairly extensive investigation some years ago using electron diffraction to determine preferred orientation in nickel deposits. Information concerning orientation is usually only of interest in fundamental investigations. Research establishments associated with the industry are more likely to be concerned with the examination of structure and the evaluation of physical and mechanical properties than with orientation which has not been found to be of importance in the practice of electroplating. Controversial theories have been put forward to relate brightening to orientation features, but these are discussed in more detail in Chapter 6.

NUCLEATION AND GROWTH OF ELECTRODEPOSITS

In order to study nucleation of electrodeposits, it is preferable to arrange for the system to be as simple as possible, i.e. to have a single crystal substrate and a pure simple salt solution. Unfortunately, very little fundamental work has been concerned with nickel deposition, but copper deposition from acid copper sulphate solution has been studied in some detail. The copper/copper sulphate system was investigated because of its apparent electrochemical simplicity, the ease with which copper cathodes could be prepared from single crystals and the technological importance of the system. In the absence of knowledge concerning the fundamental aspects of nickel deposition, the results for copper are briefly reviewed, as much of the information is likely to be relevant to the deposition of nickel from nickel sulphate solutions.

Solution purification

Since very small amounts of organic and inorganic contaminants can greatly influence the deposition process, it is essential that extremely pure solutions are used. Purification methods for commercial nickel baths are described in Chapter 7 and similar treatments are used on a laboratory scale for both copper and nickel plating solutions. The purification treatment must be more thorough for investigational work than for commercial plating baths, but carbon treatment and electrolysis at a low current density are still the main features, although a greater proportion of both per unit volume than would be justified in commercial practice are usually necessary to ensure a very high degree of purity. Standards cannot be fixed universally because the purity of the initial solution varies, and small amounts of impurity (particularly organic contamination) sufficient to influence the deposition characteristics cannot be estimated and probably cannot be detected by analytical methods. The most satisfactory method of evaluating the purity of a base solution is to carry out physical and mechanical tests and structural examination on specimen electrodeposits.

COPPER DEPOSITION

Metallurgical studies of electrodeposits have proved that the usual processes of crystal growth occur during electrodeposition. Normally, crystals develop in such a manner that they become bounded by plane crystallographic facets, the more rapidly growing being eliminated during growth. After nucleation on plane crystallographic facets, growth occurs by the spreading of layers. Pick and Wilcock[2] used the time-lapse photography technique to observe the development of growing copper deposits. A variety of deposit structures formed; these fell into groups that could be related to the atomic configuration of the original cathode surface. The surface geometry of the deposits did not change once a steady growth rate had been established, but the scale of the surface features coarsened, and this has been explained by the 'bunching' hypothesis. This is based on the supposition that growth layers do not all advance at the same rate, but that some are overtaken by those proceeding at a faster rate. In this way, multiple steps are formed in the structure. Depositing ions are thought to have appreciable mobility on copper surfaces, so that after deposition, migration can occur to positions of lowest energy. Many sites of this type occur in a developing surface and are associated with fluctuations in the deposition process and the initial-basis metal preparation. The layers spread from these sites, but since the rate of arrival of atoms at the advancing step edges is not uniform the layers do not spread at the same rate and merging of steps can occur.

Using copper single crystals having orientations near {100} planes as cathodes, Storey, Barnes and Pick[9] evaluated the effect of current density and temperature on the deposit. Four basic structures were observed, viz. ridge, platelet, block and polycrystalline. These structures are illustrated in Fig. 3.9. However, cathode overpotential was found to be of greater significance than current density. Barnes[10] investigated the effects of pH and polarisation on copper deposits.

Growth layers can originate at several different types of nuclei. Primary nucleation is not required on atomically rough substrates, but on close packed planes nucleation of growth layers is necessary. By means of the electronmicroscope Vaughan and Pick[11] studied the appearance of the surface after short plating times of the order of 15s, and observed step-like structures which had been initiated at isolated growth centres. The purity and orientation of the cathode surface strongly influences the process of nucleation. Small impurities such as oxide particles act as sites at which nucleation occurs preferentially.

Three types of nuclei (truncated pyramids, small hillocks of no obvious crystalline form and circular discs) were observed after deposition had been allowed to proceed for 15s using as the cathode a cube-textured copper sheet with {100} planes parallel to its surface. The nucleus form depends on the surface orientation of grains. It is thought that the discs, which are random and numerous, could result from primary nuclei formed at the sites of dislocations which emerge from the surface. No positive evidence has been obtained to prove that nucleation can occur at clean dislocations, but the occurrence of spirals in electrodeposits supports the view that some emergent dislocations do provide suitable growth sites.

NICKEL DEPOSITION

Cliffe and Farr[12] have investigated the deposition of nickel and cobalt from sulphate electrolytes in order to eliminate effects due to chloride ions, although these would usually be present in commercial solutions. The factors varied in their investigation were orientation of substrate, temperature of electrolyte solution, plating time and current density; the composition of the electrolyte solution and the condition of the substrate were maintained constant throughout. Nickel deposits formed on substrates having orientations near {100} planes showed a pronounced coarsening of the surface structure above 94°C when plating was carried out for 1 h at 1 A/dm^2. A more clearly defined structure became visible at 94°C, the background still being irresolvable, but at 98°C the coarse structure predominated. No further significant change in structure occurred between 98°C and the boiling point of the solution. The back-reflection X-ray technique was used to obtain information concerning the different types of deposit produced at the various plating conditions. The diffraction patterns resulting from the fine structure produced below 90°C consisted of diffuse spots. Coarse crystallographic deposits plated at 98°C and 1 A/dm^2 resulted in sharp spots indicating epitaxial growth of the deposit, and coarse nodular deposits plated at 98°C

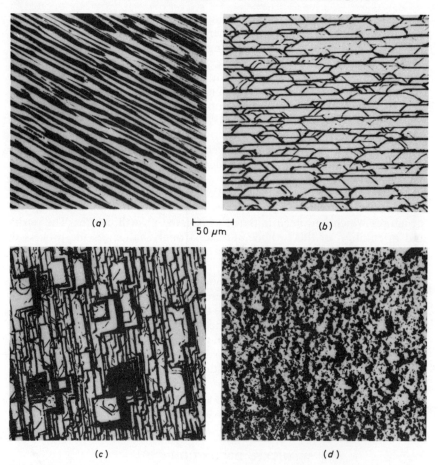

Figure 3.9. The four basic structures observed on growing copper electrodeposits. (a) *Ridge,* (b) *platelet,* (c) *block and* (d) *polycrystalline (courtesy Storey, Barnes and Pick[9])*

and current densities greater than 1 A/dm^2 resulted in rings. Photomicrographs of the surface of nickel deposits plated at various temperatures are illustrated in Fig. 3.10. Neither purification with activated carbon nor pre-electrolysis were found to have any significant effect on the surface topography of the deposits obtained. In this respect the behaviour was different from that observed for the acid copper sulphate solution.

Jones and Kenez[13] electrodeposited nickel from a conventional Watts solution onto polycrystalline nickel having large well-defined grains, the object being to investigate the influence of grain boundaries and grain orientation on nucleation and development of electrodeposits plated under conditions closely resembling those used commercially. The initial growth forms in Watts nickel deposits were much smaller than those obtained from acid copper solution, and this made it impossible to follow the early stages of growth by optical microscopy using the time-lapse photography technique. These authors were able to show that preferential coverage occurred at an early stage on substrate grains having orientations near {111} planes.

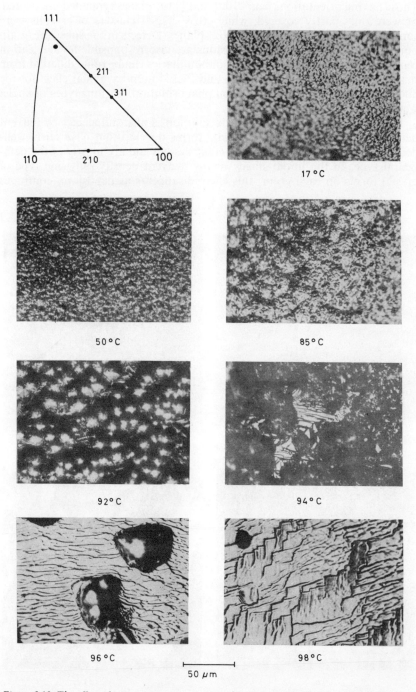

Figure 3.10. The effect of increasing temperature on the structure of nickel deposits, plated from a solution containing 240 g/l of nickel sulphate and 30 g/l of boric acid (pH 3·0) onto a single crystal nickel base, which had the orientation shown in the stereographic projection (after Cliffe)

Those having orientations near {100} and {110} planes remained uncovered, or were only partly covered, while cross-shaped clusters of crystals were sometimes formed on {103} and {102} planes. Defects in the substrate, grain boundaries and non-metallic inclusions are also responsible for nucleation centres; the form and growth at grain boundaries can be quite different from that over adjacent grains. Vaughan and Pick[11] found a similar effect for the nucleation of copper from copper sulphate solution. Various types of nuclei were found at impurity stringers.

The form of deposits from Watts solution is also influenced by current density. At less than 1 A/dm² crystal forms develop that have fairly well-defined facets; these belong to the cubic system and are mostly of the octahedral type. In the initial stages, at low current density only one type of crystal forms on each grain: this effect disappears as deposition continues

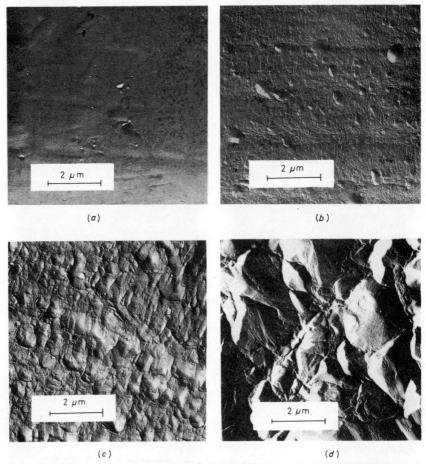

Figure 3.11. Electron micrographs showing the development of the surface topography of a Watts nickel deposit with increasing thickness. (a) Surface of steel substrate, (b) Watts nickel 0·6 μm thick, (c) Watts nickel 1·8 μm thick and (d) Watts nickel 12·5 μm thick (after Dennis and Fuggle[6])

after the formation of a coherent deposit. At current densities within the usual commercial plating range, growth centres are circular in shape and are not of external crystalline form. Once a coherent deposit has formed the surface structure has the appearance of ill-defined hillocks and this does not change much with increasing thickness of deposit.

Weil and Cook[14] demonstrated the development of nickel deposits with time of plating on a specular, fine-grained substrate. Using a negative parlodion replica technique they showed that a few grains developed preferentially at an early stage of deposition. Some evidence has been obtained using selected area diffraction to suggest that these rapidly-growing grains have a {100} type of plane parallel to the surface. Growth layers are discernible at an early stage and become coarser as deposition proceeds, probably by means of a bunching mechanism similar to that which has been shown to occur in deposits from acid copper solution. The layers are not parallel to the substrate surface but inclined to it and often result in pyramid structures. In an earlier paper, Weil and Paquin[15] showed that three structural types were formed (which corresponded to three different fibre axes) in deposits plated from Watts solution containing a selection of organic additives. Those having a $\langle 100 \rangle$ fibre axis consisted of platelets stacked parallel to each other which formed the pyramid-type structure. The deposit from the Watts bath itself is an example of this, but the thickness and spacing of platelets varies widely in deposits plated from baths containing additives. The second type which was associated with a strong $\langle 311 \rangle$ and a weak $\langle 111 \rangle$ fibre axis had a surface consisting of what appeared to be fine equiaxed crystallites. Deposits having a $\langle 110 \rangle$ fibre axis usually contained spiral growths and in most instances the deposits appeared dull.

One of the present authors[6] has investigated the effect of various substrates and their finishes on the surface topography of Watts nickel deposits. The substrates employed were of the type likely to be used for commercial plating. The electron micrographs shown in Fig. 3.11 illustrate the development of the deposit with increasing thickness on a fairly high quality rolled-steel basis metal. The general trend for the development of the structure is similar to that observed by Weil and Cook[14] on a fine-grained specular substrate and Jones and Kenez on a large-grained polycrystalline substrate.

CHROMIUM DEPOSITION

Jones and Kenez[16] have published the results of an investigation of the nucleation of chromium deposits on nickel undercoats. In the case of chromium deposition the surface condition of the nickel is of particular importance in influencing the nucleation process. Nickel passivates rapidly and it is often necessary to activate the surface to facilitate nucleation of the chromium deposit. Even though nuclei cannot be resolved by optical microscopy, changes in reflectivity of the surface can be detected and so certain information can be obtained by optical means. An increase in plating temperature, a decrease in the catalyst ratio (see Chapter 9) and an increase in current density lead to a reduction in the inhibition period (the time from the application of current to the occurrence of nuclei). The catalyst effect is the most significant and the same pattern of behaviour occurs either in the case of activated or non-activated surfaces. Anodic treatments in sulphuric acid are more effective than acid dips as a means of activation. Some areas had more nucleation centres than others, but no reason could be found for this and the distribution and orientation of growth sites appeared to be random. Electron micrographs showed the nuclei to be of well-defined cubic habit. Lateral growth occurred, but the growth centres had a limiting size and complete surface coverage occurred due to further nucleation and coalescence of the growth sites. As in the case of nickel at high current density, the growth centres have a less clearly defined crystallographic form and tend to be circular in shape. This evidence substantiates the theory that chromium is deposited in the body centred cubic form from baths consisting of chromic and sulphuric acids. No evidence of hexagonal habit could be observed to support the theories which suggest that chromium is first deposited as a hexagonal hydride which, being unstable, reverts to the stable cubic form.

Small indentations in the nickel surface, either scratches or micro-hardness indentations, act as sites for preferential nucleation. It is thought that this may be due to the fact that these are areas of low overpotential for deposition, arising from the mechanical strain induced in the substrate.

The efficiency of chromium deposition increases as the current density increases, and the rate of development of the deposit is more rapid once nucleation has taken place. This probably accounts for the lack of crystallinity at growth centres under these conditions. The second phase of deposition involves renucleation on initial crystals, and well-defined crystal forms can occur under suitable conditions at this stage. The knowledge concerning the mechanism responsible for controlling crystal size is limited. Usually high cathode potential is characterised by the formation of small grains.

STRUCTURE OF ELECTRODEPOSITS

The structure of an electrodeposit depends on the relative rates of formation of nuclei and the growth of existing ones. If the conditions favour the formation of fresh nuclei then fine-grained deposits are formed, while preferential growth of existing nuclei leads to the production of large-grained deposits. Usually fine-grained deposits are smoother, brighter, harder and less ductile than coarse-grained ones, although exceptions to this generalisation do occur. Fine-grained deposits produced by adding organic compounds to the plating solution can have a dull matt appearance instead of a smooth bright finish. Treed deposits produced by plating at a high current density appear coarse and rough, despite their small grain size, for the small grains form a coarse aggregate. In general, any change in the plating conditions that results in an increase in cathode polarisation leads to a reduction in grain size. Variation of the simple plating conditions without resort to addition to the solution, enables a certain amount of control to be exercised over the structure and physical properties of the deposit. An increase in current density causes a reduction of the metal ion concentration in the cathode film with a consequent increase in concentration overpotential and decrease in grain size. If other factors remain constant, an increase in the degree of agitation lowers the concentration overpotential and hence results in a larger grain size. An increase in temperature similarly leads to a reduction in cathode polarisation.

The type of solution and its composition also influences the characteristics of the deposit. Concentration overpotential will be high in a dilute solution and therefore the grain size will be small and the macro throwing power fairly good. However, in dilute solutions the limiting current density is so low that this type of solution is of only limited use commercially. Since fine-grained deposits are usually preferable to coarse-grained ones, it is necessary to devise solutions which provide this type of structure, but as dilute solutions are unstable other formulations are adopted. The normal procedure is to use the solutions which have a high concentration of metal compounds but a low metal ion concentration. Complex formation or the common-ion effect make this possible with the metal compounds serving as a reservoir of metal ions. Copper deposition provides an example of a metal which can be obtained by either technique. Fine-grained deposits are obtained from the double cyanide solution, much coarser deposits from the acid copper sulphate solution and even coarser deposits from solutions of copper sulphate containing no added sulphuric acid. Most of the coarse-grained deposits have a columnar structure which can be revealed easily by etching and optical microscopy. The columnar grains are formed in the direction of the current flow lines and in thick deposits, planes of weakness can arise at features such as sharp corners, just as in castings having a columnar structure.

Epitaxial growth

Growth can be defined as being epitaxial when the atomic arrangement in a crystalline substrate is perpetuated in the deposit. The effect of the substrate diminishes as the coating thickness increases. The rate at which this occurs is influenced by the type and state of the substrate and certain plating conditions. The effect of the substrate rapidly diminishes if the surface is mechanically polished and is prolonged when the substrate has a large grain size. An increase in current density, an increase in polarisation and the incorporation in the solution of surface active organic compounds are all factors which diminish the extent of epitaxial growth. Polycrystalline material is deposited at an early stage and lattice distortion occurs.

STRUCTURE AND SURFACE TOPOGRAPHY OF ELECTRODEPOSITED NICKEL

Transverse sections of Watts and semi-bright nickel deposits have a characteristic columnar structure when etched in equal volumes of nitric and acetic acid [Fig. 3.1(a)]. Etched bright deposits have an equally characteristic lamellar structure [Fig. 3.1(b)]. For comparison, scanning electron micrographs are shown in Fig. 3.3. The structure of a deposit can serve as a useful indication as to its sulphur content, a factor which has a significant bearing on its electrochemical behaviour. Usually, columnar structures are an indication that the deposit is sulphur free or at least has a very low sulphur content, while lamellar structures indicate the presence of sulphur. Metallography is a useful means of investigating the corrosion mechanism of composite nickel plus chromium coatings; the types of pits formed are illustrated in Chapter 11. Etching reveals the boundary between columnar and lamellar deposits, and so in addition to permitting the mode of corrosion to be examined, it also provides a means of determining the thickness of the two nickel layers in a *duplex* coating.

As illustrated by the work of Beacom *et al.*[17], the difference between columnar and striated structures may not be as clear cut as at one time believed. A few laminations can be detected in columnar structures and vice versa, provided that suitably selective etching reagents are used. The structure revealed by etching a Watts deposit electrolytically in a solution containing 200 g/l ammonium nitrate is shown in Fig. 3.2. The effect of surface contours of the substrate on the etching characteristics of deposits

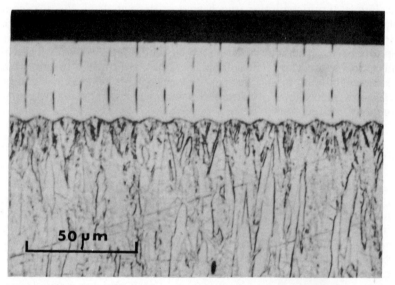

Figure 3.12. Nickel deposited on an 0·5 μm roughness standard from a Watts bath containing 2 g/l sodium allyl sulphonate and 0·004 g/l N-allyl quinaldinium bromide, and then chemically etched (courtesy Beacon, Hardesty and Doty[17]*)*

plated from Watts solution containing organic additives has also been demonstrated by Beacom *et al.*[17] (Fig. 3.12). Brenner, Zentner and Jennings[18] have carried out an extensive programme to relate variations in structure to operating variables and bath composition.

The grain structure of Watts nickel deposits can be revealed clearly by thin foil transmission electron microscopy. Fig. 3.5 shows the effect of variations in current density on the grain size of deposits plated from a solution containing 300 g/l $NiSO_4 \cdot 6H_2O$, 28 g/l NaCl and 40 g/l H_3BO_3. Not only does the grain size tend to decrease with increase in current density but it also becomes less uniform. At the highest current density (10 A/dm^2) some very large grains were formed. Fig. 3.5 also shows the problems encountered in illustrating a representative region of a deposit plated under particular conditions. The electron micrographs shown in Figs. 3.5(c) and 3.5(d) were taken from adjacent areas of the same foil. Electron microscopy definitely shows that electrodeposited nickel is twinned and that dislocations are present. The traditional Watts bath (300 g/l $NiSO_4 \cdot 6H_2O$, 36 g/l $NiCl_2 \cdot 6H_2O$ and 40 g/l H_3BO_3) produces a deposit which at 4 A/dm^2 has a less uniform grain size than that obtained from a Watts bath containing the equivalent concentration of chloride ion but present in the form of sodium chloride. Deposits plated from unpurified solutions ('AnalaR' salts dissolved in de-ionised water) have a similar structure to those plated from purified solution, except that they are more heavily twinned and contain more dislocations. This is probably due to the presence of a few 'foreign' atoms being incorporated in the nickel lattice. Transmission electron microscopy of thin foils prepared from transverse sections of electrodeposits involves many practical problems. A thick deposit must be plated, but this takes many hours and preparation of a thin foil from this deposit is even more difficult than in the case of bulk metals. The results so far obtained provide little new information over that which can be obtained using the rapid technique described earlier, except to illustrate the presence of columnar grains in Watts nickel[19]. If the location from which the specimen had been cut could be determined accurately it should be possible to follow the deviation from epitaxial growth with increasing thickness of deposit.

Even the large grained electrodeposits have a relatively small grain size compared to most bulk metals and for comparison the structure of annealed wrought nickel is shown in Fig. 3.13. Dislocations in the wrought material

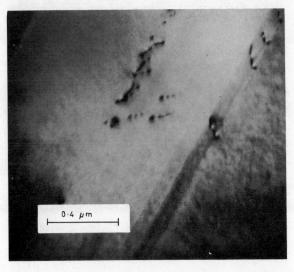

Figure 3.13. Transmission electron micrograph of annealed wrought nickel (after Dennis and Fuggle; see Reference 25 later)

were observed to move while the specimen was being examined in the microscope. In electrodeposited nickel, no movement of dislocations could be detected while the foil was in the microscope. This suggests that the dislocations are pinned by impurity atoms.

Several authors have published electron micrographs illustrating the structure and surface topography of electrodeposits plated from nickel solutions containing organic addition agents[6, 20, 21, 22]. Some compounds result in a very-fine grain size even when present at fairly low concentration, and deposits plated from commercial bright nickel baths have such an extremely small grain size that this cannot be resolved satisfactorily using the electron microscope (Fig. 3.7). Less active compounds do not have such a dramatic effect on the structure. For example, coumarin, which is the best known compound added to the Watts bath to provide a semi-bright levelled deposit, causes only a slight reduction in grain size but the dislocation density is increased considerably and the grain boundaries cannot be clearly distinguished (Fig. 3.6).

Electron diffraction patterns can be used to provide a certain amount of information even when the grain size is too fine to resolve. Complete rings instead of discrete spots are an indication of fine grain size. Diffraction patterns characteristic of f.c.c. metals are obtained from large grains using selected area diffraction (Fig. 3.14). Diffraction patterns also indicate the occurrence of twinning and enable the twinning plane to be determined. In the example shown in Fig. 3.15, twinning has taken place on {111} planes as is usual for f.c.c. metals.

Figure 3.14. Electron diffraction pattern obtained from a large grain in a Watts deposit (after Dennis and Fuggle[6])

Crossley, Kendrick and Mitchell[23] have investigated the structure of deposits plated from 'all-chloride' nickel plating solutions and have shown that the appearance and properties can be changed by using a square wave supply as the current source. Depending on the characteristics of the current source, the structures of the deposits obtained could resemble either those

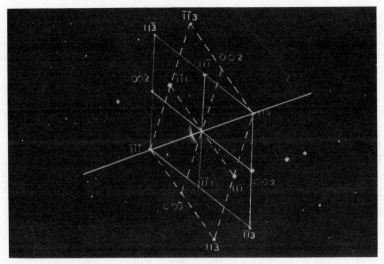

Figure 3.15. Electron diffraction pattern obtained from twinned region in a Watts deposit (after Dennis and Fuggle)

obtained by plating from the Watts bath or those containing various organic additives.

Deposits from sulphamate solutions (600 g/l $Ni(SO_3NH_2) \cdot 4H_2O$, 15 g/l $NiCl_2 \cdot 6H_2O$ and 40 g/l H_3BO_3) have been shown by Saleem, Brook and Cuthbertson[24] to have similar structures to those of Watts deposits. There was extensive evidence of twinning but the grain size was not markedly dependent on current density over the range 8 to 60 A/dm^2, and it was only about a quarter the size of that of Watts nickel plated at 4 A/dm^2.

Figure 3.16. Electron micrograph (replica) showing the effect on surface topography of the presence of 0.1 g/l coumarin in a Watts nickel solution. Plating conditions: 4 A/dm^2, pH 4, 60°C, deposit thickness of 12.5 μm on rolled mild steel substrate (after Dennis and Fuggle[6])

It has been indicated earlier that some nickel deposits exhibit preferred orientation, but the extent of this is dependent on the plating conditions and the particular bath used. Watts deposits plated at current densities within the normal operating range (e.g. 4 A/dm^2) have a fibre axis in the [100] direction but at a low current density this has been shown to change to the [211] direction[25]. The orientation can be influenced either by organic or inorganic additions to the plating solution, but many bright nickel deposits do not have a preferred orientation. Coumarin does not have too drastic an effect on the physical properties of deposits from this bath which exhibit the same preferred orientation as deposits from Watts nickel solution free from additions.

The surface of Watts nickel has a dull matt appearance to the naked eye, but at high magnification this is shown to be due to the pyramid shaped formations which develop (Fig. 3.11). A bright finish can only be produced on this type of deposit by mechanical polishing, and this, of course, was the earliest method of obtaining a decorative nickel coating. Electropolishing of the electrodeposited metal is unsuccessful as it usually results in pitting, but in any case, from an economic point of view it would be a wasteful processing sequence to deposit a fairly expensive metal by electrodeposition and then to remove a large proportion of it by a dissolution process, particularly when alternative methods are available for depositing a bright coating

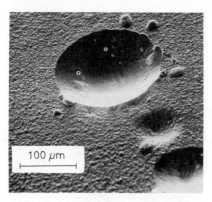

Figure 3.17. Scanning electron micrograph of a gas pit in a Watts deposit

Figure 3.18. Scanning electron micrograph of a 'coumarin' pit in a deposit plated from a Watts solution containing 0·1 g/l coumarin

straight from the bath. Semi-bright deposits, as the name implies, are less dull and matt than Watts deposits, but not as bright or microscopically smooth as fully bright deposits. The electron micrograph illustrated in Fig. 3.16 shows that the pyramid growths have become quite rounded as a result of adding coumarin to the Watts solution. In the case of a fully bright deposit the surface is so smooth that facets and surface growths cannot be detected even at the highest magnifications available. The only features which can occasionally be detected are a few small pits. In a high-quality deposit these would be so small that they would not detract from the appearance of the coating, since they would not be visible to the naked eye. Pitting in any type of deposit is often due to gas bubbles adhering to the developing surface; gas pits can be recognised by their characteristic shape (Fig. 3.17). This type of defect can be prevented by using adequate agitation. Deposits plated from Watts solution containing coumarin are liable to contain large pits which again are easy to recognise by their shape (Fig. 3.18).

STRUCTURE AND SURFACE TOPOGRAPHY OF ELECTRODEPOSITED CHROMIUM

Decorative chromium coatings are too thin (0·25–1·25 μm) to be examined in cross-section by optical microscopy, but thick chromium deposits can be sectioned to reveal the structure and discontinuities. Jones et al.[26] have etched chromium deposits in a number of reagents and have shown that the features revealed are dependent on the etching technique used. Electrolytic etching in a mixture of hydrochloric acid and methyl alcohol provides a fairly uniform etch and shows a fibrous structure; bands due to changes in plating conditions are also revealed. The fibrous columnar structure is not the true grain size, since it is known from other investigations[27,28] that chromium deposits consist of very small crystals. Chemical etching in oxalic acid reveals striations in chromium deposits plated under such conditions that cracking occurs. However, this etch severely attacks the crack lines and consequently interferes with observation of the banded structure. Other etching techniques have been used which are more satisfactory[26] and which to a certain extent reveal both columnar and striated structures. Optical microscopy is useful for examining the surface of all types of decorative chromium deposits in order to study discontinuities, either pores or cracks. A low magnification is required to assess macro cracking and a magnification of the order of ×200 is adequate to assess the frequency of micro cracking. Photomicrographs are shown in Chapter 9 to illustrate the influence of plating parameters and the underlying metal on crack formation.

Replica techniques have been used to examine surface features of chromium deposits but are not entirely satisfactory, since the plastic replica may be damaged if keyed too securely in the cracks. Scanning electron microscopy is now usually more suitable, particularly for examining corroded nickel +chromium coatings as illustrated by the electron micrographs shown in Chapter 11.

Electrodeposited chromium in common with bright nickel has an extremely fine grain size, and attempts to examine the structure by transmission electron microscopy have not revealed much detail. Cleghorn and West[7] have used the technique to examine chromium deposits stripped from electropolished nickel substrates and have shown that epitaxial growth occurs. The chromium deposit adopted different orientations on different grains. Although the grain size was estimated to be approximately 100 Å, 'single crystal' spot patterns were obtained by electron diffraction since all the grains in the selected area had nearly the same orientation.

REFERENCES
1. ARROWSMITH, D. J., DENNIS, J. K. and FUGGLE, J. J., *Electroplating & Metal Finishing*, 22, 19 (1969)

2. PICK, H. J. and WILCOCK, J., *Trans. Inst. Metal Finishing*, **35**, 298 (1958)
3. JONES, M. H. and SAIDDINGTON, J., *Proc. Amer. Electroplaters' Soc.*, **48**, 32 (1961)
4. WEIL, R. and READ, H. J., *Metal Finishing*, **53** No. 11, 60 (1955)
5. BRAMMAR, I. S. and DEWEY, M. A. P., *Specimen Preparation for Electron Microscopy*, Blackwell Scientific Publications (1966)
6. DENNIS, J. K. and FUGGLE, J. J., *Electroplating & Metal Finishing*, **20**, 376 (1967) and **21**, 16 (1968)
7. CLEGHORN, W. H. and WEST, J. M., *Trans. Inst. Metal Finishing*, **44**, 105 (1966)
8. FINCH, G. I. and LAYTON, D. N., *J. Electrodepositors' Tech. Soc.*, **27**, 215 (1951)
9. STOREY, G. G., BARNES, S. C. and PICK, H. J., *Electrochim. Acta.*, **2**, 195 (1960)
10. BARNES, S. C., *Electrochim. Acta.*, **5**, 79 (1961)
11. VAUGHAN, T. B. and PICK, H. J., *Electrochim. Acta.*, **2**, 179 (1960)
12. CLIFFE, D. R. and FARR, J. P. G., *J. Electrochem. Soc.*, **111**, 299 (1964)
13. JONES, M. H. and KENEZ, M. G., *Plating*, **53**, 995 (1966)
14. WEIL, R. and COOK, H. C., *J. Electrochem. Soc.*, **109**, 295 (1962)
15. WEIL, R. and PAQUIN, R., *J. Electrochem. Soc.* **107**, 87 (1960)
16. JONES, M. H. and KENEZ, M. G., *Proc. Amer. Electroplaters' Soc.*, **51**, 23 (1964)
17. BEACOM, S. E., HARDESTY, D. W. and DOTY, W. R., *Trans. Inst. Metal Finishing*, **42**, 77 (1964)
18. BRENNER, A., ZENTNER, V. and JENNINGS, C. W., *Plating*, **39**, 865 (1952)
19. WARING, R. S., Project Report, University of Aston in Birmingham, Department of Metallurgy (1968)
20. FROMENT, M. and OSTROWIECKI, A., *Metaux*, **42**, 83 (1966)
21. MAURIN, G. and FROMENT, M., *Metaux*, **42**, 102 (1966)
22. CROSSLEY, J. A., BROOK, P. A. and CUTHBERTSON, J. W., *Electrochim. Acta.*, **11**, 1153 (1966)
23. CROSSLEY, J. A., KENDRICK, R. J. and MITCHELL, W. I., *Trans. Inst. Metal Finishing*, **45**, 58 (1967)
24. SALEEM, M., BROOK, P. A. and CUTHBERTSON, J. W., *Electrochim. Acta.*, **12**, 553 (1967)
25. DENNIS, J. K. and FUGGLE, J. J., *Trans. Inst. Metal Finishing*, **46**, 185 (1968)
26. JONES, M. H., KENEZ, M. G. and SAIDDINGTON, J., *Plating*, **52**, 39 (1966)
27. WOOD, W. A., *Trans. Faraday Soc.*, **31**, 1248 (1935)
28. SNAVELY, C. A. and FAUST, C. L., *J. Electrochem. Soc.*, **97**, 99 (1950)

BIBLIOGRAPHY

FISCHER, H., *Elektrolytische Abscheidung und Elektrokristallisation von Metallen*, Springer, Berlin (1954). A comprehensive survey of earlier theories on the electrocrystallisation of metals (in German)

FISCHER, H., 'Electrocrystallisation of Metals Under Ideal and Real Conditions', *Agnew. Chem. Internat. Edn.*, **8**, 108 (1969)

FISCHER, H., 'The Nucleation Dependent Growth Layer—A Structure Element in Electrocrystallisation', *Plating*, **56**, 1229 (1969)

WIEGAND. H. and SCHWITZGEBL. K., 'Eigenspannun, Kristallitgrösse und Texturen in Galvanischen und Chemischen Nickelschiten und Ihr Zusammenhang mit Mechanischen und Technologischen Werkstoffkennwerten'. *Metall.*, **21**, 1024 (1967). Compares grain structure and texture of electroplated and electroless nickel deposits (in German)

Chapter 4

Plating Baths and Anodes Used for Industrial Nickel Deposition

PLATING BATHS

Watts nickel bath

Most commercial nickel plating solutions are based on the one named after Watts who first introduced a bath having the formulation:

> Nickel sulphate $NiSO_4 \cdot 7H_2O$ 240 g/l
> Nickel chloride $NiCl_2 \cdot 6H_2O$ 20 g/l
> Boric acid H_3BO_3 20 g/l

The name *Watts Bath* is now used to cover a range of solutions whose compositions vary within the range shown in Table 4.1, the chloride ion sometimes being introduced in the form of sodium chloride. Sodium

Table 4.1 CONCENTRATION RANGES OF INGREDIENTS OF WATTS BATH

Chemical	Concentration range (g/l)
Nickel sulphate, $NiSO_4 \cdot 6H_2O$	150 to 400
Nickel chloride, $NiCl_2 \cdot 6H_2O$	20 to 80
or	
Sodium chloride, NaCl	10 to 40
Boric acid, H_3BO_3	15 to 50

chloride is cheaper than nickel chloride and is satisfactory for most purposes, although it has been reported that sodium ions are detrimental in the presence of some organic addition agents; this cannot be so in the majority of cases since many organic compounds are added in the form of their sodium salts.

Nickel sulphate is the principal ingredient; it is used as the main source of nickel ions because it is readily soluble (570 g/l at 50°C), relatively cheap, commercially available and is a source of uncomplexed nickel ions. However, it is known that a certain amount of ion association occurs in concentrated solutions due to ions of opposite charge being held together by coulombic forces. This reduces the effective concentration of free ions and the activity coefficient is a measure of the extent to which association takes place. In nickel plating solutions the activity of nickel ions is governed by the concentration of nickel salts in solution, their degree of dissociation and the nature and concentration of other components of the solution. If the concentration of Ni^{2+} available for deposition is low, burnt deposits will be produced at a relatively low current density, and in addition the limiting current density will be low (see Chapter 2). For these reasons the concentration of nickel sulphate must be high.

The presence of chloride ions has two main effects; it assists anode corrosion and increases the diffusion coefficient of nickel ions thus permitting a higher limiting current density. Saubestre[1] quotes values for the diffusion coefficients of nickel ions in sulphate and chloride solutions at specified conditions and shows that the limiting current density at a cathode in a chloride bath is approximately twice that in a sulphate bath, other factors being equal. Earlier, Wesley et al.[2] had calculated the limiting current densities in 1M $NiSO_4$ solution and 1M $NiCl_2$ solution and obtained similar results.

Boric acid is used as a buffering agent in Watts nickel solution in order to maintain the pH of the cathode at a predetermined value. Boric acid solutions of the strength used in Watts nickel solutions have a pH of about 4·0 when interfering factors are taken into consideration. From this, it would appear that boric acid should be most suitable as a buffer at about pH 4, which is rather convenient, since most nickel solutions are operated near this value. However, it is satisfactory over the range of pH 3 to 5, probably due to the formation of complexes of boric acid and nickel. The buffer action of boric acid is particularly important in solutions of low pH (high activity of hydrogen ions), since hydrogen discharge occurs and consequently the pH increases in the cathode film with the possibility of co-deposition of nickel hydroxide. Other buffers such as acetate and formate can be used successfully, particularly at the lower pH values.

Cathode efficiency of nickel deposition

The standard electrode potentials $E^0_{Ni^{2+}/Ni} = -0.25$ V and $E^0_{H^+/\frac{1}{2}H_2} = 0.00$ V indicate that thermodynamically hydrogen discharge should take place in preference to nickel ion discharge when the ions are present at unit activity. However, the cathode efficiency for nickel deposition from a Watts bath is $\approx 95\%$, and this is due to the much higher activity of the nickel ion (≈ 1 g ion/l) compared to that of the hydrogen ion (10^{-3}–10^{-6} g ion/l corresponding to pH 3–6), which affects the reversible potentials and the rate of diffusion of the two species into the cathode layer. In addition, account must be taken of the respective overpotentials.

Saubestre[1] estimated values for deposition potentials that are in good agreement with actual values by taking into account activities of the discharging species and their overpotentials. On the basis of these factors it can be shown that cathode efficiency increases with increase in activity of nickel ions, pH, temperature and current density. Saubestre has also shown that a mixed sulphate/chloride solution gives a similar cathode efficiency to a sulphate solution.

Hard Watts bath*

This is mainly used for engineering purposes. The increase in hardness is achieved at the expense of other properties. The incorporation of ammonium ions or organic additions in the plating solution results in modification of structure and certain properties such as ductility are adversely affected. On the other hand, this solution does provide a means of producing a hard deposit when this is the most important feature required for a particular purpose.

Nickel sulphate bath

A simple solution of nickel sulphate in water has little commercial application, but sound deposits can be produced at a reasonable efficiency if boric acid is added as a buffer. However, for commercial applications, the sulphate plus chloride solution of the Watts type is superior in performance, except with inert anodes (see page 68).

All-chloride and high-chloride baths

Watts type solutions having chloride/sulphate ratios higher than that in the conventional Watts bath, most often with the proportions almost reversed, are used for certain purposes such as high-speed bright plating in vats and in barrel-plating operations (Table 4.2). In the latter case, the higher conductivity of the solution is its most important feature. In the extreme case,

Table 4.2 CONCENTRATION RANGES OF HIGH- AND ALL-CHLORIDE BATHS

Chemical	Concentration range (g/l)
Nickel chloride, $NiCl_2 \cdot 6H_2O$	150 to 300
Nickel sulphate, $NiSO_4 \cdot 6H_2O$	0 to 200
Boric acid, H_3BO_3	20 to 50

Note. Total concentration of nickel salts does not often exceed 300 g/l

only chloride is present and this type of bath is used for heavy nickel plating. All the solutions in this category have good conductivity, good throwing power, and high current densities are permissible[2]. The deposits obtained from them have a higher tensile stress and lower ductility than those obtained from Watts solutions, Kendrick[3] and Crossley, Kendrick and Mitchell[4] have investigated the effect of plating from an all-chloride nickel bath using square-wave a.c. At high frequencies (400 Hz or above) and with relatively

---

* Compositions of baths are given in Chapter 5.

large amounts of deplating during each cycle, matt, soft, ductile deposits are produced having low internal stresses. At low frequencies (100 Hz or below) and with only small amounts of deplating during each cycle, lustrous, hard, brittle highly-stressed deposits are produced. At the present time this development has no commercial application, but it does provide another variable which may yet prove useful in controlling the properties of electrodeposits.

Nickel sulphamate bath

Small volumes of plating solutions based on nickel salts other than nickel sulphate and nickel chloride are used for certain purposes and the commonest of these is the nickel sulphamate [$Ni(NH_2SO_3)_2$] solution. The advantages of this solution are the high rates of deposition possible and the low stress in the deposit. It is a more expensive plating solution than sulphate, but is used where the aforementioned properties are important, for purposes such as electrotypes in the printing industry and for making gramophone record stampers. In instances such as these the low stress is essential to prevent distortion. The throwing power of sulphamate solutions is rather better than that of sulphate solutions but the reported results vary somewhat.

The process is mainly used for heavy nickel deposition and electroforming, but these aspects will be discussed more fully in the next chapter. The essential features of the solution are similar to those for nickel sulphate, the pH of the bath usually being between 3·0 and 5·0 and most often between 3·5 and 4·5, with boric acid being a satisfactory buffer when used at about the same concentration as in the sulphate solution. Sulphamate solutions tend to hydrolyse at elevated temperatures, but if their pH is kept above 4, this occurs only at a very slow rate. Nickel chloride is present to assist anode corrosion and its concentration is rather more critical than in the sulphate solution. The necessity for the presence of chloride ions has been a point of controversy, but most formulations include a minimum of 5 g/l nickel chloride as this quantity appears essential for good anode efficiency except when using nickel anodes containing some sulphur. An excessive amount of chloride should be avoided because this increases the magnitude of tensile stress in the deposit. Bromide additions can be used as alternative anode depolarisers and if nickel bromide is used at the same concentration as nickel chloride, it produces only about two thirds of the stress produced by nickel chloride, i.e. 80 N/mm^2 instead of 125 N/mm^2 at 50°C and 5·5 A/dm^2.

As the internal stress in a deposit from a sulphamate solution is less tensile than that in a deposit from a sulphate solution, claims have been made that it is far superior for repairing worn components where good fatigue properties are required. This aspect is frequently over-emphasised, as will be discussed in the next Chapter. The low tensile stress can be improved marginally by addition of a stress reducer to the solution so that the electrodeposit has a compressive internal stress. Sodium naphthalene 1,3,6-trisulphonic acid is often used for this purpose. A recent development by Kendrick[5] involves the use of a more concentrated nickel sulphamate solution (600 g/l). This enables much higher current densities to be used, e.g. at 70°C satisfactory deposits can be obtained at 85 A/dm^2. At this concentration of 600 g/l of nickel sulphamate, the internal stress is a minimum

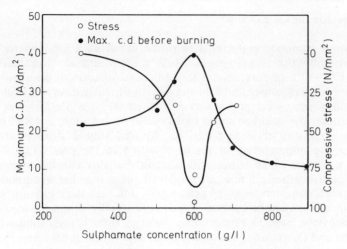

Figure 4.1. Effect of concentration of nickel sulphamate on maximum current density and stress in deposits 25 μm thick plated at 5·4 A/dm², 60°C (after Kendrick[5])

at 5·5 A/dm², and at the same concentration the maximum current density that can be utilised before burning occurs is also a maximum (Fig. 4.1).

Nickel fluoborate bath

Nickel fluoborate[6] is the basis of the only other bath of significant commercial importance. In addition to nickel fluoborate this solution contains boric acid and chloride ions. The solution is intrinsically highly buffered and so pH changes on plating are very small, the usual operating value chosen being pH 3. The deposit has good ductility and low internal stress[7]. This bath has certain specialised uses for electroforming, particularly for the production of electrotypes. The solution is easy to operate; it has high conductivity, good anode corrosion characteristics and tolerates relatively high metallic contamination, but is much more expensive than a Watts solution.

ANODE PROCESSES

Nickel dissolution is the main anode process in commercial practice but gas liberation (normally oxygen) occurs if the anode efficiency is low or if inert anodes are employed for special purposes. Gaseous chlorine is unlikely to be liberated under normal operation, since the small quantities discharged will dissolve in water forming hypochlorous acid. Chlorine can have deleterious effects in certain bright-nickel plating solutions since chlorination of some organic brighteners can occur.

Anodes for nickel plating

The purity of anode material is of prime importance for modern plating requirements; the nickel (plus cobalt) content should be at least 99% (BS 588: 1970[8]). Impure anodes lead to contamination of the solution and inferior physical properties of the deposit. In the early days of nickel plating technology the nickel purity was as low as 90%; while this was hardly adequate for the cold low speed baths in use at that time, it would be completely useless in modern solutions. Anodes should dissolve smoothly without the undercutting of grains resulting in the production of small nickel particles, which is not only wasteful but can also have detrimental effects on the cathode. If fines are produced, these, together with anode stubs, constitute a considerable scrap production, although the anode stubs can be used up in inert anode baskets. Anodes are usually contained in cotton or polypropylene bags to prevent fine particles from being dispersed in the solution and thus causing roughness by incorporation in the cathode surface.

Depolarised anodes

Depolarised anodes were the first type to be deliberately developed to have improved dissolution characteristics; these anodes are rolled and consist of 99% purity nickel together with 0.5% of nickel oxide[9]. They have a fine grain size and the nickel oxide is segregated at the grain boundaries. This type of anode is suitable for use in Watts-type solutions, particularly those without brightener additions. Dissolution occurs smoothly and a brownish film forms on the surface, but this is not of the same type as that which forms on cast carbon-containing anodes; it does not discourage the formation of fine material and bags are therefore essential.

Carbon-containing anodes

Cast and rolled nickel anodes (99%) containing a controlled amount of carbon and silicon (0.25% of both) are now in common use. These additions reduce the amount of fine material formed and increase the electrochemical reactivity of the nickel so as to prevent anodic passivation. The carbon-containing anodes are particularly suitable for use in nickel baths containing organic brighteners, as these produce a most uneven dissolution of depolarised anodes, which are now mainly used only for 'dull' plating. A carbon-silica sheath forms as the anode dissolves; this is porous but adheres tenaciously to the surface so that dissolution takes place through the film and formation of fines is reduced. In commercial electroplating, this type of anode is generally contained in a bag, since even a small number of particles in the bath is undesirable.

Electrolytic nickel

Electrolytic nickel[10-13] has been used to a certain extent for a number of years but its dissolution characteristics are not particularly desirable due to its fairly high purity (99·9%). In order to obtain reasonably satisfactory anodic dissolution the pH of the solution must be less than 4·5 and it must contain at least 6 g/l of chloride ion; even then the anode becomes spongy and nickel particles fall out. Nevertheless, pure electrolytic nickel is functioning satisfactorily in numerous commercial installations. However, its anodic behaviour can be improved by incorporating a small amount of sulphur in the electrolytic nickel. A sulphur-containing organic compound is added to the bath at the end of the refining cycle so that a controlled amount of sulphur (0·02%), usually in the form of sulphides, is incorporated in the final product. The incentive to make use of electrolytic nickel has always existed due to its cost advantage over the more expensive cast or rolled anodes. The present technique of incorporating sulphur does not add greatly to the manufacturing costs, as it involves only a minor modification of one of the final stages in the process. This sulphur-bearing material dissolves smoothly and at a uniform rate in modern electroplating solutions independently of their chloride content; it can be obtained either in sheet form or as small rectangles or discs. Di Bari[14] states that a small quantity (0·1%) of fine powder is produced during their dissolution but this is non-metallic nickel sulphide.

Di Bari[14] also investigated the effect of sulphur and other additives on the electrochemical reactivity and type of corrosion of nickel anodes. Using

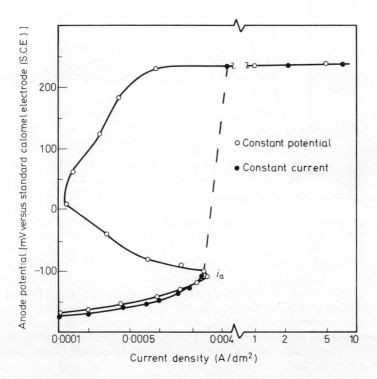

Figure 4.2. Anode potential/current density curves of regular electrolytic nickel obtained by two methods in a Watts bath containing 14 g/l of chloride maintained at pH 4 and 55°C (after Di Bari[14])

constant current and constant potential techniques he obtained curves relating anode potential with current density. However, the constant-potential technique is more useful since it provides a curve with a well defined transition from the active to the passive state. Fig. 4.2 shows a typical curve for commercial grade electrolytic nickel in a Watts bath. The point i_a, the critical current density, is a criterion of anode activity, since it defines the current density above which the anode becomes passive. If this occurs at a low value, the material is not suitable for use as commercial anodes. Di Bari found that the point i_a is the limit of smooth uniform dissolution, and if an anode is to corrode in a satisfactory manner in service, its value must be greater than the anode current density used in commercial plating practice. Although the presence of chloride in the bath breaks down the passive region, it does not eliminate the active to passive transition, as shown in the potentiostatic E versus I diagram. Di Bari correlates this behaviour to the pitting of stainless steel and suggests that an oxide film is

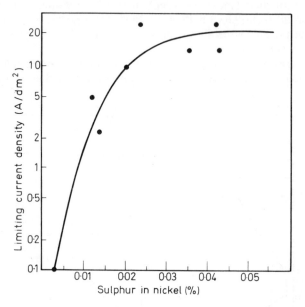

Figure 4.3. Approximate relationship between limiting anode current density and the sulphur content of nickel anodes in nickel sulphate baths (low chloride, less than 0·5 g/l) at 55°C (after Di Bari[14])

formed which results in the spongy nature of commercial electrolytic nickel during anode corrosion. The incorporation of sulphur in electrolytic nickel improves its performance as an anode because it increases the anode limiting current density and dissolution takes place uniformly at 100% anode efficiency. The relationship between anode limiting current density and % sulphur is illustrated in Fig. 4.3.

Sulphur has been found to be the only addition of practical value. With other additions investigated, problems arose, such as large wastage of metal in the form of fines, low anode efficiency, change of bath pH and incorporation of additive from the anode in the deposit.

Anode baskets

Small pieces of nickel can only be used as an anode if retained in inert containers to facilitate dissolution. Various plastic or plastic-covered metal containers have been used, but these have a relatively short life due to mechanical damage inflicted in service and of course need some current carrier to effect electrical connection, usually an inserted nickel anode. With the advent of a cheaper and ready supply of titanium, this has been used to a considerable extent in the plating industry for anode baskets, anode hooks and clamps; it also serves as an insoluble conductor to the nickel metal which still performs as a soluble anode. Anode hooks manufactured in titanium are tapped into conventional anodes in the usual manner but have an advantage in that the hooks do not dissolve even if they become submerged. Obviously, nickel hooks do corrode in such circumstances, thus allowing anodes to fall to the bottom of the vat. Titanium clamps are used to hold electrolytic sheet and as in the case of titanium hooks the clamp itself is never attacked so that there is little tendency for the sheet to fall out of the clamp, although periodic inspection is advisable.

Titanium is an ideal material for the construction of the aforementioned components[15] since it has an adherent oxide film. Its low conductivity, i.e. high resistivity (48 $\mu\Omega$ cm compared with nickel's 7 $\mu\Omega$ cm and copper's 1·7 $\mu\Omega$ cm, at 20°C) has not proved too disadvantageous, although larger cross-sections of titanium compared to copper must be used to carry a given current. The oxide film is not destroyed when titanium is used under anodic conditions provided that the applied voltage does not exceed a certain limiting value, which depends to some extent on the chloride content of the bath. However, the use of titanium is not advised in sulphamate or fluoborate baths. In all-chloride solutions titanium will corrode at potentials of approximately 12 V, while in all-sulphate baths the metal forms an anodic oxide film up to much higher voltages, when spark breakdown occurs. Fortunately, where the sulphate/chloride ion ratio exceeds 3:1, as in the Watts bath, the oxide film on titanium does not break down until a high voltage is reached[16]. The oxide film is mechanically weak and is readily fractured by the pressure resulting from the weight of the nickel contained in the basket[17], thus permitting a flow of current to the nickel. Primary forms of nickel, other than electrolytic pieces, are used in titanium baskets in large quantities at the present. Nickel shot or pellets produced by the carbonyl process are frequently employed; problems due to the outer skins being left undissolved have been encountered but have been largely overcome by experience and the realisation that a high chloride ion content appears essential. Carbonyl shot containing sulphur has been investigated for use as anode material; it is claimed that this has improved dissolution characteristics compared with the conventional type.

Basket design is most important[18]; a typical one is shown in Fig. 4.4. These should be constructed to avoid 'bridging' so that the formation of large voids is prevented, good contact maintained and the basket kept full of nickel. Well designed baskets, if properly maintained, ensure that the anode/cathode ratio is constant throughout the whole period of operation and that the current densities and plate distribution on the cathodes remain constant. Conventional anodes decrease in size and become tapered in use so that it is difficult to maintain a constant anode/cathode ratio. Since baskets are constantly replenished with nickel, the accounting and costing of particular

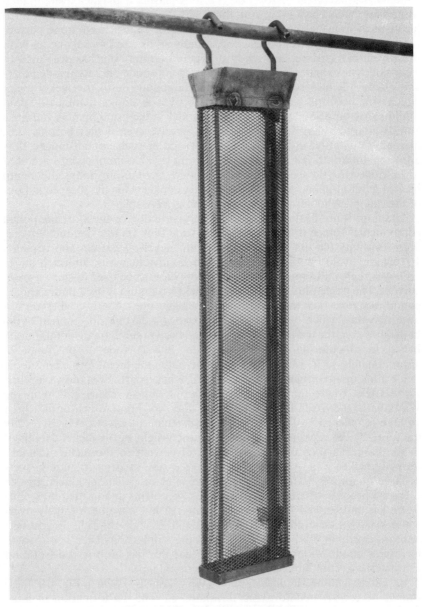

Figure 4.4. Typical titanium anode basket

jobs is simplified, whereas with conventional anodes large periodic expenses are incurred when replacing a number of anodes. Lack of dissolution of the fine material produced in baskets has proved troublesome in some instances, but this has been overcome and many large automatic and manual plants now use primary nickel as anode material. The whole basket is contained in a bag. Since the bag is usually in service for a long time without being removed, the process of 'double-bagging' is frequently employed. in which a cotton bag is slipped over a tightly fitting polypropylene bag. However, in many cases either material by itself has proved to be adequate.

Inert anodes

In certain circumstances some insoluble anodes are employed in solutions which are otherwise kept in balance by soluble anodes. This need arises if the anode efficiency is greater than the cathode efficiency or if auxiliary anodes are required to achieve uniform thickness of coating on a complex object. Auxiliary anodes may be soluble, in which case a special type of nickel containing both carbon and silicon is sometimes used[19]. If high purity nickel is used for this purpose, even dissolution is said to be assisted by the superimposition of a.c. current[20]. Bouckley and Watson[21] have discussed the problems that arise when using inert auxiliary anodes if a large proportion of the surface area of a component is plated by means of the auxiliary anodes. They describe the successful use of bags of sulphur-containing electrolytic nickel for use as internal anodes to plate the inside of kettles. The cost of consumable anodes is high if they are of complex shape, since inevitably their useful life is quite short. When assessing the economic advantages to be achieved by using conforming auxiliary anodes to plate articles of complex shape with a uniform coating, it is essential to take into account the extra cost of auxiliary anodes (either inert or soluble), the cost of adding salts if inert anodes are used, the problems associated with the use of inert anodes and the time taken to place auxiliary anodes in position. The main advantages are the reduced plating time to obtain a particular thickness in a recess and the saving of electrodeposited metal. The latter is particularly important if the metal is either expensive or in short supply, as has occurred in the case of nickel from time to time.

Inert anodes are frequently made from platinised titanium, which is titanium coated with a thin, not necessarily continuous, layer of electrodeposited platinum approximately 2·5 μm thick. Platinised titanium functions essentially as a platinum electrode and in nickel plating solutions has a reasonable working life. In acid solutions, titanium forms a protective oxide coating on being anodically polarised. The main anode process is the evolution of oxygen, but slight dissolution of titanium occurs concurrently. In many plating solutions the rate of dissolution of metal is negligible, but in certain baths it can be appreciably faster and can result in the quite rapid deterioration of platinised titanium anodes. Dissolution of titanium leads to undermining of the platinum and subsequent flaking off. Warne and Hayfield[22] have studied the behaviour of platinised titanium in several nickel plating solutions, including two commercial bright plating baths, which result in quite rapid dissolution of titanium. In most cases failure is definitely caused by the dissolution of titanium and the undermining of platinum, and not by the poor adhesion of platinum to the titanium substrate. On the other hand, certain organic compounds, possibly sulphur-containing first-class brighteners, form complexes with platinum during anodic polarisation, and thus cause its dissolution to occur at an accelerated rate. However, in spite of the shorter working life of platinised titanium in bright nickel baths it is still economically advantageous to use it for the construction of auxiliary anodes, and it is, at the present time, widely used for this purpose.

Where insoluble anodes are used to plate only specific parts of an article, the reduction in the concentration of nickel ions in the solution will not be rapid. On the other hand, if the majority of the anode area is inert, provision must be made for fairly frequent additions of nickel salts to the solution in order to maintain the correct concentration. In bright plating solutions, inert anodes can lead to rather special conditions as far as organic brightener stability is concerned. If chlorine is liberated, as will occur unless the chloride ion concentration is less than 25% of that of the sulphate ion, conditions conducive to chlorination of some organic compounds are established. Accordingly, the type and concentration of brightener must be selected carefully when there is a large through-put of work involving the use of inert anodes. Chlorine liberation can also cause anode bags to rot[12], resulting in the inevitable release of particles into the solution. Therefore, when it is essential to employ only insoluble anodes of lead, as is the case for some specialised engineering applications, baths entirely devoid of chloride must be used. As stated previously, these baths contain nickel sulphate and boric acid, possibly plus a little sodium sulphate. They are operated in much the same manner as are Watts baths, except for their much lower pH, which is maintained at values between 1.5 and 2.5, and the unusual techniques that are required to replenish their nickel content. Details of one method which uses freshly precipitated nickel hydroxide are given by Hothersall and Gardam[23]. Wesley and his co-workers[24] describe an electrolytic regeneration cell, which has a high anodic and a low cathodic current efficiency.

Black nickel

No description of the field of nickel plating would be considered complete if it did not include a section on black nickel plating. Nevertheless, the use made by industry of this process is very limited. Obviously, black as a relief to the shiny blue appearance of standard nickel plus chromium plate can have an aesthetic attraction, for example on business machines. It also has certain technical merits particularly for scientific and photographic equipment. Unfortunately, black nickel coatings have little abrasion or corrosion resistance. For this reason, they are usually deposited over an undercoat of dull or bright nickel, but on ferrous substrates sometimes over zinc or cadmium plate, and subsequently dipped in oil, wax or lacquer. If the latter topcoat is used for the best corrosion protection, little advantage is gained compared with the use of nickel plate plus a black-pigmented lacquer.

Fishlock[25] has discussed the production of black nickel electrodeposits and the mechanism responsible for their formation. Two types of bath are used for this process, one based on nickel sulphate[26] and the other on chloride[27]. Both contain large quantities of zinc and thiocyanate ions, as will be seen from Table 4.3. It is these additives that are responsible for the black colouration, for the deposit contains large quantities of zinc and

Table 4.3 BLACK NICKEL PLATING PROCESSES

	A	B
Nickel sulphate, $NiSO_4 \cdot 6H_2O$	100 g/l	–
Nickel chloride, $NiCl_2 \cdot 6H_2O$	–	75 g/l
Ammonium sulphate, $(NH_4)_2SO_4$	15 g/l	–
Ammonium chloride, NH_4Cl	–	30 g/l
Zinc sulphate, $ZnSO_4$	22 g/l	–
Zinc chloride, $ZnCl_2$	–	30 g/l
Sodium thiocyanate, NaCNS	15 g/l	15 g/l
pH	5·5–6·0	3·5–5·5
Temperature of bath	26–32°C	Room
Cathode current density	0·2 A/dm^2	0·15 A/dm^2

sulphur. The sulphate bath (*A*) was the first to be introduced, but requires very close control of the bath pH and temperature in order to obtain a constant colour. Even then, the normal variance of current density over a cathode can lead to lack of colour uniformity. This makes the process difficult to operate on a mass production basis, where colour matching is important. The more recent bath (*B*) based on nickel chloride is claimed to be much more tolerant to changes in concentration of its ingredients and acidity. Variation in current density is also said to have far less effect than in the sulphate bath. A comparative study of these two baths and a similar black plating solution based on nickel sulphamate has been published by Shenoi and Indira[28].

It will be noted that the current densities employed in both baths are much lower than those used for normal dull or bright nickel plating. Thus only a thin deposit is produced in reasonable periods of time, usually 30 min being recommended, but this is no major disadvantage since thicker coatings from these baths are liable to spall if damaged, because of their low ductility and adherence. It is said to be possible to deposit black nickel coatings using barrel plating techniques.

Barrel nickel plating

The plating of small articles in bulk is most economically performed by placing them as a mass inside a non-conducting and revolving container, in which some form of electrical contacts are present[29–31]. As the container is usually based on a cylindrical shape, it is commonly termed a *plating barrel*. The main advantage of barrel plating is that the cost of jigging or wiring of each individual item is thereby avoided. An additional benefit is that each article should have almost the same thickness and distribution of nickel plate as any other. A small increment in thickness over the average can thus be allowed, based on statistical calculations, to ensure that all the articles comply with the minimum specified, with much greater certainty than with conventional vat plating[32]. As the articles are continuously moving in the

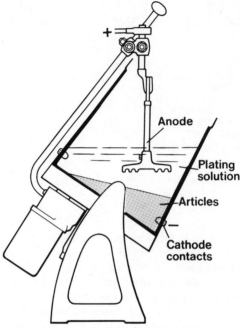

Figure 4.5. Schematic diagram of an open-ended barrel (alternative names are oblique or inclined barrel)

barrel there are no small bare patches of substrate, where electrical contact would otherwise have to be made, i.e. the wire or jig 'marks' which are unavoidable on vat plated work.

Articles to be barrel plated must not be too heavy, certainly not more than 500 g, otherwise as they tumble over each other they may 'peck' or

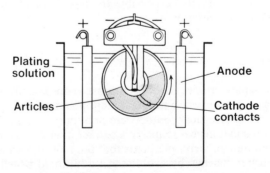

Figure 4.6. Totally-immersed horizontal barrel (a partially-immersed barrel is less than half immersed)

otherwise damage themselves. They must be of the type that roll freely and do not stick to each other or to the sides of the barrel; also they must not 'lock' or 'nest' together. Providing that their shapes and weight comply with these limitations, very many items can be successfully barrel nickel plated. It has been found that the definition that they should be of 'small' size can be quite loosely applied; articles having a diameter greater than 100 mm and even longer rods have been satisfactorily plated. It is frequently thought that only thin nickel deposits of unpredictable thickness can be applied in barrels, but under accurately calculated and properly controlled conditions, plating to specifications is readily achieved[33].

The older type of barrel is that known as the open-ended oblique or inclined type (Fig. 4.5). This open-ended barrel has solid walls and acts as both the container for the components to be plated and the plating solution which are removed by tilting the barrel. While being used for plating, it revolves at an angle of 35°–45° to the vertical on its own axis, with a cast nickel disc anode suspended in the solution above the work, which makes electrical contact through stainless steel studs inserted in the base and/or at the bottom of the sides. Although this type of barrel has not the output of the more modern type, it is still used for pointed articles or those having dimensions such that they would pass through the perforations or holes of the immersed type.

These more modern barrels are usually cylinders or hexagons operated in a horizontal position. Most commonly they are constructed from plastics, often Perspex or polypropylene, with holes drilled in them (Fig. 4.6). They are partially or fully immersed in a vat containing conventional anodes, placed along two sides of the tank. Current is fed to the work via the so-called 'danglers', which are insulated leads having bare knobs at their ends, or from cones, stars or discs placed at each end of the barrel. Since the plating solutions can be kept hot, and the anode area is much greater, higher cathode current densities can be used than in the open ended barrel. These are limited by the resistance offered by the insulating barrel walls, which necessitates the use of a high-voltage current source of 12–20 V compared with the 5–8 V normally adequate for standard vat plating. The flow of the solution is also restricted by the number and size of the perforations, and to ensure as free a passage as possible for both current and electrolyte solution, without seriously weakening the barrel walls, many ingenious modifications have been made to the basic idea of this immersed plating barrel[34]. In its various forms, this type of barrel is by far the most common now employed and is almost universal on automatic plating plants.

One different type of barrel, which is suitable for manual or automatic plants, is intermediate between the two main kinds being perforated but also open-ended, and so it has to be inclined and is only moved to a horizontal position when being transferred from tank to tank. This inclined barrel is immersed in a conventional type of plating vat and external anodes are required. Another horizontal type has no lid being rocked not revolved.

Barrel nickel plating, as first conceived, was not only a convenient means of dealing with a multitude of small parts but also had the ability to produce plate having some lustre from solutions normally depositing only dull nickel. This brightness is due to the self-burnishing action of the parts as they tumble in the barrel. This movement results in each very thin increment of nickel plate being polished before the next layer is deposited. In order that this burnishing action be appreciable, it is essential that sufficient time per unit thickness be allowed. This necessitates the use of a low current density (0·3 A/dm² or less) even for immersed barrel plating. If appreciable thicknesses of nickel are required, long plating times are obviously needed. Electrolyte solutions employed for this are based on the Watts bath but with somewhat higher pH values than normally used for vat plating, as will be seen from Table 4.4. The inorganic additions are claimed to provide improved lustre and also extra conductivity. Even then the voltage applied

Table 4.4 TYPICAL DULL NICKEL BARREL-PLATING SOLUTIONS

(a) Nickel sulphate, $NiSO_4 \cdot 6H_2O$	150 g/l
Ammonium chloride, NH_4Cl	30 g/l
Boric acid, H_3BO_3	30 g/l
pH	5·0–5·6
Temperature of solution	Room ($\not< 15°C$)
(b) Nickel sulphate, $NiSO_4 \cdot 6H_2O$	250 g/l
Nickel chloride, $NiCl_2 \cdot 6H_2O$	50 g/l
Boric acid, H_3BO_3	40 g/l
Magnesium sulphate, $MgSO_4 \cdot 7H_2O$	180 g/l
or	
Sodium sulphate, Na_2SO_4	50 g/l
pH	5·2–5·8
Temperature of solution	40°–55°C

will have to be between 6 and 10 V, although only low current densities are used.

Barrel plating now commonly utilises true bright nickel deposition. Since the brightness of the deposit is not only much greater but also no longer depends on sufficient time being allowed for burnishing to take place, the current density can be increased to approximately 1 A/dm² with a corresponding reduction in plating time. The base solutions used for this are very similar to those used for bright vat plating, as given in Tables 4.1 and 4.2. The Watts bath is often used as the base solution, particularly for open-ended barrels, while the high-chloride solution is frequently preferred for immersed barrels because of its greater conductivity. Wallbank[35] has discussed their relative merits. In both cases the brightener systems used are much the same as those found satisfactory for bright nickel deposition in ordinary vats. Nevertheless, although the combinations may be the same, the proportions of the different organic compounds may vary. Often it is found best to increase the concentration of the brightener of the first class, especially if it also is a stress-reliever, and decrease the content of the brightener of the second class, i.e. the leveller (see Chapter 6). This is to decrease the ever-present tendency of the barrel nickel plate to exfoliate or flake-off in a manner often referred to as *tinselling*. This can happen with dull barrel nickel plate but bright nickel is much more prone to it because the brightener of the second class adsorbs onto the fresh nickel surface and passivates it, thus preventing satisfactory adherence of the next layer. If this layer is brittle and tensile stressed, it can crack and flake off spontaneously.

Sometimes if it is rubbed either by the burnishing action in the barrel or as an inspection test after plating, myriads of tiny particles of nickel are detached and have the appearance of tinsel. For this reason, an excess of brighteners, particularly of the second class, must be avoided even more carefully than when vat plating. If drip feeding of these is impossible, then small and frequent additions are necessary. Experience has shown that the best results are obtained when additions are made between every barrel load being plated. To obtain the optimum utilisation of bright barrel nickel plating solutions, they should be heated to the same temperatures as for vat plating. Although the average current density is quite low, the actual current density on the articles that are both in contact and opposite the anode at any one time is much higher. On occasions this can lead to the shape of the perforations being imprinted on the nickel deposit, if correct operating conditions, including speed of barrel movement, are not employed. While continuous filtration is not essential for successful barrel nickel plating, periodic filtration is advisable and so are purification treatments for the removal of any impurities that may accumulate.

Subsequent barrel chromium plating is frequently performed on parts that have been barrel nickel plated. This is done in one of two types of barrels—either inclined and open-ended but perforated, i.e. the 'intermediate' type, or the horizontal partially immersed kind[36]. The latter can have either solid or perforated walls; those with solid walls can be constructed to give a continuous output, in which case the interior of the cylinder is fitted with a helical track giving a continual feed-through of articles. The usual material of construction is steel, with the stainless variety being preferred, the helix being of suitable non-conducting and chemically-resisting plastics. In all cases, the chromic acid bath contains a high proportion of silicate or fluosilicate ions and little or no sulphate ions, in order to avoid the 'milky' appearance that otherwise occurs when chromium is repetitively deposited. These barrels always have central internal anodes and also external ones in the case of the 'intermediate' type, lead-tin alloy being normally used.

It is difficult to apply more than very thin coatings of chromium by this method. To ensure that these coatings adequately cover and adhere to the nickel plate, if a delay between barrel nickel and chromium plating is unavoidable, this must either be kept in an active condition by immersing the nickel plated parts in a weak sodium cyanide solution or else activated by immersion in 10% (v/v) sulphuric acid solution or other more vigorous dips, such as those described in Chapter 6.

REFERENCES
1. SAUBESTRE, E. B., *Plating*, **45**, 927 (1958)
2. WESLEY, W. A., SELLERS, W. W. and ROEHL, E. J., *Proc. Amer. Electroplaters' Soc.*, **36**, 79 (1949)
3. KENDRICK, R. J., *Trans. Inst. Metal Finishing*, **44**, 78 (1966)
4. CROSSLEY, J. A., KENDRICK, R. J. and MITCHELL, W. I., *Trans. Inst. Metal Finishing*, **45**, 58 (1967)
5. KENDRICK, R. J., *Trans. Inst. Metal Finishing*, **42**, 235 (1964)
6. LAINER, V. I. and PANCHENKO, I. I., *Russian Engineering Journal*, No. 5, 56 (1959). Translation of *Vestnik Mashinostroeniya*, No. 5, 65 (1959)
7. ROEHL, E. J. and WESLEY, W. A., *Plating*, **37**, 142 (1950)
8. *Nickel Anodes for Electroplating*, BS 558: 1970

9. WESLEY, W. A., *Trans. Inst. Metal Finishing,* **33**, 1 (1956)
10. SELLERS, W. W. and CARLIN, F. X., *Plating,* **52**, 215 (1965)
11. DI BARI, G. A. and PETROCELLI, J. V., *J. Electrochem. Soc.,* **112**, 99 (1965)
12. SELLERS, W. W., *Electroplating and Metal Finishing,* **17**, 415 (1964)
13. CHATTERJEE, A. N. and RAY, S. K., *Electroplating and Metal Finishing,* **20**, 244 (1967)
14. DI BARI, G. A., *Plating,* **53**, 1440 (1966)
15. COULSON, I. B., *Electroplating and Metal Finishing,* **17**, 418 (1964)
16. DUGDALE, I. and COTTON, J. B., *Corrosion Science,* **4**, 397 (1964)
17. WARNE, M. A. and MOORE, D. C., *Electroplating and Metal Finishing,* **18**, 224 (1965)
18. WATSON, S. A., *Metal Finishing Journal,* **18**, No. 205, 36 (1972)
19. INTERNATIONAL NICKEL CO. LTD., U.S. Pat. 3 449 224 (10.6.69)
20. INTERNATIONAL NICKEL CO. LTD., U.K. Pat. 1 059 899 (22.2.67)
21. BOUCKLEY, D. and WATSON, S. A., *Electroplating and Metal Finishing,* **20**, 303 (1967)
22. WARNE, M. A. and HAYFIELD, P. C. S., *Trans. Inst. Metal Finishing,* **45**, 83 (1967)
23. HOTHERSALL, A. W. and GARDAM, G. E., *J. Electrodepositors' Tech. Soc.,* **27**, 181 (1951)
24. WESLEY, W. A., CARR, D. S. and ROEHL, E. J., *Plating,* **38**, 1243 (1951)
25. FISHLOCK, D., *Metal Colouring,* Robert Draper Ltd., Teddington (1962)
26. POOR, J. G., *Metal Finishing,* **41**, 694, 769 (1943)
27. ANON, *Metal Finishing Journal,* **4**, 436 (1958)
28. SHENOI, B. A. and INDIRA, K. S., *Metal Finishing,* **61**, 65 (1963)
29. OLLARD, E. A. and SMITH, E. B., *Handbook of Industrial Electroplating,* 3rd edn. Butterworths, London (1964)
30. LA MANNA, F. J., Chapter in *Metal Finishing Guidebook,* Metal and Plastics Publications Inc., Westwood, 434 (1971)
31. ALLEN, R. (Editor), *Canning Handbook on Electroplating,* 21st edition, W. Canning & Co. Ltd., Birmingham, 127 (1970)
32. WALLBANK, A. W. and LAYTON, D. N., *Trans. Inst. Metal Finishing,* **32**, 308 (1955)
33. *Electroplated Coatings on Threaded Components—Nickel or Nickel Plus Chromium on Steel, Copper or Copper Alloy Components,* BS 3382: 1965 (Parts 3 and 4)
34. JACKSON, W. H. and GRAHAM, A. K., Chapter 25 of *Electroplating Engineering Handbook,* 3rd edn, Graham, A. K. (Editor), Reinhold Publishing Corporation, New York (1971)
35. WALLBANK, A. W., Section in *Nickel Plating, Techniques and Applications,* 2nd edn., International Nickel Co., London (1967)
36. MORRISET, P., OSWALD, J. W., DRAPER, C. R. and PINNER, R., *Chromium Plating,* Robert Draper Ltd., Teddington, 444 (1954)

BIBLIOGRAPHY

BRUGGER, R., *Nickel Plating,* Robert Draper Ltd., Teddington (1970)
The INCO Guide to Nickel Plating, International Nickel, London (1972)

Chapter 5

Engineering Applications

APPLICATIONS OF THICK ELECTRODEPOSITS

The corrosion-resistant properties of nickel electrodeposits are often thought of as being of use only for protecting consumer items, large or small, where decorative embellishment is the most important factor. However, nickel plate has many applications in the engineering field where its functional behaviour, rather than its appearance, is the main criterion. When nickel is electroplated for this purpose, the coatings deposited are usually thicker than for decorative corrosion-protective uses, and so these are termed *heavy* nickel coatings, which may be arbitrarily defined as those greater than 50 µm thick. They were first used to reclaim components which had worn or corroded in service, or which had been incorrectly machined during manufacture. Nickel was used to build up either the whole or just the affected portion of the unserviceable article to a size greater than that actually required. The nickel coating was then machined so that the plated article had the desired dimensions. Turning, milling or grinding operations are all suitable for this. Clearly, the thickness of nickel required for this reclamation work depends on the depth of the damage to be repaired, but experience indicates that approximately 12·5 mm is probably the economic, if not the technically possible limit, although this limit must obviously depend on the value of the part being salvaged.

Heavy nickel coatings are now often applied to new iron or steel components to prevent their corroding or otherwise suffering damage caused by the normal wear and tear experienced in certain uses, the thickness deposited varying from 50 to 500 µm according to the service condition. Machining of such coatings is frequently not necessary. These nickel coatings prevent the basis metals from being corroded, and by preventing this attack they therefore reduce the danger of corrosion products of these substrates being produced which would contaminate materials being processed in various types of equipment. This ability to prevent metallic contamination, together with their non-toxicity, renders nickel electrodeposits ideal for food-handling plant. Other industries whose products must not be contaminated by metallic impurities, particularly iron, also make use of thick electrodeposits of nickel. Certain cylinders which are subject to wear have their service life greatly extended in this manner. Two other properties of nickel are also found of benefit for specialised purposes, i.e. its resistance to scaling at high temperatures and its good performance when used as a bearing surface in contact with bare steel, in conditions where fretting corrosion would otherwise be liable to occur. A further use of nickel in the engineering field is in the electroplating of steel sheet and pipes, which are available in a range of sizes. These partly-fabricated bimetallic parts complement steel plates that are produced in clad form by conventional mechanical methods with thicker coatings of nickel than are deposited by the electroplating technique.

For all these applications, good adhesion between coating and substrate is even more essential than for decorative plating, both because of the mechanical finishing operations often required on the nickel plated components and also because of the more demanding functions for which they will be used. Therefore, their preparation prior to plating must be very thorough, and usually for ferrous metals, after the conventional removal of soils, it is necessary to etch anodically in strong sulphuric acid solution (typically 50% w/w) instead of using a dilute acid dip (see Chapter 6). This anodic treatment at ≈ 22 A/dm^2 for times typically of 1 to 5 min obviously removes far more metal than can a short immersion in dilute acid. Thus the outer fragmented layer of metal, which often includes oxides and imbedded particles of other metals or abrasives, can be totally removed. This produces the maximum adhesion obtainable, with bond strengths greater than either the tensile strength of the substrate or nickel.

TYPES OF NICKEL PLATING SOLUTIONS USED

Watts bath

While the Watts nickel solution is by far the most popular one used for deposition of heavy nickel coatings, other baths are employed when their specialised advantages are desired. For example, the sulphamate solutions are becoming more widely used both where nickel plate having a low internal stress is required and where a high rate of deposition is desired.

A number of baths have been used for heavy nickel deposition; these are listed in Table 5.1. A range of concentration is given for most ingredients, since many are non-critical in content, provided that they do not depart too far from the limits stated.

In Table 5.2, the mechanical properties of the nickel deposits obtainable from them are summarised. It will be noted that the nickel plated from any type of solution can apparently have a fairly wide range of properties. This is due partly to the paucity of information that is available for some types of solution, but also because for well documented processes, such as the Watts and sulphamate baths, conflicting values have been reported by different authors. There is some possibility that erroneous results may have been obtained from impure solutions, but in most cases these were purified before testing. However, factors such as variations in solution composition, acidity, temperature and current density can all modify to some extent the mechanical properties of the plate deposited. Fig. 5.1, which is taken from the extensive work of Brenner and his colleagues[1], illustrates how these variables affect deposits from the Watts solutions. Changes in current

Table 5.1. COMPOSITION OF BATHS USED FOR NICKEL DEPOSITION

Type of bath	Constituents of solution (typical concentrations in g/l)					
	Nickel sulphate ($NiSO_4 \cdot 6H_2O$)	Nickel chloride ($NiCl_2 \cdot 6H_2O$)	Nickel sulphamate ($Ni(SO_3NH_2)_2 \cdot 4H_2O$)	Nickel fluoborate ($Ni(BF_4)_2$)	Boric acid (H_3BO_3)	Other
Watts	240–330	37–52	—	—	30–45	—
Hard Watts (containing ammonia)	180–230	—	—	—	30	Ammonium chloride 25
All chloride	—	250–300	—	—	25–30	—
Conventional sulphamate	—	0–15	300–450	—	30–45	—
Conventional sulphamate plus organic stress-reliever	—	30	300–450	—	30	Sodium naphthalene trisulphonate 7·5
Concentrated sulphamate	550–650	5–15	—	—	30–40	—
Fluoborate	—	—	—	300–450	22–37	—

Table 5.2. RANGE OF PHYSICAL PROPERTIES OF NICKEL ELECTRODEPOSITED FROM THE BATHS LISTED IN TABLE 5.1

Type of bath	Ultimate tensile strength (N/mm^2)	Yield strength (N/mm^2)	Elongation (%)	Hardness (HV)	Internal stress (N/mm^2)	Limiting current density before burning (A/dm^2)
Watts	380–450	220–280	20–30	150–200	140–170	20
Hard Watts (containing ammonium ions)	1000*	750*	5–8	350–500	280–340	20
All chloride	750–900	650*	8–13	200–250	280–340	40
Conventional sulphamate	500–800	500*	10–20	160–240	7–70	20
Conventional sulphamate plus organic stress-relievers	1500*	800–1000	2–5	400–600	−40 to +14	20
Concentrated sulphamate	750–1000	—	10–15	200–300	−100 to +140	40
Fluoborate	380–550	—	17–30	170–220	100–170	40

* Typical values

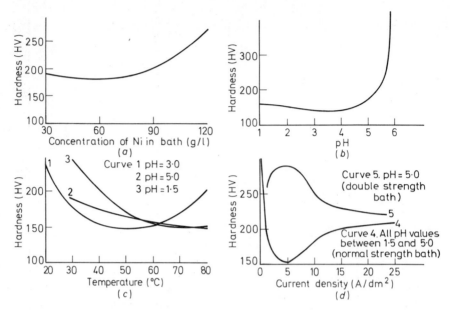

Figure 5.1. Effect of solution and operating conditions on the hardness of Watts nickel deposits. (a) Concentration of nickel ions, (b) pH of solution, (c) temperature of solution and (d) current density of deposition. Plating conditions: $5 A/dm^2$, $55°C$, unless otherwise indicated (after Brenner et al.[1])

density cause far greater alterations in the properties of deposits from the sulphamate bath as will be seen from Fig. 5.2.

With such a range of mechanical properties available from the one metal,

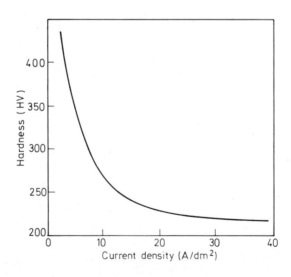

Figure 5.2. Effect of current density on the hardness of nickel deposited from a concentrated sulphamate solution (after Kendrick)

the engineer has a wide choice from which to make his selection. The Watts bath is so simple to control and maintain in its optimum condition that it is still predominant in the heavy nickel field, although deposits obtained from it are fairly soft and have a moderately high tensile stress. The stress is not usually deleterious since the ductility is so good. When harder plate is desired than that obtainable from a straight Watts bath, additions such as ammonium salts or organic compounds, such as saccharin or *p*-toluene sulphonamide, can be made to this bath; the latter compounds also reduce the tensile stress in the deposits. Unfortunately their ductility is lowered, for the usual inverse relationship between the tensile strength and ductility of metals holds good. While possibly the lower elongation may not be objectionable, the higher tensile stress caused by the ammonium ions limits the use of this solution more than those containing the organic compounds, their deposits having a low tensile or even compressive stress.

Sulphamate baths

The standard sulphamate solution[2] gives deposits having a lower stress than those from a Watts bath, but with somewhat lower ductility. The first property makes this process suitable for electroforming, but when harder and stronger deposits having little or no internal stress are required, compounds such as a naphthalene sulphonic acid are added to the solution. The sulphamate bath containing such an organic compound was at one time favoured for electroforming because deposits from it have a very low stress. There are some difficulties in controlling the exact stress obtained when using organic stress relievers, just as is the case when these are added to a Watts bath. This lack of precise control can cause difficulties in commercial electroforming, unless frequent stress measurements are performed. Also, the presence of small quantities of sulphur in the nickel plate renders it notch sensitive. For this reason, the high-density sulphamate bath was developed. Not only can this bath be operated at double the current density of a conventional sulphamate bath without the occurrence of burning, but it will also deposit harder deposits with little loss of ductility. Probably its most important property is its ability to deposit nickel with either zero or a compressive stress, without any organic chemical being added to the bath. Thus, it is free from the difficulties which attend the introduction of such a chemical. An inorganic compound, which acts as a stress reliever is, in fact, responsible for the compressive stress in the sulphamate plate, but this compound is formed *in situ* by the anodic behaviour of non-activated nickel at low current densities (0·5–1 A/dm^2). This electrolytic oxidation is done in a separate tank connected to the main plating vat. By varying the anodic current density, the deposit's hardness and internal stress can be adjusted to values whose limits are given in Table 5.2.

All-chloride bath

All-chloride solutions have some application for heavy nickel plating because it is possible to employ double the current density that can be used in a

Watts bath before burning results, and their throwing power is also superior. The deposits from such baths have a high tensile stress and low ductility.

Fluoborate bath

Fluoborate baths have not achieved much popularity for deposition of nickel, although the deposits from them are slightly less stressed than those from a Watts bath. The mechanical properties of the electroplate from fluoborate solutions are excelled by those from a sulphamate bath, and so, except for some uses in the printing industry, the former bath is but little used.

MECHANICAL PROPERTIES

The mechanical properties of the deposits that can be obtained from these various baths are summarised in Table 5.2. The limiting current density figures given in the Table are included since they have some relevance to the choice of a bath for application to any particular purpose. These values are only approximate, although relative, depending not only on the solution chosen but on the geometry of the cathode and the plating cell and to a large extent on the type and degree of agitation used; to obtain the figures quoted

Figure 5.3. A number of 'diamond-coated' tools fabricated by plating with nickel from baths in which diamond particles were suspended (courtesy Diagrit Electrometallics Ltd.)

violent air agitation would be essential. It must be emphasised that these are not average current densities as usually quoted; they are maxima on edges, etc.

It will be seen from the Table that gains in tensile strength and hardness are achieved only at the expense of ductility. While the former may sometimes be more important, it may be unwise to attempt to use deposits of greater hardness than 400 HV, for although even 600 HV is obtainable by addition of organic compounds to the bath, the nickel plate so produced is brittle. Although nickel plate of about 200 HV is softer than many steels onto which it is deposited, this is not the case for other basis metals which are suitable for certain engineering purposes, e.g. copper and its alloys or aluminium and its alloys. The mechanical properties shown in Table 5.2 are only obtainable provided the solutions are pure, since impurities such as those to be discussed in Chapter 7 decrease ductility and increase stress.

The prevention of pitting and roughness by the standard means of agitation and filtration are even more vital when heavy nickel deposition is being performed, for any tendency to produce these defects will be accentuated by the great thicknesses of plate applied. Wetting agents are often added to reduce the interfacial tension, these materials often being termed *anti-pit agents*. These are of benefit in preventing the formation of pits, which can result in either the ruining of a component which has been plated for many hours, or at least a greater allowance having to be made for the nickel to be machined off. When machining is to be done, it might be thought that rough deposits would not be detrimental, but this operation may tear out a nodule, leaving a hole.

COMPOSITE COATINGS

Although it might be considered that the range of properties available in nickel electrodeposited from the various aforementioned solutions would satisfy most purposes, the requirements for engineering materials are becoming more and more demanding. Therefore, efforts have been made to improve the wear resistance of nickel coatings, particularly at high temperatures, by including in them a second phase, usually particles having an abrasive or refractory nature, such as carbides or borides[3]. These composite coatings thus form a type of cermet.

The technique adopted to produce such coatings is to deposit nickel under conventional conditions from conventional plating baths in which are contained solids kept in suspension by some form of agitation. Mechanical stirring is most often used, but air agitation or continuous circulation via a pump are other methods that are sometimes employed. The solutions are usually of the Watts type, although the sulphamate and chloride baths have also been tried with success.

Diamond dust and grit were the first solid materials to be incorporated into a nickel electrodeposit, in order to manufacture diamond tools, and this technique is still used (Fig. 5.3). Of course, these coatings are rough and must be so in order to function correctly. For most other industrial purposes, smooth composite coatings are desired and so non-metallic particles of 1–12 μm size are generally used. If certain particles of this size are suspended in nickel plating baths, electrodeposits can be produced which are indistinguishable in appearance from conventional nickel plate, but can have very different mechanical properties.

These particles can be silicon, or tungsten or zirconium diboride for greater abrasion resistance, mica or graphite to form a self-lubricating surface or alumina or silica to act as dispersion-hardening materials. The proportion of incorporated particles in the nickel coatings can vary between 1 and 50% by volume, but is usually in the range of 5 to 30%. This content is achieved by adding the particulate matter to the nickel bath in concentrations between 25 and 100 g/l, more than this latter quantity being found to have little effect on the volume included in the nickel deposits. In fact, too large a content is said to produce highly tensile stressed electroplate. The most important factor which controls the extent of incorporation appears to be the degree of agitation. This must be vigorous enough to maintain the solids in suspension so that there is a sufficient number of them immediately adjacent to the cathode surface but not so violent as to displace the particles before they are trapped by inclusion in the growing electrodeposit. As already mentioned, the actual bath used seems to be of secondary importance as far as nickel coatings are concerned.

However, for other metal/non-metal combinations where the nature of the plating solution can be varied considerably, e.g. copper from acid or cyanide baths, this has not been found to be the case and so it has been postulated[4] that there is some attractive force holding the particles on the cathode and that these are not just included in the deposit by chance entrapment. This attractive force can be influenced by the interaction between the particle and the solution, and so unconventional nickel plating baths might give quite different results from those in common use.

Little data has been published on the mechanical properties of these so-called electrodeposited composite coatings. Most of the work done on them has been in the nature of direct comparisons against engineering components coated with some standard material and functioning either in their normal manner or else in some accelerated test. These comparisons have shown that the electrodeposited cermets containing abrasives have greater wear resistance than conventional nickel plate, particularly at temperatures greater than 300°C. The hardness of such combinations has been reported to be between 400 and 550 HV.

The dispersion-hardened nickel/alumina composites have been claimed to have lower tensile strengths than Watts nickel. Figures of 200[5] and 280[6] N/mm^2 have been quoted. The hardness of such combinations is said to be approximately 500 HV. Gillam, McVie and Phillips[7] have examined the structure of this type of deposit.

Electrodeposited nickel can also be reinforced by fibres incorporated into it by winding a continuous filament or yarn of them onto a slowly revolving mandrel onto which the nickel is being deposited[8,9]. Fibres of tungsten, stainless steel and carbon have been incorporated into electrodeposited nickel by this technique. The majority of the experimental work has been aimed at producing much stronger electroforms than otherwise obtainable. However, the production of such fibre-reinforced composites is still only in the development stage and many difficulties have been encountered in their manufacture, notably in obtaining good adhesion at the fibre-to-matrix interface and in the avoidance of gross porosity, both of which reduce the strength of the reinforced nickel.

EFFECT OF PLATED COATINGS ON FATIGUE STRENGTH

The use of a nickel electroplate has often been shown to be responsible for a loss in fatigue strength of plated steel articles[10], this effect obviously being greatest when the ratio of nickel thickness to substrate thickness is highest. This feature is of importance for parts which are subjected during service to alternating stresses of a magnitude that approaches the endurance limit for the unplated steel. As this reduction in fatigue strength can often be as much as 30% and sometimes even 50%, this effect can be catastrophic on components which, for weight reasons, are designed to allow little safety margin, particularly as the percentage reduction becomes greater as the strength of the substrate increases. The internal tensile stress present in some nickel deposits is one of the major causes of this accentuation of fatigue failure, for although their tensile and fatigue strengths have an important influence on the fatigue behaviour of the plated article, there is a linear relationship

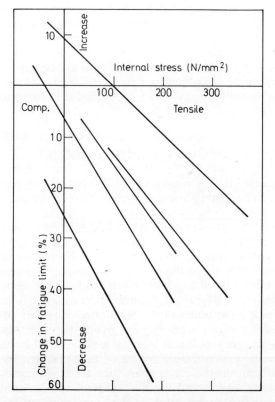

Figure 5.4. Linear relationship between the change in fatigue limit of steel after nickel plating and the internal stress of the nickel deposit (after Williams and Hammond[11])

between the internal stress of the nickel deposit and the percentage change in fatigue strength that they induce[11], as shown in Fig. 5.4. Those nickel deposits that have internal tensile stresses can be shown to reduce drastically the fatigue strength of components on which they are applied, while those compressively stressed will have little or no effect. If the substrate's surface is compressively stressed by shot peening and then plated with a nickel coating having a tensile stress, the consequent reduction in fatigue strength is much less than it would have been without the shot peening[12]. This beneficial effect is even greater if the nickel itself is shot peened. Therefore, there is some justification for the addition of a stress-reducer such as saccharin to a Watts nickel bath in order to produce a compressively-stressed deposit for engineering purposes. It is fortunate that the addition of organic stress relievers to nickel plating solutions has a dual effect, both raising the endurance limit and reducing the stress in the nickel plate. Also, the use of the dense nickel sulphamate solution is likely to be beneficial because of its low tensile or compressive stress and higher tensile strength.

The detrimental effects of some nickel electrodeposits on fatigue properties have usually been demonstrated by tests done in air. However, the combined effect of fatigue and corrosion can seriously reduce the normal endurance limit of the substrate and as the nickel coating will often protect the underlying metal from corrosion, the plated metal may have almost as good a fatigue strength as when unplated, if the comparative tests are performed in a corrosive environment[13, 14]. However, if the nickel becomes cracked at a relatively low stress level, as can occur because of its low intrinsic fatigue strength, the stresses on the substrate at the base of the crack will be enhanced because of the notch effect and a stress-corrosion phenomenon. The latter occurs if the coating is cathodic to the basis metal. e.g. nickel on steel. Consequently, nickel plate is only of limited value in preventing corrosion fatigue.

HYDROGEN EMBRITTLEMENT

When high-tensile steels (stronger than 1250 N/mm^2 tensile strength) are the substrate, their embrittlement by absorbed hydrogen can seriously detract from their anticipated mechanical strength and give rise to delayed failure[15]. During the process of nickel plating, hydrogen can be introduced into the steels from two sources[16], either during the pretreatment processes of cathodic alkaline cleaning or acid pickling or during the plating process itself, the former often being the most potent source. The seriousness of its presence depends on the strength of the steel and the stresses to which it will be subjected in service. Hydrogen has little effect on a steel's fatigue strength or its tensile properties under rapid strain rates. However, at slow rates of strain the ductility of the steel may be greatly reduced. Also, fracture of high-tensile steels may occur at stresses well below their normal tensile strength, after an unpredictable period of time, i.e. delayed failure. The effects of hydrogen on these two properties are made use of in two tests used for assessing and comparing the effects of various treatments which may introduce hydrogen into steel, i.e. the 'slow-bend' test for ductility and the 'static fatigue' test for delayed failure.

Hydrogen can be removed or redistributed by heat treatment ('baking'), this 'stress-relieving' having a time/temperature relationship. The hydrogen is concentrated in the surface layers and at 'traps' within the metal from where it is either released to the atmosphere or diffused into interstitial sites within the steel, where it is not deleterious. In this manner, a uniform hydrogen concentration is obtained, which must be reduced below the critical level for that particular steel.

With uncoated steel components or those which are not completely enveloped in electroplate, this 'outgassing' can be readily carried out by heating at temperatures of $\approx 200°C$, even if the time necessary to do so is lengthy, as is the case for ultra-high-tensile steels[17]. This time will not be prolonged by a coherent, continuous nickel coating, since this metal is permeable to hydrogen. Even so, it is best to avoid preplating treatments that generate hydrogen at the steel's surface, or if this is impossible, the component must be baked to remove hydrogen before it is nickel plated.

ELECTROFORMING

Electrodeposition of nickel for engineering purposes has been discussed in the previous section as if it were used solely to combine some superior property of this metal with that of a substrate. However, electrodeposited nickel can have an existence in its own right in the form of articles that are manufactured by electroforming it onto mandrels, permanent or destructible, which are subsequently removed. The reasonable hardness, tensile strength, ductility and good corrosion resistance of electrodeposited nickel make it most suitable for the production of complex shaped components, which are difficult and expensive to manufacture by conventional means[18, 19]. These nickel electroforms are then frequently used as moulds from which the exact original shape of the mandrel is reproduced by moulding or casting. Thus, this technique is ideal where perfect reproduction of the surface of a certain object is required. For example, the use of nickel electroforming is well established in the electrotyping industry[20] and the gramophone record field. Another common application is for the making of moulds for plastics[21]. Such traditional uses are for electroforms of which a few only are required. However, recent developments have lead to the electroforming process being used as a manufacturing operation in its own right. Functional or domestic consumer items are being produced in quantity by this method. Examples of the first class are the nickel foil and fine mesh[22] that are produced by plating onto rolls, from which they are then peeled off. In the second class are decorative items that either possess such a shape that manufacture by other techniques is difficult or else are embellished with considerable intricate and detailed patterning. Electroforms of both classes are included in Fig. 5.5. In either case, the items can be produced more cheaply and also give better service than when produced from more corrodible metals by conventional techniques, with subsequent polishing and plating then being necessary.

Obviously, for all these applications no warping of the electroform must occur when it is removed from the mandrel, and therefore the internal stress in the electrodeposited nickel must be very low. Again, this can be achieved by the addition of organic chemicals to a Watts or nickel sulphamate bath.

Figure 5.5. A group of articles that have many different functions, but all of which have been electroformed, illustrating the versatility of the technique (courtesy Electroformers Ltd.)

An alternative is to use the conventional nickel sulphamate solution without organic additions but with a low chloride content. If such a solution is kept in a pure condition by attention to cleanliness and continuous purification, low tensile stresses can be maintained in the nickel deposits. The dense nickel sulphamate bath has some advantages for nickel electroforming since even lower stresses can be obtained from it and at a faster deposition rate.

The importance of cleanliness when electroforming has already been mentioned, and indeed, the same dissolved and solid contaminants—both organic and inorganic—that adversely affect nickel plate which is deposited for decorative purposes, can be most deleterious in nickel electroforming solutions. The same precautions as in conventional electroplating must be used to keep any inadvertent contamination to a minimum. However, the maximum tolerable concentrations of any specific impurities are likely to be lower, for many of their harmful effects are accentuated in these thick nickel deposits (see Table 7.1).

By using continuous operating techniques and the choice of a particular nickel plating solution, it is possible to manufacture articles having not only close tolerances but also possessing controlled mechanical properties[23]. These properties can vary from those of a very strong metal having a tensile strength of 1400 N/mm^2, a yield strength of 950 N/mm^2 and an elongation of 2%, to a more ductile one having 10% elongation but still possessing a tensile strength of 820 N/mm^2 and a yield strength of 600 N/mm^2. When their excellent corrosion resistance is also considered, it will be seen that many components could be fabricated by electroforming, with advantage. Engineers thus have available another production tool which should not be regarded as being applicable only when the use of more conventional methods is impossible.

CHROMIUM ELECTRODEPOSITS

Although the emphasis in this text is on nickel deposits, some mention must be made of chromium applied for engineering purposes. This is the so-called *hard chrome plate* (the name being a misnomer, for while these coatings are certainly thicker than those used for decorative purposes they are no harder and are obtained in quite a lustrous condition). These 'hard chrome' coatings are normally between 8 and 250 μm thick (see page 189).

Thick chromium deposits are not used nearly so often for salvage as nickel coatings, particularly where a very heavy build-up is required, as the use of nickel is more technically practicable and economical. Not only is it more difficult to apply chromium than nickel electrodeposits, because of the poor throwing power of the former plating process, but any excess chromium plate is too hard to be turned off, unlike nickel, and instead it must be ground away. Nevertheless, chromium plate in spite of it being very brittle and having an elongation that is much lower than 0·1%, is employed for recovery of worn or scrap engineering components. This brittleness, together with the inherently high tensile stress normally present in chromium electrodeposits, have prevented their being used for electroforming, the metal being too cracked to have any strength. However, some work has been reported[24] in which the electroforms are sintered at very high temperatures after dissolution of the mandrel, and it has been claimed that this heat treatment will produce chromium having a high ductility (up to 17%) and a tensile strength of ≈ 200 N/mm^2. This type of stress-relieved chromium may have certain specialised applications which justify the cost of this procedure.

The properties of chromium plate which render it of most value for engineering purposes are its intrinsic hardness, which is between 800 and 1000 HV, and its low coefficient of friction. The coefficient of friction for chromium on steel is 0·16 compared with 0·30 for steel on steel. Most frequently these thick chromium coatings are plated directly onto a steel substrate, but can be applied onto other metals such as aluminium and its alloys, or over an undercoat of nickel. Use of the latter two-coat system enables the toughness of nickel to be combined with the wear resistance of chromium, with a concurrent gain in corrosion resistance. These benefits can be conferred on nickel deposits applied to new or salvaged components or electroforms. The high hardness and resistance to abrasion of chromium electrodeposits are retained largely unchanged at temperatures of up to 200°C, suffer some reduction between that temperature and 400°C, but are markedly lower at higher temperatures.

The same careful attention to preparation of the substrate is required as for heavy nickel deposition, and anodic etching of the basis metal is again utilised. The chromium plating bath is sometimes also used as the etch instead of a strong sulphuric acid solution.

Because of its tendency to crack, 'hard' chromium electroplate, when it is the sole metal applied, confers little corrosion protection to its substrate until 80–120 μm have been deposited. Chromium coatings of lower thicknesses are considerably inferior to nickel layers of equivalent thicknesses. However, this cracking, so detrimental to corrosion performance, can be put to a beneficial use. If the surface crack pattern is enhanced by modifications to the plating bath so that it becomes more marked and uniform, it confers a much improved oil wettability and retention to the chromium. A micro-cracked deposit is an example of this type of chromium. Cracks that have already been produced can be enlarged by making the chromium plated component anodic in a suitable solution, e.g. the chromium plating bath itself or another chromic acid solution. An alternative technique to produce channels in a chromium plated surface is to etch it through a plastic mesh screen applied tightly to it[25].

Due to their high tensile stress, thick deposits, i.e. those greater than 1 μm of decorative bright chromium, are invariably cracked. However, by adjustment of the composition of the plating solution and the use of modified operating conditions, in particular solution temperatures greater than 65°C, crack-free deposits can be obtained at the thicknesses required for engineering purposes. These alterations in the process result in the chromium plate produced being dull and white in appearance and also being softer than conventional deposits. The hardness of this crack-free plate is in the range of 425 to 700 HV. This lack of lustre and relative softness of the crack-free chromium deposits can be acceptable for certain applications because of the greater protection it confers to a substrate, which is not exposed through any discontinuities.

Thicker chromium coatings are normally deposited from much the same chromic acid based solutions as are decorative coatings. However, they are usually rather more dilute, containing between 150 and 300 g/l of CrO_3, partly to reduce physical loss by drag-out and spray of the chromic acid, but mainly because of the greater cathodic efficiency obtained at these lower concentrations. A large proportion of hard chromium plating is still done in the solution containing 250 g/l of CrO_3 and 2·5 g/l of H_2SO_4, the same composition as originally recommended by Sargeant, which has a cathodic efficiency of $\approx 12\%$. However, baths which contain fluorides in addition to sulphate as catalyst ions have a higher efficiency of $\approx 18\%$. A typical solution of this type contains 250 g/l of CrO_3, 1 g/l of H_2SO_4 and 2·5 g/l of H_2SiF_6. In both baths, the efficiency of deposition increases as the current density is raised. Because of this and also to obtain the required thickness as rapidly as possible, higher current densities and solution temperatures are used for

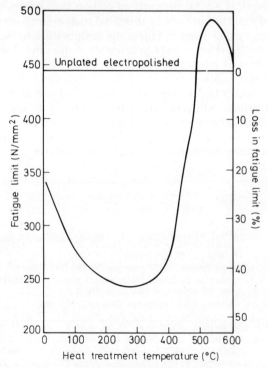

Figure 5.6. Effect of heat treatment on the fatigue limit of a high tensile steel plated with chromium (after Hammond and Williams[10])

hard chromium plating than for decorative applications. As the current densities are most frequently between 40 and 75 A/dm^2, solution temperatures are usually in the range of 55° to 62°C. Use of these conditions results in chromium deposition taking place at the rate of 50 to 75 μm/h, much the same as is obtained in conventional heavy nickel electroplating.

The large amounts of hydrogen evolved during chromium plating inevitably embrittle a steel substrate and it is necessary to heat-treat high-strength steels at temperatures up to 200°C to remove or to redistribute the absorbed hydrogen. However, heat-treatment at this temperature has a dire effect on the fatigue strength of the plated component, reducing this even more than would be the case in its as-deposited condition[10] (Fig. 5.6). It has been shown that chromium plate has a greater deleterious effect on fatigue strength than has nickel plate, both because of its often higher internal tensile stress and its lower mechanical strength. The fatigue limit of chromium-plated steel can be shown to be directly related to the internal stress in the chromium and the density of cracks on its surface[26]. Whatever this reduction in fatigue strength compared with the unplated steel may be, it can be doubled by heat treatment at temperatures between 200° and 300°C, where the baking effect is greatest. However, if the chromium-plated steel is heated at 450° to 500°C for 1 h, the fatigue strength of the plated article then becomes the same as before plating, i.e. it is better than in the as-deposited condition (Fig. 5.6). Therefore, if fatigue strength is a vital property of certain steel components, it is now recommended that they are not heat treated to alleviate hydrogen embrittlement or for any other reason, unless the temperature of heat treatment is greater than 440°C. This is only practicable if the steel does not suffer an unacceptable loss of temper or if a reduction in chromium plate hardness to about 600 HV is allowable. If this is not the case, shot peening of the substrate to induce a high compressive stress in its surface before chromium plating is performed, will give satisfactory results[12]. In spite of these detrimental side effects, chromium plating is just as important to the engineering industry as it is to decorative finishing, and can be regarded as complementary rather than competitive to heavy nickel plating.

REFERENCES

1. BRENNER, A., ZENTNER, V. and JENNINGS, C. W., *Plating*, **39**, 865 (1952)
2. HAMMOND, R. A. F., *Metal Finishing Journal*, **16**, 169, 205, 234, 276 (1970)
3. KEDWARD, E. C., *Metallurgia*, **79**, 225 (1969)
4. BRANDES, E. A. and GOLDTHORPE, D., *Metallurgia*, **76**, 195 (1967)
5. BROWNING, M. E., et al., U.S. Technical Documentary Report No. ML TDR 64–26. *Deposition Forming Processes for Aerospace Structures* (1964)
6. SAFRANEK, W. H., *Metalworking Production*, 19th July, 55 (1967)
7. GILLAM, E., MCVIE, K. M. and PHILIPS, M., *J. Inst. Metals*, **94**, 228 (1966)
8. DONOVAN, P. D. and WATSON-ADAMS, B. R., *Metals & Materials*, **3**, 443 (1969)
9. BAKER, A. A., ALLERY, M. P. B. and HARRIS, S. J., *J. Materials Science*, **4**, 242 (1969)
10. HAMMOND, R. A. F. and WILLIAMS, C., *Metallurgical Review*, **5**, 165 (1960)
11. WILLIAMS, C. and HAMMOND, R. A. F., *Trans. Inst. Metal Finishing*, **34**, 317 (1957)
12. ALMEN, J. O., *Product Eng.*, **22**, 109 (1951)
13. FORSMAN, G. and LUNDIN, E., *Proc. 1st World Metallurgical Congress*, Cleveland. Amer. Soc. Metals, 606 (1951)
14. ALLSOP, R. T., *Coil Spring Journal*, **32**, 5 (1953)
15. SMIALOWSKI, M., *Hydrogen in Steel*, Pergamon Press. Oxford (1962)
16. READ, H. J., *Hydrogen Embrittlement in Metal Finishing*, Reinhold Publishing Corp., New York (1961)
17. *The Pretreatment and Protection of Steel Parts of Tensile Strength Exceeding 90 tonf/in^2*, Min. of Defence Spec. DEF-162, H.M.S.O., London (1968)
18. *Applications, Uses and Properties of Electroformed Metals*, Symposium on Electroforming, A.S.T.M., Philadelphia (1962)
19. SPIRO, P., *Electroforming—A Comprehensive Survey of Theory, Practice and Commercial Applications*, 2nd edn, Robert Draper, Teddington (1971)

20. SQUITERO. A. D.. *Development of All-nickel Shell Electrotype Printing Plates*. International Nickel Symposium on Nickel Deposition in the Engineering Industries. London. Oct. (1963)
21. SPIRO. P.. *Nickel Electroforming of Moulds and Dies for the Plastics and Mass Production Industries*. see ref. no. 20
22. VAN DER WAALS. J.. *Electroforming of Nickel Screens*. see ref. no. 20
23. CHRISTIAN. J. L.. SCHECK. W. G. and COX. J. D.. 'Mechanical Properties of Electroformed Nickel at Room and Cryogenic Temperatures'. *Advances in Cryogenic Engineering*. edited by TIMMERHAUS. K. D.. **11**. 409, Plenum Press. New York (1966)
24. BRANDES. E. A. and WHITTAKER. T. A.. *The Engineer*. **220**. 929 (1965)
25. MONOCHROME. LTD.. U.S. Pat. 2620296 (30.1.53)
26. STARECK, J. E., SEYB. E. J. and TULUMELLO, A. C., *Proc. Amer. Electroplaters' Soc.*. **42**. 129 (1955)

BIBLIOGRAPHY

Electroplated Coatings of Chromium for Engineering Purposes, BS 4758: 1970
Electroplated Coatings of Nickel for Engineering Purposes, BS 4758: 1971

Chapter 6

Bright Nickel Plating

The introduction of nickel plating solutions, from which inherently bright plate could be deposited, was a major innovation, for since then post-plating polishing and 'colouring' operations have rarely been essential. The early bright nickel deposits had no ability to selectively 'fill in' scratches and other surface defects, and thus their brightness depended largely on that of the substrate, for they did not possess a property that is now known as *levelling*. The application of a bright but non-levelling coating exaggerates the effect of defects in the substrate surface because the eye is more sensitive to imperfections in a bright surface than in a dull, matt surface.

At the present time the term *bright nickel* is usually assumed to imply that the deposit is fully bright and levelled. An addition of a single organic compound will not produce a deposit possessing sufficient of both these characteristics, and a combination of several specific compounds, usually organic, is required to produce a commercially acceptable deposit. Semi-bright deposits which have good levelling properties but which, as the name implies, are only partially bright, are obtained from much simpler baths often containing only one organic compound. Although the behaviour of organic compounds is complex in electrolyte solutions from which fully bright electrodeposits can be obtained, certain groups of compounds do have a particular use or uses. The main categories are classified as *brighteners of the first class, brighteners of the second class, levellers, stress relievers, depolarisers* and *wetting agents*. Often, certain chemicals serve a dual purpose and fulfil two of these functions. For example, a compound may be added to serve as a stress reliever but may also assist in the brightening action. Another example of an addition having more than one effect is when a brightener of the first class also produces levelling. Classification of any particular compound into one of these main groups can therefore be used only as an indication of its main effect.

BRIGHTENERS

Brighteners of the first class when used by themselves do not produce fully bright deposits unless the substrate has been polished thoroughly so as to be lustrous itself. The concentration which may be used is not critical and a relatively high concentration may be permitted. A value of 15 g/l is common for compounds of the naphthalene polysulphonic acid type and 1·5 g/l for the aromatic sulphonamides or sulphonimides. Brighteners of the first class do not have such a marked effect on the physical properties of deposits as

Table 6.1. BRIGHTENERS OF THE FIRST CLASS

Type of compound	Examples*
Aromatic sulphonic acids	Benzene sulphonic acid
	1.3.6 naphthalene sulphonic acid
Aromatic sulphonamides	p-toluene sulphonamide
Aromatic sulphonimides	o-benzoic sulphonimide (saccharin)
Heterocyclic sulphonic acids	Thiophen-2-sulphonic acid
Aromatic sulphinic acids	Benzene sulphinic acid
Ethylenic aliphatic sulphonic acids†	Allyl sulphonic acid $CH_2=CH-CH_2-SO_3H$

* Although the free acids are listed above, their water-soluble salts with non-detrimental cations have equal benefits in nickel plating solutions. The most common cations are the alkali metals, particularly sodium, but nickel has been employed.
† These compounds also act as brighteners of the second class and so are included in Table 6.2.

Table 6.2. BRIGHTENERS OF THE SECOND CLASS

Active group	Type of compound	Example
C=O	Aldehydes	Formaldehyde $H-\overset{\overset{\displaystyle O}{\|\|}}{C}-H$
	Chloro and bromo substituted aldehydes	Chloral hydrate $CCl_3-C\overset{(OH)_2}{\underset{H}{\diagdown}}$
	Sulphonated aryl aldehydes	o-sulpho benzaldehyde (benzene ring with CH=O and SO$_3$H)
C=C	Allyl and vinyl compounds	allyl sulphonic acid $CH_2=CH-CH_2-SO_3H$
O=C–O, C=C	1,2 Benzo pyrones	Coumarin
	Unsaturated carboxylic acids and their esters	o-hydroxy cinnamic acid (benzene ring with OH and $-CH=CH-COOH$)
		diethyl maleate $CH_3-CH_2-O-\overset{\overset{\displaystyle O}{\|\|}}{C}-CH=CH-\overset{\overset{\displaystyle O}{\|\|}}{C}-O-CH_2-CH_3$

	Acetylenic compounds	
C≡C	Alcohols	2-butyne-1,4-diol HO—CH$_2$—C≡C—CH$_2$—OH
	Carboxylic acids	phenyl propiolic acid C$_6$H$_5$—C≡C—COOH
	Sulphonic acids	2-butyne-1,4-disulphonic acid HO$_3$S—CH$_2$—C≡C—CH$_2$—SO$_3$H
	Amines	3-dimethylamino-1-propyne (CH$_3$)$_2$—N—CH$_2$—C≡CH
	Aldehydes	propargyl aldehyde HC≡C—CHO
	(Numerous other water-soluble acetylenic compounds have also been patented)	
C≡N	Nitriles	ethyl cyanohydrin HO—CH$_2$—CH$_2$—CN
		succindinitrile NC—CH$_2$—CH$_2$—CN
	Thionitriles	β-cyanoethyl thioether NC—CH$_2$—CH$_2$—S—CH$_2$—CH$_2$—CN
C=N	Quinolinium, quinaldinium and pyridinium compounds	quinoline methiodide (*N*-methyl quinolinium iodide)

[structure: N-methyl quinolinium iodide]

[structure: triphenyl methane dye — Magenta]

Amino polyaryl methanes   triphenyl methane dyes, e.g. Magenta

Table 6.2 (*Continued*)

Active group	Type of compound	Example
	Azine, thiazine and oxazine dyes	Azine dye—methylene blue $(CH_3)_2N\text{—}[\text{phenothiazine ring}]\text{—}N(CH_3)_2$ $Cl^{\ominus}$
	Alkylene amines and polyamines	tetraethylene pentamine $NH_2\text{—}(CH_2)_2\text{—}NH\text{—}(CH_2)_2\text{—}NH\text{—}(CH_2)_2\text{—}NH\text{—}(CH_2)_2\text{—}NH_2$
$N{=}N$	Azo dyes	*p*-amino azo benzene [phenyl–N=N–phenyl–NH₂]
$N\text{—}C{=}S$	Thiourea and derivatives	thiourea $NH_2\text{—}\overset{S}{\underset{\|}{C}}\text{—}NH_2$ allyl thiourea $CH_2{=}CH\text{—}CH_2\text{—}NH\text{—}\overset{S}{\underset{\|}{C}}\text{—}NH_2$ *o*-phenylene thiourea (2-mercapto benzimidazole) [benzimidazole-2-thiol structure with —SH]
$\text{—}(CH_2\text{—}CH\text{—}O)\text{—}$	polyethylene glycols	$HO\text{—}(CH_2\text{—}CH_2\text{—}O)_n\text{—}H$ where *n* is an integer from 10 to 40

brighteners of the second class, which however, can only be used at a low concentration. Often the effects of the two types of brighteners when used in conjunction are synergistic; the effect of the two together is greater than the sum of the effects of each when used separately. When used in conjunction with brighteners of the second class they may enable a lower concentration of the latter to be used and so reduce their deleterious effects on mechanical properties. Compounds of the first class have the group $=\!C\!-\!SO_2$ in the molecule; this class can be subdivided and various advantages are claimed in the literature and in patents for specific types of compounds. e.g. tolerance to metallic contamination, superior brightness, benefits in high chloride solution, etc. The types of compound found most useful in commercial practice are listed in Table 6.1.

Brighteners of the second class are responsible for the ability to obtain fully bright deposits, but they cannot be used as the sole additive since they cause brittleness and induce tensile stress in the deposits. This class is characterised by the presence in the molecule of an unsaturated group. Any of the following may be active in this respect:

$C\!=\!O$, $C\!=\!C$, $C\!=\!C\!-\!C\!-\!O$, $C\!\equiv\!C$, $C\!\equiv\!N$, $C\!=\!N$, $N\!=\!N$, $N\!-\!C\!=\!S$ and
$-\!(CH_2\!-\!CH_2\!-\!O)\!-$

Examples of patented compounds containing these groups are given in Table 6.2. The concentrations of these compounds, particularly their upper limits, are critical. For example, an acetylenic compound will usually be present only at a concentration of the order of 0·01 g/l, higher concentrations giving better levelling but producing excessive brittleness. 'missing' and possibly flaking plate. Depolarisers, such as thiocarboxylic acids are added to counteract these detrimental effects and so allow more of these brighteners to be used. Thiomalic acid is an example of such depolarisers.

LEVELLERS

Levelling agents are usually non-sulphur containing compounds. coumarin being the best known. These are the active ingredients of the commercial solutions used to deposit semi-bright deposits, where the ability to hide surface defects is the prime objective. Semi-bright deposits are harder than those plated from Watts solution but can readily be buffed to a bright finish. Their surface topography is smoother than that of Watts deposits as shown by the electron micrographs illustrated in Figs. 3.11 and 3.16; this is in addition to the levelling phenomenon. Their grain size is smaller than that of Watts deposits but not as small as that of bright deposits[1].

STRESS RELIEVERS

Stress relievers are added to counteract the deleterious effect of second class brighteners. Usually compounds causing compressive stress are required to 'neutralise' the detrimental tensile stress caused by brighteners of the second class. Commercial solutions are often formulated so that the deposit plated from a fresh solution has a fairly low compressive stress (35–100 N/mm^2).

Saccharin (*o*-benzoic sulphonimide) is probably the commonest stress reliever used and this is an excellent example of a compound which can also serve as a brightener of the first class.

WETTING AGENTS

The importance of wetting agents should not be minimised since no brightening system can be commercially successful unless it produces pit-free plate. To ensure this, a wetting agent compatible with the other compounds is required. Unsuitable wetting agents can result in loss of brightness, loss of levelling and even fine surface pitting. The ability of any compound to lower surface tension is by no means the only criterion.

Organic compounds chosen from some or all of these main groups are present in all successful semi-bright or bright nickel plating baths and examples of satisfactory combinations are given in Appendix 1 at the end of the book.

PROPERTIES OF ELECTRODEPOSITED BRIGHT NICKEL

Mechanical properties of most importance for a bright nickel deposit are the same as those of any bulk metal, but in addition several other properties peculiar to electrodeposited metal must be included. To simplify the discussion, the properties of the bath and deposit will be dealt with separately. First Eckleman[2] and later Saltonstall[3] listed the requirements of both deposit and electrolyte solution. These have changed little in the intervening years except that standards have become more exacting.

BRIGHTNESS

A fully bright deposit must be obtained over a wide current density range and irrespective of thickness. Dark deposits must be avoided; the 'depth' of colour must be good. The latter is particularly difficult to evaluate and can really only be assessed by the human eye. Brightness itself is only a subjective property and cannot be determined in a quantitative manner. Most measurements are dependent on the allied property of reflectivity, but the levelling properties of the deposit are also important and it is the combination of these that it is sought for in a commercial process. The bright plating range can be investigated in a qualitative manner using the Hull cell, in which the anode and cathode are not parallel, as mentioned in Chapter 2. Since in this cell one end of the cathode is much nearer to the anode than the other, the appearance of the deposit over a wide range of current densities can be observed by means of one plating test. The Hull cell[4] is a versatile device and is useful for both development and control of electroplating solutions. Agitation, either by mechanical stirring or air bubbles, can be arranged and the cell can also be modified in several ways. Different sizes are available and holes may be cut in the sides or bottom to allow circulation of solution. All these factors influence the current distribution, but in each case the test panel can be compared to calibration charts or standards. Partial or total absence of deposit at low current densities, e.g. 'missing' or 'skipping', due to the effect of brighteners, is a typical defect that can be revealed by a Hull cell test.

REFLECTIVITY

This factor is associated with brightness but its estimation does not provide a means of evaluating brightness. A bright smooth surface would have a high reflectivity but a bright surface need not necessarily have a high specular reflectivity. This property is not often evaluated in the plating industry; it is only of significance for certain applications such as reflectors. It is influenced by other factors such as basis metal imperfections, pitting and roughness which are taken into consideration by other tests. The general impression on the eye of a smooth bright surface is usually of greater significance than the rather academic measure of reflectivity. A Guild Reflectometer[5] is the instrument recommended in a British Standard for the estimation of reflectivity of anodised aluminium.

ROUGHNESS AND PITTING

The former defect should not occur in any nickel deposit provided that cleanliness of operation is observed; entrapment of foreign matter in the deposit should not then occur. Roughness should therefore arise only when the filtration rate is inadequate or when the filter cloths are damaged and so allow carbon, filter media or other particles to pass into the plating tank. A punctured anode bag causes the same trouble, as this permits anode sludge to enter the bath. Often it is difficult to distinguish between fine pitting and roughness even under the optical microscope, but the scanning electron microscope is extremely useful for this purpose. Magnetic particles and slivers of metal resulting from the manufacture of the component are sometimes attached to its surface and are then preferentially plated upon and so produce roughness.

Pitting is usually caused by gas bubbles adhering to the cathode surface. The electrodeposit grows around the attached gas bubble and hence a hole is produced in the coating (Fig. 3.17). This necessitates the use of a wetting agent to lower the interfacial tension so that gas bubbles are more easily detached from the cathode surface. Pitting also results from other sources, the most obvious being residual grease on the cathode surface and inclusions in the cathode surface. It can be initiated by tiny particles lodging on the surface which are too small to cause roughness.

As noted earlier, the selected wetting agent must be compatible with the brightener system and the anionic type are used almost exclusively. Air agitation of vats is favoured in most industrial installations, but cathode rod movement is used in some instances to reduce the need for filtration, since any particles present tend to settle to the bottom of the vat. In the latter case very active wetting agents are essential in order to depress the surface tension as much as possible. On the other hand, if air agitation is used with such high foaming wetting agents, an excessive foam blanket will be produced on the solution. This could result in various troubles, of which the most serious are the dangers of hydrogen explosions caused by sparks igniting the entrapped gas and the possibility of the whole plating vat being engulfed in foam!

POROSITY

In present day nickel plating practice, porosity should not occur, except in coatings of less than approximately 20 µm. Various tests have been developed which are claimed to illustrate porosity in deposits, but in many instances these are aggressive and lead to the formation of pores at the more active regions in the coating; discretion is therefore needed when interpreting such results. Two types of porosity occur, i.e. intrinsic and that due to bad housekeeping, the latter being a microscopic form of the pitting defects caused by the same effects described for Roughness and Pitting. Porosity in some thin coatings is inevitable; chromium deposits are an example of this. Intrinsic pores can be troublesome in thin nickel deposits used for certain purposes such as an undercoat for gold.

Interest in intrinsic porosity has lessened since it has been found that the corrosion of nickel plated metals occurs primarily at points where external corrosive attack has penetrated the coating rather than at previously existing pores. Nevertheless, in the past, much effort was devoted to attempts to establish reasons for this intrinsic porosity in nickel electroplate. Watts nickel baths being usually employed for ease of reproducibility. The American Electroplaters' Society has been particularly active in this respect and Ogburn and Benderly have produced a comprehensive report[6]. An earlier A.E.S. Research Report by Thon and his co-workers[7] included results on the varying intrinsic porosity of nickel on copper pretreated in different ways, the porosity being measured by the gas permeability method devised by that team. Brook[8] later carried out similar work using an autoradiographic technique to reveal pores. It is possible that their results were influenced by the so-called *zoning* effect found at different microscopic distances from the surfaces of metals, depending on their type and previous metallurgical history, as first reported for steels by Clarke and Britton[9] and subsequently also for copper and brass by Clarke and Leeds[10].

CORROSION RESISTANCE

For decorative purposes it is usually the corrosion resistance of the composite nickel plus chromium coating that is of importance rather than that of the nickel alone. However, single nickel coatings are satisfactory for some purposes and good corrosion resistance is not always a criterion for acceptance of a nickel plus chromium coating.

Corrosion resistance of modern nickel plus chromium systems is discussed fully in Chapter 11. Most bright nickel deposits contain sulphur and are less electropositive and hence more electrochemically reactive than sulphur-free deposits[11]. This apparent disadvantage is usefully employed in double layer nickel coatings, where a difference in potential between the two nickel deposits is essential if the system is to function correctly. The chromium overlay also performs an important role in corrosion behaviour of the coating; its purpose is not just to prevent tarnishing as was thought originally. The effects of the various chromium coatings will also be dealt with in Chapters 10 and 11.

CHROMABILITY

As already stated, bright nickel deposits are almost always part of a nickel plus chromium coating and it is therefore essential that the surface of the nickel should be receptive to the electrodeposition of chromium. Certain brightener systems result in an adsorbed layer of organic compounds on the nickel surface and this can prevent the deposition of chromium or cause defects in it. These defects occur in the form of white patches in the chromium coating and are caused by its structure being locally affected by the underlying nickel. Thorough swilling may not be sufficient to remove this adsorbed layer and an activation treatment may be necessary. A dip in dilute acid or alkali sometimes removes the film and so enables subsequent satisfactory chromium plating of nickel to be carried out. Cathodic treatment in an alkali-metal cyanide solution is even more effective. Nickel passivates fairly quickly and if a delay occurs between nickel and chromium plating, particularly if the nickel is allowed to dry, the chromium coating may be unsatisfactory. Activation processes of the aforementioned type or other suitable activating dips are then necessary before chromium plating. The following solutions are frequently effective as activating dips:

(a) 20 g/l of oxalic acid.
(b) 3·5% w/w sulphuric acid containing 0·04% w/w potassium iodide and 0·001% w/w iodine.

This phenomenon of passivity is so erratic and elusive that it is impossible to be dogmatic about cause or cure.

ADHESION AND SURFACE PREPARATION

Lack of adhesion is usually the result of inadequate or incorrect pretreatment of the basis metal surface, but can be caused by excess of brighteners of the second class. The two objects of the cleaning cycle are to remove soils (which are usually grease) and to remove oxides and sulphides, etc. Special problems exist in the case of light metals, low melting point metals, some copper alloys and cast iron. Components of complicated shape lead to the entrapment of compacted polishing composition in recesses, folds and blind holes. Cleaning solutions should not be allowed to dry onto the surface of a component between two stages in a plating sequence; this occasionally occurs on automatic plants if the cleaner is too hot or a long dwell period in the air is necessary. Metal cleaning and degreasing and the subsequent acid treatments for oxide removal are extensive subjects and are dealt with adequately in other publications[12-14]. This topic will therefore only be outlined here to indicate the main techniques which are employed and the pitfalls which cause poor adhesion. The physical and chemical processes are dissolution, saponification, emulsification and colloidal chemical processes. However, these processes can be assisted by the passage of current through suitable solutions (electrolytic cleaning) or by ultrasonic vibration. Dissolution in organic solvents is a satisfactory means of removing many oils and greases, but some of the modern die lubricants are silicone based and hence are difficult to remove by the cleaning process. The disadvantage of solvent degreasing is that a film of soil always remains on the surface, since it is impossible to continually use fresh solvent for the last stage. This can be partially overcome by using vapour degreasing as an alternative; the surface is continually in contact with clean solvent until it attains its boiling temperature. Emulsification and saponification of greases in alkaline solutions are assisted by electrolytic cleaning action, either anodic or cathodic, whichever is appropriate. For a particular current density, twice as much hydrogen as oxygen is liberated, so that cathodic treatment provides a better scrubbing action, but it also causes hydrogen embrittlement of steel and can result in the deposition of films and other metals onto the workpiece. If copious amounts of hydrogen are adsorbed by the substrate it may cause the nickel deposit to blister at a later date. Removal of contaminating soils is encouraged, since hydrogen evolution can cause the hydrogenation, and oxygen evolution the oxidation, of these materials.

Simple non-electrolytic cleaning processes are made more effective by mechanical means such as hand cleaning, forced solution circulation in soak cleaning, spray cleaning or by the washing action of the high pressure exerted on the surface in vapour cleaning. However, ultrasonic vibration is the most effective of all due to cavitation and the severe turbulence induced. The final stage in a cleaning cycle is concerned with the removal of oxide by acid dipping, or with neutralisation of the film of alkaline cleaning solution on the surface, if the plating solution concerned is an acid one. It is particularly important to ensure that the correct sequence is observed through the cleaning cycle, double swilling in counter-flow rinses between each operation being often advisable. Failure to carry out these precautions is likely to result in poor adhesion. Typical cleaning cycles for steel, brass, copper and zinc castings are listed below.

Process sequence suitable for cleaning steel

1. Anodic treatment in a hot alkaline solution.
2. Rinse in running water.
3. Dip in hydrochloric acid solution (50% v/v concentrated acid).
4. Rinse in running water.
5. Anodic treatment in a hot alkaline solution.
6. Rinse in running water.
7. Dip in sulphuric acid solution (5% v/v concentrated acid).
8. Rinse in running water.
9. Nickel plate.

Process sequence suitable for cleaning copper and brass

1. Anodic treatment in a hot alkaline solution.
2. Rinse in running water.
3. Cathodic treatment in a cold or warm alkaline solution.
4. Rinse in running water.
5. Dip in sulphuric acid solution (5% v/v concentrated acid).
6. Rinse in running water.
7. Nickel plate.

Process sequence suitable for cleaning zinc alloy die-castings

1. Cathodic treatment in hot alkaline solution.
2. Rinse in running water.
3. Anodic treatment in warm alkaline solution.
4. Rinse in running water.
5. Dip in hydrofluoric acid solution (2·5 v/v concentrated acid) or sulphuric acid solution (0·5% concentrated acid).
6. Rinse in running water.
7. Copper plate in cyanide solution.
8. Rinse in running water.
9. Dip in sulphuric acid solution (5% v/v concentrated acid).
10. Rinse in running water.
11. Nickel plate.

N.B. If the surface of any of these metals is grossly contaminated by soils such as polishing compositions, oils or greases, it is preferable to remove the bulk of this extraneous matter by non-electrolytic techniques as mentioned previously, before commencing to apply the appropriate electrolytic cleaning sequence as outlined above.

Poor adhesion is exaggerated if the electrodeposit has a high tensile stress since this tends to lift the coating off the substrate; a compressive stress is not so detrimental in this respect. When nickel is correctly bonded to the substrate, the bond strength is often greater than the tensile strength of either the substrate or the coating and failure occurs in one or the other. Linford[15] has carried out a series of contaminating treatments on copper basis metal and evaluated their effects on the initial stages of nickel being electrodeposited from a Watts bath and also on the resultant bond strength.

DUCTILITY

Ideally this property should be as high as possible in order that the coating can withstand small amounts of deformation in service without cracking; it becomes less important when a rigid basis metal such as a zinc alloy die-casting is used. If the ductility is high, a relatively high internal stress can be tolerated before there is any danger of spontaneous cracking. Since chromium electrodeposits usually have a high tensile internal stress, cracking of the nickel coating can occur when the chromium overlay is deposited, if the nickel has a low ductility. If cracks of this type penetrate straight through to the substrate, corrosion of this soon takes place causing undermining of the coating. Internal stress and ductility are not interrelated. but the effects of these phenomena are interdependent.

INTERNAL STRESS

This has already been mentioned in connection with adhesion and ductility, but high stress has another serious effect in that it can cause distortion of thin plated components. This is usually of greater significance in the case of electroforms than for decorative coatings, but delicate components could be distorted even by thin coatings if the stress is high. Bright nickel deposits produced commercially may have either tensile or compressive stresses depending on the brightener system employed and the condition of the bath. As stated earlier, deposits plated from fresh solutions usually have a compressive stress, but as the solution ages and organic breakdown products build up, the stress may gradually become tensile, if the solution has not been formulated correctly.

HARDNESS

In the same way that stress and ductility can be controlled by judicious choice of additives, so the hardness can be controlled. The relationship between hardness and ductility follows the same trend as for conventional metal in that as the hardness increases the ductility decreases. Semi-bright nickels have a hardness of about 300 HV, while most commercial bright nickels have a hardness of about 600 HV.

EFFECT OF HYDROGEN ABSORPTION

Hydrogen absorption can arise from several sources in metal finishing, from preliminary pickling, electrolytic cleaning, and deposition processes accompanied by hydrogen ion discharge. The substrate can be affected by all three operations and intermediate layers may be affected by the second and third. For example, since chromium deposition is such an inefficient process (10–20%), copious amounts of hydrogen are liberated at the nickel surface. As a prerequisite for hydrogen evolution, the discharge reaction must occur:

$$H_3O^+ + e + M \rightarrow M\text{---}H + H_2O$$

followed by:

$$M\text{---}H + M\text{---}H \rightarrow 2M + H_2$$

and a certain proportion of the adsorbed hydrogen (M---H) may enter the metal. This readily diffuses into the nickel layer, and as will be shown in Chapter 9, affects the apparent stress in the chromium. Bright nickels are much more sensitive to the effect of absorbed hydrogen than Watts and semi-bright deposits.

Ductility values are rarely determined in the presence of a chromium overlay since the type and thickness of chromium affects the behaviour of a composite coating when deformed. If the ductility of the nickel deposit is evaluated after stripping off the chromium it is possible that the value will be lower than for an unchromed specimen due to hydrogen embrittlement. Conflicting evidence has been published concerning hydrogen embrittlement of electrodeposited nickel, but this is almost certainly due to the variations possible in the electrodeposited metal. The behaviour after stripping chromium is erratic and is dependent on the type of nickel undercoat and the type of test used. As in the case of the aforementioned stress effect, bright nickels are affected to a greater extent than dull nickels.

The effect of hydrogen absorption on the substrate can have serious consequences, and baking may be required to diffuse out the gas. This is usually of greater significance for such deposits as zinc or cadmium plate, but may be necessary for certain hard chromium deposits. It has been shown by Rollinson[16] that plating solutions which are regarded as having a high cathode efficiency (98%) can still cause serious hydrogen embrittlement. However, this section is primarily concerned with the effect of hydrogen on nickel coatings and not with the effect on the substrate, the latter being discussed in Chapter 5.

PROPERTIES OF BRIGHT NICKEL BATHS

STABILITY

A solution should be stable so that deterioration does not occur during extended use. The inorganic constituents of a Watts bath are unlikely to deteriorate, but the organic constituents may take part in chemical and electrochemical reactions. The products of these reactions are usually referred to as *organic break-down products* and they can have a detrimental effect on the deposit, affecting appearance, ductility, internal stress, hardness and corrosion resistance. Brightener systems should be chosen so that these effects are a minimum; this usually means that a compromise is reached. It is unlikely that optimum values could be achieved simultaneously for all the properties of the electrolyte solution and electrodeposit. For example, the normal trend would be that if the brightness and sulphur content increase, then the nobility and ductility decrease, while the hardness increases. A high concentration of addition agent required to give good brightness usually leads to faster accumulation of break-down products. The ideal situation which has been achieved with modern bright nickel baths is that these degradation products are of a harmless nature. However, if the break-down products are deleterious, the point will be reached at which an unacceptable deposit is produced. This situation is avoided by carrying out appropriate purification treatments, which should be as simple and cheap to carry out as possible. Purification methods are described in detail in Chapter 7, but the commonest method is to adsorb the break-down products on activated carbon. Batch treatment on an industrial scale is a dirty, time-consuming operation which involves closing down the plant for several hours. However, a much more serious problem arises if break-down products are not adsorbed on carbon or other similar adsorptive media. Continuous carbon treatment is a more elegant method of treating the solution; this is feasible when the relative adsorption coefficients are such that the break-down product is removed from the solution in preference to the organic chemical used as addition agent.

A solution should be tolerant to inorganic impurities. These can arise from several sources, e.g. spray or splash from adjacent vats and by carry-over (in holes or recesses) from one solution to another. Dissolution of the basis metal is also responsible, zinc contamination being a common feature when plating zinc alloy die-castings. The acid nickel solution readily attacks any exposed zinc in recesses where the copper undercoat is thin and porous or even completely absent. Early bright nickels were usually more susceptible to inorganic impurities than dull nickels, but modern solutions have been formulated to obviate this difficulty.

CATHODE AND ANODE EFFICIENCIES

Cathode efficiency should be as near to 100% as possible to avoid wastage of power and time. A process which wastes power is obviously economically inefficient, but a lengthy plating time is also expensive; the throughput of the plant is reduced and the cost of overheads per unit weight of metal deposited is increased. Most bright nickel solutions operate at a cathode current efficiency of 95–97%, and so the problem does not really arise as far as this metal is concerned but the process is sufficiently inefficient for considerable quantities of hydrogen to be liberated at the cathode. Removal of hydrogen ions from the cathode film causes a subsequent rise in the pH of the bath. Periodic or continuous pH control is therefore necessary to ensure that plating is carried out within the recommended range. As indicated in the section in this chapter concerned with pitting, gas bubbles should not adhere to the cathode surface but be free to flow up the metal surface. Low efficiency and hence vigorous gas liberation can cause vertical channelling in the developing cathode surface.

Where soluble anodes are used, anode efficiency should also be as near to 100% as possible, and this is usually achieved in nickel plating baths.

OPERATING RANGE

On large installations certain plating conditions are often controlled automatically, i.e. temperature and pH. Brightener additions can be controlled by means of an ampere-hour meter, but as yet this technique is employed only on a limited number of plants. Manual control is used most frequently, probably because greater flexibility is possible so as to cope with fluctuations in output; extra additions may be necessary to compensate for any loss by volatilisation or decomposition during idle periods. If such loss occurs, small periodic additions are preferable to large infrequent additions in order to ensure that optimum conditions are maintained, and therefore drip feeding of the brighteners, etc., is employed. However, commercial solutions should be formulated to tolerate a wide variation of plating conditions, but yet still be capable of producing a uniform appearance over a complex article. If excessive additions are made, or if the brightener addition is added too near to the cathode, temporary defects such as passivation leading to 'skipping', poor chromability or loss of adhesion may occur.

SIMPLICITY OF OPERATION

The inorganic constituents of most bright nickels are similar, being based either on the Watts solution or its high-chloride version. Nickel sulphamate baths have, however, been occasionally utilised as a base for producing fully bright or semi-bright deposits. Ideally, for ease of operation of any bright nickel solution the additions should be readily water soluble or miscible and non-volatile. The solution itself should be tolerant to organic and inorganic impurities and easily purified if grossly contaminated. The final two points have already been discussed, but the form of the brightener warrants a brief mention. Brightener in liquid form can be conveniently added to plating solutions and is then usually miscible. If one liquid consisting of several compounds can be added, this is even simpler, provided that the various ingredients are present in the correct ratio for maintenance, which is usually not the same as their ratio in the bath. If an additive is not readily water soluble or miscible, defective plating can result. Coumarin forms an oily liquid on the surface of the solution if the temperature of the latter is above the melting point of that compound. If work is introduced into the vat through this film, pitting may occur due to some of the undissolved coumarin adhering to the cathode surface.

A volatile brightener is difficult to control in small scale plating operations as the brightener is lost by incorporation in the deposit and volatilisation. On a large scale, the correct addition rates can be evaluated more easily but added expense arises if the vat stands idle at working temperature, particularly if air agitation is not turned off, for the brighteners would still be volatilised. However, use of a volatile brightener has one compensation, i.e. if an excessive addition is added by mistake, its concentration can be decreased by raising the temperature of the bath.

THROWING POWER

Throwing power and covering power should be good so that the minimum thickness obtained on significant surfaces of complex shaped articles is sufficient to give reasonable protection in the relevant environment. Additives have only a minor effect on throwing power; this is influenced more by the base solution. However, additives can have considerable effects on covering power, 'skipping' or 'missing'.

THE INCORPORATION AND EFFECT OF ORGANIC ADDITION AGENTS

MECHANISMS OF INCORPORATION OF ORGANIC COMPOUNDS IN ELECTRODEPOSITS

Numerous organic compounds have been cited in patents and other literature that are claimed to modify the properties of nickel deposits[17-21]. The general classes of these have already been indicated, and it is only possible in the present text to describe the behaviour of a few compounds in detail. The mechanisms of incorporation of certain compounds have been investigated by Edwards and Levett[22] using radioactive tracer techniques. Their work was based on the use of organic compounds in which particular carbon or sulphur atoms in the molecule had been labelled by employing radioactive isotopes for their synthesis, so that the decomposition of the organic compound and/or its incorporation in the deposit could be studied.

Diffusion control

Thiourea is typical of compounds whose rate of incorporation is controlled by the rate of diffusion of the organic compound from the bulk of the electrolyte solution to the developing electrodeposit. Diffusion control is the easiest mechanism to evaluate mathematically, but it is unlikely that many compounds behave solely in this manner. The sulphur atom in thiourea is so reactive that it can cause sulphidation of the cathode surface even in the absence of current flow. Consequently, thiourea molecules react with the surface as soon as they arrive and the diffusion process is slow in comparison with the rate of reaction. Since the rate of incorporation is fully diffusion controlled, the sulphur content of the deposit is proportional to the con-

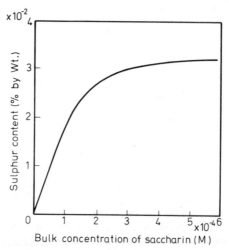

Figure 6.1. Relationship between bulk concentration of saccharin in Watts bath and sulphur content of deposit. Plating conditions: 55°C, pH 4, 4 A/dm² (after Edwards[23])

centration of the compound in the plating solution and inversely proportional to the current density. At a high bulk concentration of thiourea in the solution or at low current density, the sulphur content of the deposit can reach a high value. This has catastrophic effects on the physical properties of the deposit which becomes extremely brittle and highly stressed. Depending on the plating conditions, its appearance may be either bright or black. Thiourea is an extremely active compound and has no commercial applications in modern commercial nickel plating solutions.

Edwards[23] has demonstrated that with variation of temperature, the sulphur content of the deposit varies as would be expected if the diffusion process obeyed the Arrhenius equation (the logarithm of the rate of diffusion plotted against the reciprocal of the absolute temperature measured in kelvins is a linear relationship). Further work on the estimation of carbon content has revealed that much less carbon than sulphur is present in the deposit, which indicates that most of the molecules are incorporated in the deposit in a decomposed form.

Diffusion and adsorption control

Saccharin and p-toluene sulphonamide are examples of organic compounds from which the rate of incorporation is controlled by diffusion and adsorption. Both compounds are of use in commercial nickel plating solutions, and they do not result in such a high sulphur content in the deposit as does thiourea.

The concentration of organic molecules in the cathode film is determined by the balance between the rate of replenishment by diffusion and the rate of consumption by incorporation in the deposit. Unlike the case of thiourea, when the bulk concentration is above a certain value the molecules do not react with the cathode surface as soon as they arrive. Fig. 6.1 shows the relationship between the sulphur content of the deposit and the concentration of saccharin in the solution. A linear relationship is maintained up to a certain concentration, indicating diffusion control, but subsequently the sulphur content does not increase at such a high rate with an increase in the concentration of saccharin in solution. In this region adsorption control is most significant. The surface concentration of adsorbed molecules is a function of the concentration of unadsorbed molecules in the cathode film. By use of the Langmuir isotherm, it is possible to show that the rate of incorporation of sulphur from saccharin and p-toluene sulphonamide is an adsorption controlled process.

Cathodic reduction

Many organic compounds are electrolytically reduced at the cathode during electrodeposition. This complicates the mechanism of incorporation since the original compound, the reduction product or both of them may be incorporated in the deposit. Coumarin is one of the most well-known compounds which is cathodically reduced and the rate of carbon incorporation from this has been studied by having a carbon atom radioactively labelled (^{14}C isotope) in the position marked by an asterisk in the formula.

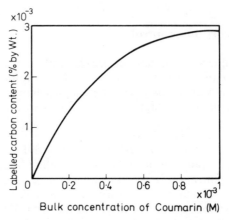

Coumarin   Melilotic acid

The relationship between labelled carbon content of deposit and bulk concentration of coumarin in the solution is shown in Fig. 6.2. The curve is similar to that shown in Fig. 6.1 for saccharin. but the curvature extends over a wider range of concentration. It has been intimated that at pH 4. 90% of the coumarin consumed at the cathode is electrolytically reduced to melilotic acid; at lower pH values an even greater proportion is reduced to

Figure 6.2. Relationship between bulk concentration of coumarin in Watts bath and labelled carbon content of deposit. Plating conditions: 55°C. pH 4, 4 A/dm^2 (after Edwards[23])

melilotic acid. Only 10% of the consumed coumarin is therefore incorporated in the deposit at pH 4. The labelled carbon content of nickel electrodeposited from solutions containing melilotic acid alone is much less than from solutions containing equivalent concentrations of coumarin. However. the carbon content of the deposit increases at a greater rate than the melilotic acid concentration in solution. This is an example of yet another mechanism which affects the rate of incorporation. In this case. the incorporation of some carbon from melilotic acid in some way increases the surface density of sites at which it can itself be adsorbed. Quinoline methiodide is a further example of a compound which exhibits the same type of behaviour.

The study of the behaviour of melilotic acid shows that the reduction product formed from coumarin does not greatly affect the carbon content of a deposit plated from a solution which initially contained only coumarin. In an aged solution the melilotic acid concentration builds up and at a certain level leads to unsatisfactory deposits. The ductility decreases and the

BRIGHT NICKEL PLATING

stress becomes more tensile. At this stage, the solution must be purified by carbon treatment to remove the undesirable reduction product. The controlling mechanism for the incorporation of carbon from a solution containing coumarin is an adsorption-diffusion mechanism similar to that which applies to the incorporation of sulphur from saccharin. Superimposed is the reduction process, but initially this does not result in very much modification for the amount of carbon incorporated from melilotic acid at low concentrations is so small.

INTERACTION OF ORGANIC ADDITIONS

In the examples considered so far, the effect of single additions only has been considered, except in the case where unavoidably a reduction product is formed. In commercial solutions a number of compounds are normally used so it is necessary to investigate the effect of multiple additions. The simplest case is that in which there is no interaction and each compound is incorporated at exactly the same rate as it would have been if it had been the sole addition. This usually occurs at low concentration where the

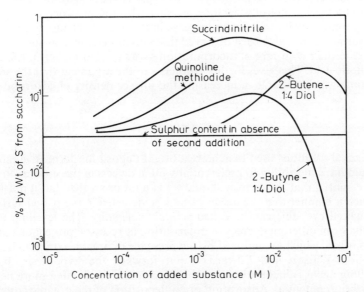

Figure 6.3. Effect of added substances on the rate of incorporation of sulphur from saccharin in a Watts bath. Plating conditions: 55°C, pH 4, 4 A/dm², concentration of saccharin 2×10^{-3} M (after Edwards[23])

impurity content of the deposit is proportional to the bulk concentration of additive in the deposit. Saccharin and *p*-toluene sulphonamide are examples of compounds which behave in this way. Above the critical concentrations, the two additives compete for incorporation. The rate of incorporation of sulphur from compounds such as thiourea (fully diffusion controlled) is unaffected by quite high concentrations of other additives, e.g. saccharin. However, the rate of incorporation of sulphur from saccharin is reduced by the presence of thiourea. Presumably very few adsorption sites are available for the saccharin. The most interesting case of interaction is that in which the presence of a second compound results in an increase in the rate of impurity incorporation from the first compound. Edwards[23] has shown the effect of succindinitrile, quinoline methiodide, 2-butene-1:4 diol and 2-butyne-1:4 diol on the rate of incorporation of sulphur from saccharin (Fig. 6.3). At the optimum concentration of succindinitrile the amount of sulphur incorporated from saccharin is eighteen times greater than in its absence. The most likely explanation of the maxima in the curves is that even in the instances where a second compound initially results in an increase in impurity content in the deposit from saccharin, a stage is reached at which competition for adsorption sites occurs and the amount resulting from saccharin decreases. Du Rose[19] has put forward an explanation of this synergistic effect based on an increase in the rate of reduction of sulphur compounds at more cathodic potentials, while Beacom and Riley[24] have explained such behaviour on the basis of close packing of charged particles in the presence of other particles of opposite charge. However, Edwards[23] has obtained experimental evidence to support the suggestion that it is due to a change in deposit structure produced by the presence of the second compound, which results in an increase in the number of adsorption sites on the surface. It has also been observed that compounds such as succindinitrile increase their own rate of incorporation as was noted earlier in the case of melilotic acid. This gives further emphasis to the argument that the solution in some way is capable of increasing the surface density of adsorption sites.

LEVELLING

All nickel solutions used to achieve a bright deposit for decorative purposes should have the ability to preferentially fill in defects in the surface. Not only is it essential that the deposit should fill in a pit or scratch, but it must do so in such a manner that the defect cannot be detected in the plated surface by the naked eye; this can be called *scratch hideability*. The levelling usually results from differing degrees of polarisation at recesses, projections and the flat surface, which are induced by the presence of organic compounds in the solution. Watson and Edwards[25] put forward the hypothesis that the levelling agent is incorporated in the deposit and results in a more negative deposition potential. Adsorption or co-deposition of the organic compound in the deposit leads to the establishment of a diffusion layer with a reduced concentration at the cathode surface. Since the concentration of additive in the bulk solution is quite low, the rate of diffusion to peaks is much faster than to recesses. A greater quantity of additive is incorporated at the peaks and the cathode potential is more negative in these regions than at the recesses. The main points of their hypothesis have now been confirmed by experimental results. The influence of certain variables on levelling for a number of compounds have been investigated in detail.

Cathode potential

It has been shown that levelling usually occurs when an organic compound results in a more negative deposition potential, as postulated in the above hypothesis. The relationship between current density and cathode potential for a selection of concentrations of saccharin and coumarin is shown in Fig. 6.4. Coumarin is well known as an effective leveller while saccharin does

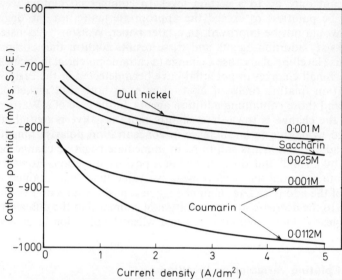

Figure 6.4. Relationship between cathode potential and current density for Watts solutions containing either coumarin or saccharin (after Watson and Edwards[25])

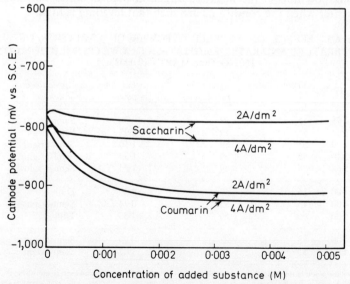

Figure 6.5. Relationship between cathode potential and concentration of either coumarin or saccharin in Watts solutions at current densities of 2 and 4 A/dm² (after Watson and Edwards[25])

not cause levelling. The curves clearly indicate that of the two compounds only coumarin leads to a shift of the cathode potential in the negative direction. Fig. 6.5 shows the relationship between cathode potential and concentration for the same compounds at current densities applicable to commercial plating. These curves show that in the case of saccharin an increase in concentration has little effect on cathode potential at a particular current density and that in the case of coumarin an increase in concentration is significant only up to a certain level. In commercial plating it would therefore be pointless to exceed the appropriate figure, for the degree of levelling would not be improved. In a later paper, Watson[26] has listed the effects of sixty addition agents and these results confirm that compounds which cause levelling also cause a change in cathode potential in the negative direction. Small changes in potential have been detected in the early stages of deposition (plating times of up to 1 min) in both dull nickel plating solution and those containing addition agents. In the case of a Watts nickel solution, the change is towards a slightly more negative potential; this is considered to be due to an increase in concentration polarisation. If coumarin is added to a Watts solution, an immediate negative change occurs, followed by a small and slow change in a positive direction. However, the steady state potential has a more negative value than that attained in the absence of the additive. The small change in a positive direction is thought to be due to the decrease in concentration of coumarin in the diffusion layer and the steady state potential is reached when the diffusion layer is established.

Effect of plating variables on levelling

Agitation of nickel solutions containing coumarin causes a greater change of cathode potential in the negative direction than in static solutions and this is greater when the conditions are such that levelling is high.

Table 6.3. EFFECT ON LEVELLING POWER OF VARIATION OF pH, TEMPERATURE AND RATE OF AGITATION IN A WATTS SOLUTION CONTAINING COUMARIN (0·0002M)†

pH	Temperature (°C)	Oscillation rate (cycles/min)	Levelling power*	Appearance of deposit
2·98	55	59	0·38	Semi-bright
3·44	55	59	0·97	Semi-bright
4·00	55	59	1·10	Semi-bright
5·40	55	59	1·39	Semi-bright
5·64	55	59	1·26	Semi-bright
4·00	36	59	0·39	Dull
4·00	45	59	0·74	Semi-bright
4·00	65	59	1·53	Dull
4·00	55	Zero	0·93	Semi-bright
4·00	55	27·3	1·07	Semi-bright

* Levelling power in this instance was determined by dividing the difference between average groove depths, before and after plating, by the average deposit thickness. The value obtained by this means is influenced by the thickness of deposit and the cathode profile and so is only a comparative measure of levelling.
† Data taken from Watson[26].

BRIGHT NICKEL PLATING

Fig. 6.6 shows the effect of additive concentration on levelling for several compounds at constant conditions of pH (4·0), temperature (55°C), current density (2A/dm^2) and time (15 min). Watson (Table 6.3) has illustrated the effect of variations in pH, temperature and agitation on levelling of deposits

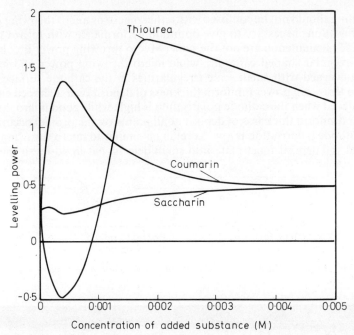

Figure 6.6. Experimental graphs of levelling versus addition agent concentration (after Watson and Edwards[25]). For definition of levelling power, see Table 6.3

plated from a Watts nickel solution containing coumarin (0·0002M). Levelling is influenced by the depth and shape of recesses, but exact rules cannot be laid down to state the effect of variables on the degree of levelling for all solutions. However, in general, a concentration of leveller near to the optimum value is required and a high concentration of the dischargeable complex (hydrated metal ion, etc.). An increase in current density and a decrease in temperature usually result in an increase in the thickness of the diffusion layer and a decrease in levelling. An increase in coating thickness and an increase in the degree of agitation normally lead to an improvement in levelling.

Incorporation of levelling agents in the deposit

It has been shown earlier in the chapter that levelling compounds are incorporated in the deposit. By using a radiotracer technique, it can be demonstrated that the concentration of carbon atoms from coumarin is much higher at peaks than in recesses, thus confirming that the concentration of the compound in the diffusion layer is much greater in the region of the peaks than the recesses.

Filling in of surface defects

Levelling should not be confused with either macro or micro throwing power; the conditions necessary to give optimum performance with regard to each of these phenomenon are not the same. Macro throwing power is concerned with irregular shaped cathodes, while micro throwing power and levelling are concerned with small scale irregularities in the cathode surface. Good macro throwing power (uniform thickness of deposit over a shaped cathode) is achieved when the cathode polarisation is high, while good micro throwing power (uniform thickness of deposit at all points on the profile) occurs when the cathode polarisation is low. Levelling is similar to micro throwing power in that the deposit must plate into small defects, but in addition the recess

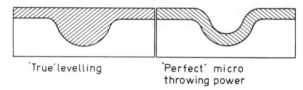

'True' levelling   'Perfect' micro throwing power

Figure 6.7. Diagrammatical representation of 'true' levelling and 'perfect' micro throwing power

25 μm   NICKEL

Figure 6.8. A cross section of a nickel electroform which has been removed from a plastics mandrel and then backed with a levelling copper deposit prior to sectioning. A void has occurred in the nickel deposited in a micro depression in the plastics surface

must be preferentially filled in (Fig. 6.7). In the case of irregular shaped defects, the mouth of the pit can become bridged over, thus showing apparent levelling although a cavity exists below the surface. This is likely to occur if the mouth of the pore is narrow or the micro throwing power poor. In the filling in of 'V' or 'U' notches, of the type popular for experimental purposes, a plane of weakness or a void can form down the centre similar to that shown in Fig. 6.8, which is actually a nickel electroform. The central

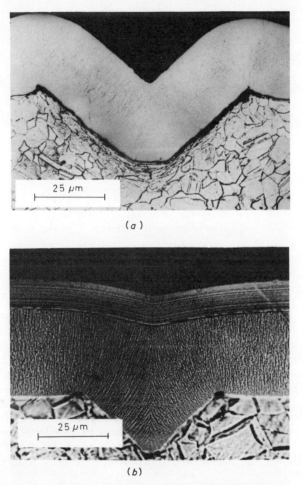

Figure 6.9. Photomicrographs showing (a) *absence of levelling (Watts deposit) and* (b) *levelling provided by double nickel system (semi-bright plus bright)*

space may contain entrapped plating solution and in practice this type of defect, as also the cavity already discussed, could result in corrosion. The degree of levelling provided by typical nickel plating solutions is shown in Fig. 6.9, while the effect of the surface roughness of the substrate on this levelling is illustrated in Fig. 6.10. Kardos and Foulke[27] have published

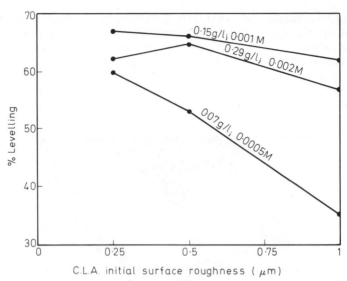

Figure 6.10. Levelling powers of nickel deposited from Watts baths containing coumarin at constant concentrations onto test plaques abraded to have various degrees of roughness. (Measured on Talysurf surface roughness meter)

interesting work and discussion on the effect of the shape of the cathode micro-profile.

EFFECT OF ADDITIVES ON STRUCTURE

The structure of Watts nickel as revealed by optical and electron microscopy has been shown in Chapter 3. Incorporation of organic material in the deposit results in a change in grain size and mechanical properties. Sulphur and carbon, etc. co-deposited with the nickel usually result in a restriction of the grain size with a subsequent increase in hardness and loss of ductility. The change in magnitude of the internal stress, the appearance of the surface, the degree of levelling and the occurrence of a preferred orientation in the structure depends on the particular compound or combination of compounds added. It has been suggested by some authors that preferred orientation is an essential feature of bright plated deposits but in more recent years it has also been shown that bright deposits can have a random orientation.

GRAIN SIZE, ORIENTATION AND BRIGHTNESS OF ELECTRODEPOSITS

Effects of plating variables on structure and properties have been described but no definite explanation has been put forward to account for brightness and high reflectivity. Two theories were published a number of years ago to account for brightness. The first[28-30] proposed that if the grain size of the deposit were very small (less than the wavelength of light) then the deposit would be bright, while the second[31-33] proposed that the deposit should have a high degree of preferred orientation so that the crystal faces would be essentially parallel in the surface. Subsequent research[34-36] has shown that some bright deposits may have these features, but certainly they are not the only reason for the existence of bright deposits. Many electrodeposits, whether dull or bright, have a fibre texture (Watts and Watts plus coumarin are examples) but equally many have a random structure. Denise and Leidheiser[37] have reported that brittle nickel deposits have a fibre orientation other than ⟨100⟩ while ductile deposits have a ⟨100⟩ orientation. The same authors found that most bright deposits did not have a preferred orientation; this is consistent with results obtained recently by one of the present authors[38]. Modern theories assume that the addition agents in the solution result in modification of the growth mechanism of the cathode, but it is now agreed that brightness cannot be related to structure by any general rule.

EFFECT OF ADDITIONS ON STRESS, DUCTILITY AND HARDNESS

Inevitably, considerable reference has already been made to the effect of additives on mechanical properties so it is not intended to deal in detail with the vast number of compounds that have been studied. Sufficient experience and information has now been accumulated to enable the effects of most of the useful compounds to be predicted. However, combinations of compounds are normally used in practice and this leads to further complexity of behaviour. To conclude this chapter the following general remarks will be made, but almost certainly exceptions to the generalisations can be found.

Incorporation of non-metallic material results in harder deposits; the grain size is smaller and the tensile strength higher than in the absence of organic additives in the bath.

The relationship between internal stress and concentration of addition agent is rarely simple. Watson[39] has observed the effect of many additions on the stress in nickel deposits and has concluded that stress is more susceptible than cathode potential to the influence of addition agents. The effects of additions fall into three classes:

1. Stress increases with increasing concentration.
2. Stress decreases to a value which is then unaffected by further increase in concentration.
3. Stress first decreases then rises as concentration is increased.

Compounds classified as brighteners of the first class have a much less severe effect on stress and ductility than brighteners of the second class. Unacceptable properties are usually associated with a high concentration of non-metallic material in the deposit, usually sulphur or carbon. Kendrick[40] has shown the relationship between stress and the concentration of sulphur incorporated from naphthalene sulphonic acids (Fig. 6.11). Compounds giving rise to a low sulphur concentration result in a moderate compressive stress while those giving a higher sulphur content result in small compressive stresses or tensile stresses. Compounds such as thiourea in which the sulphur is present in a very active form can lead to a very high sulphur content in the deposit, in which case it is extremely brittle and highly stressed. It is shown in Chapter 7 that there is little definite evidence to indicate that brittle deposits are caused by metallic impurity unless the concentration is very high (some alloys are brittle), therefore most brittle deposits are due to the presence of organic compounds in the plating solution.

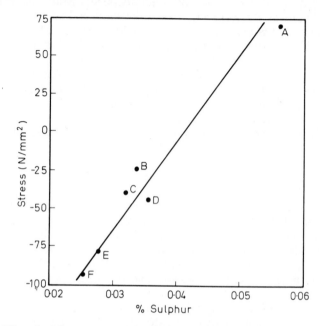

Figure 6.11. Effect of sulphur incorporation on the internal stress in deposits 50 μm thick plated from Watts solutions containing naphthalene sulphonic acids at a concentration of 4×10^{-3} M. Plating conditions: 40°C, pH 4, 4 A/dm²

A naphthalene-1-sulphonic acid
B naphthalene-1,3,6-trisulphonic acid
C naphthalene-1,5-disulphonic acid
D naphthalene-2,7-disulphonic acid
E mixed mono-naphthalene sulphonic acids
F naphthalene-2-sulphonic acid

[after Kendrick[40]]

Bright plating solutions are usually formulated to give the best compromise for all properties, brightness, levelling, ductility and stress.

REFERENCES

1. WILLSON, K. S. and ROGERS, J. A., *Proc. Amer. Electroplaters' Soc.*, **51**, 92 (1964)
2. ECKLEMAN, L., *Monthly Review, Amer. Electroplaters' Soc.*, Nov., 18 (1934)
3. SALTONSTALL, R. B., *Trans. Inst. Met. Fin.*, **31**, 223 (1954)
4. NOHSE, W., *The Investigation of Electroplating and Related Solutions with the Aid of the Hull cell*, R. Draper, Teddington (1966)
5. *Anodic Oxidation Coatings for Aluminium*. BS 1615: 1961
6. OGBURN, F. and BENDERLY, A., *Plating*, **41**, 61, 169 (1954)
7. THON, N., ADDISON, E. T., KELEMAN. D. and LING YANG, *Porosity of Electrodeposited Metals*, A.E.S. Research Reports Serial Nos. 5 (1947), 10 (1948) and 17 (1950)
8. BROOK, P. A., *Trans. Inst. Met. Fin.*, **35**, 251 (1958)
9. CLARKE, M. and BRITTON, S. C., *Trans. Inst. Met. Fin.*, **37**, 110 (1960)
10. CLARKE, M. and LEEDS, J. M., *Trans. Inst. Met. Fin.*, **43**, 50 (1965)

11. HOSPADARUK, V. and PETROCELLI, J. V., *Plating*, **48**, 479 (1961)
12. POLLACK, A. and WESTPHAL, P., *An Introduction to Metal Degreasing and Cleaning*, R. Draper, Teddington (1963)
13. SPRING, S., *Metal Cleaning*, Reinhold, New York (1963)
14. STRASCHILL, M., *Modern Practice in the Pickling of Metals and Related Processes*, R. Draper, Teddington (1963)
15. LINFORD, H. B., *Proc. 7th International Metal Finishing Conference*, Hannover, 14 (1968)
16. ROLLINSON, J. J., *Electroplating and Metal Finishing*, **14**, 323, 356, 396 (1961); **15**, 6, 228 (1962)
17. BROWN, H., *Plating*, **55**, 1047 (1968)
18. BROWN, H., *Trans. Inst. Met. Fin.*, **47**, 63 (1969)
19. SAUBESTRE, E. B., *Plating*, **45**, 1219 (1958)
20. DUBPERNELL, G., *Plating*, **46**, 599 (1959)
21. DUROSE, A. H., *Proc. 7th International Metal Finishing Conference*, Hannover, 54 (1968)
22. EDWARDS, J. and LEVETT, M. J., *Trans. Institute Met. Fin.*, **39**, 33, 45, 52 (1962); **41**, 140, 147, 157 (1964); **44**, 27 (1966); **45**, 12 (1967)
23. EDWARDS, J., *Trans. Inst. Met. Fin.*, **41**, 169 (1964)
24. BEACOM, S. E. and RILEY, B. J., *J. Electrochem. Soc.*, **108**, 758 (1961)
25. WATSON, S. A. and EDWARDS, J., *Trans. Inst. Met. Fin.*, **34**, 167 (1957)
26. WATSON, S. A., *Trans. Inst. Met. Fin.*, **37**, 144 (1960)
27. KARDOS, O. and FOULKE, D. G., *Advances in Electrochemistry and Electrical Engineering*, **2**, 145, Interscience, New York (1962)
28. MACNAUGHTON, D. J. and HOTHERSALL, A. W., *Trans. Faraday Soc.*, **31**, 1168 (1935)
29. KOHLSCHUTTER, V., *Trans. Electrochem. Soc.*, **45**, 229 (1924)
30. HOTHERSALL, A. W. and GARDAM, G. E., *J. Electrodepositors' Tech. Soc.*, **15**, 127 (1939)
31. PALATNIK, L. S., *Trans. Faraday Soc.*, **32**, 939 (1936)
32. WOOD, W. A., *Trans. Faraday Soc.*, **31**, 1248 (1935)
33. BLUM, W., BECKMAN, A. O. and MEYER, W. R., *Trans. Electrochem. Soc.*, **80**, 249 (1941)
34. CLARK, G. L. and SIMOUSEN, S. H., *J. Electrochem. Soc.*, 98, 110 (1951)
35. HOAR, T. P., *Trans. Inst. Met. Fin.*, **29**, 302 (1953)
36. SMITH, W., KEELER, J. H. and READ, H. J., *Plating*, **36**, 355 (1949)
37. DENISE, F. and LEIDHEISER, H., *J. Electrochem. Soc.*, **100**, 490 (1953)
38. DENNIS, J. K. and FUGGLE, J. J., *Trans. Inst. Met. Fin.*, **46**, 185 (1968)
39. WATSON, S. A., *Trans. Inst. Met. Fin.*, **40**, 41 (1963)
40. KENDRICK, R. J., *Trans. Inst. Met. Fin.*, **40**, 19 (1963)

Chapter 7

Control and Purification of Nickel Plating Solutions

In previous chapters, the properties of nickel plating solutions and the electrodeposits that can be obtained from them have been dealt with, assuming ideal conditions prevailed. Unfortunately, the conditions existing in many industrial plating shops are far from ideal. While any commercially viable process must be non-critical in its composition and robust enough to work satisfactorily in such conditions, steps must be taken to keep the components of the bath within the specified limits.

CONTROL OF INORGANIC CONSTITUENTS

Although in many cases the ranges of composition for inorganic constituents are fortunately wide, slow changes in their concentrations as a result of 'drag-out' (removal of plating solution due to a film of liquid clinging to the cathode when it is taken out of the bath) are inevitable. Often this loss is greatly multiplied by the cupping effect of hollow articles. The correct jigging of the work, combined with tipping of the jigs during removal from the bath can help to minimise the volume lost, but a steady, if slow, loss of the inorganic salts from the bath is inevitable. In the case of the nickel ions, this loss is partly compensated for by the greater anodic than cathodic efficiency, so that the nickel content of the solution falls at a rate proportionately slower than that of the other constituents. Because of this, in many instances, the nickel salt needs only infrequent replenishment, with the other inorganic ingredients being added 'as required'. To define 'as required' as being frequent small additions of salts rather than occasional large quantities, is easier to recommend as general practice than to state in a quantitative manner. Analytical control of nickel baths is thus highly desirable, although by no means essential.

The simple measurement of the density of a plating solution by a hydrometer is a crude but effective means of ensuring that major changes in the constituents of the bath, such as might be caused by leakage or over-dilution, have not occurred. It is obviously possible to make additions to restore the solution to approximately its correct composition by using salts in the same ratio as were present in the original bath. This is often sufficient to maintain the solution in a satisfactory state, for the limits for basic constituents are very wide, as can be seen from Table 4.1 in Chapter 4. Chemical analysis of the solution prior to its replenishment is clearly a superior technique. The inorganic constituents can be estimated readily by simple titrimetric methods, and although the results obtained are not always exact or reproduceable to a few per cent, they are more than adequate for control purposes. The details of these methods are available in many papers and textbooks[1-3]. Of the alternative ones available the authors prefer those given in Appendix 2.

Fortunately, one of the most vital features of a nickel plating slution, i.e. its pH, is the most easily controlled. This continually changes in spite of the solution being buffered, and the difference between the cathodic and anodic efficiency that helps to maintain the concentration of the nickel ions results in a gradual increase in the pH of the solution (see Chapter 4). The pH of the bath should be kept within the limits of ± 0.3 to 0.5 units of the optimum; it does not need much use of a nickel plating solution to alter its pH so that it is out of the required range. This rise in pH is counteracted by an addition of acid, usually sulphuric but occasionally hydrochloric or other acids depending on the nature of the major anion present. These changes in pH can be ascertained easily by using the standard methods for the estimation of pH. A pH meter can be employed to give the most accurate results, but the use of pH papers is common. The latter are simple and cheap, and providing the correct pH range paper is chosen, will produce results which are quite satisfactory for shop-floor control purposes. Even if the colorimetric pH values given by the papers are corrected for salt error they will often differ from the electrometric results of the pH meter by 0.1 or even 0.2 units, but this is of no significance when checking commercial plating baths.

While a monthly determination of the basic inorganic constituents is often quite sufficient, the pH changes so rapidly that a once-daily check is the least that should be made. After some little experience has been gained with the operation of any particular process, the approximate amount of acid needed, based on the quantity of electrical charge per unit volume of the solution, can be added as a daily routine. However, the periodic estimation of pH should still be carried out, since failure to do so may produce dire effects when plating either for decorative or engineering purposes.

CONTROL OF ORGANIC CONSTITUENTS

The control of the inorganic constituents has been shown above to be relatively straightforward, and so it is unfortunate that the equally important —and often even more important—control of organic constituents of bright nickel plating solutions is sometimes neglected due to the difficulties involved in their chemical estimation. The appearance of the electrodeposit and its mechanical properties are so dependent on the organic content of a nickel bath that it is regrettable that their concentration cannot always be determined by simple analytical techniques. However, in many modern bright nickel solutions, at least one of the organic chemicals present can be estimated without difficulty, and sometimes a simple titration is all that is necessary. With other processes the organic material must first be separated from its aqueous solvent by distillation or liquid/liquid extraction, using a water-immiscible solvent such as chloroform or ether in order to obtain it in a concentrated form. Following this procedure, titration is most frequently used as the method of estimation.

The simplest technique involves the use of a spectrophotometer having an ultra-violet range. If one of the organic brighteners absorbs ultra-violet light, and interference from the other chemicals present is negligible at a particular wavelength, then, with prior calibration of the spectrophotometer, a very rapid analysis can be performed on the plating solution. This can be done on a sample either taken direct from the vat, or after dilution if the absorption is too strong; filtration may be necessary if the sample contains suspended matter.

In most analyses for organic compounds the estimation is based on the reaction of a particular substituent group or atomic linkage in the organic molecule. Any chemical having these same structural features will react as if it were the one to be estimated, and will be recorded as such in the results. The chances of similar organic chemicals being accidentally added to a plating bath are rather remote; it is much more probable that the chemical or electrochemical degradation of the organic brighteners will produce other compounds having similar molecular configurations but without beneficial activity. These would be estimated as their precursors and thus the analytical results, which indicate that sufficient brightener is present, and concurrent unsatisfactory plating results may be contradictory. The likelihood of this occurring is usually minimised by use of the analytical methods published for proprietary nickel plating solutions, but the difficulty must be recognised by those developing a process containing novel brighteners. Here the budding analyst must bear in mind that he is dealing with perhaps 0·01 g/l of an organic compound in an aqueous solution containing up to 400 g/l of inorganic salts.

COMMON CONTAMINANTS OF NICKEL PLATING BATHS

During industrial operation of nickel plating solutions, it is almost impossible to prevent the accidental introduction of unwanted matter and so the effect of inorganic and organic materials that commonly find their way into these baths must be considered. These two types of impurity will be discussed separately, although often they are both present in commercial plating baths and their effects are also sometimes similar. Nevertheless, their modes of introduction, estimation and the techniques for their removal do differ greatly as will be seen in the following sections.

TYPES OF INORGANIC CONTAMINATION

It is obviously difficult to prevent the nickel plating solution coming into contact with deeply recessed areas of complex shaped components which are remote from the anodes. Although cathodic, the current on these portions is not sufficiently great to result in deposition or cathodic protection and so the underlying metal is attacked by the solution. Sometimes the situation is

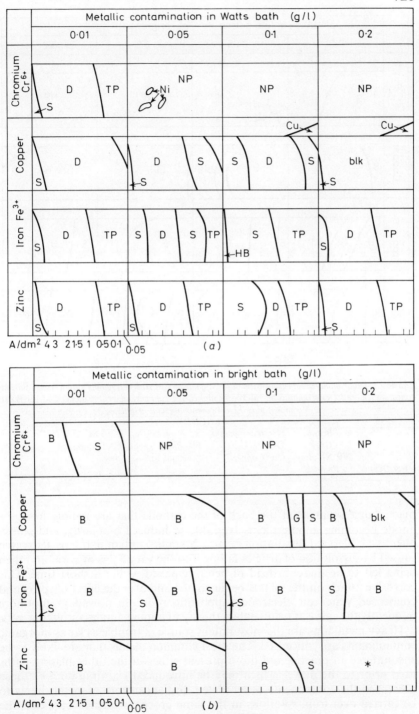

Figure 7.1 (Continued overleaf)

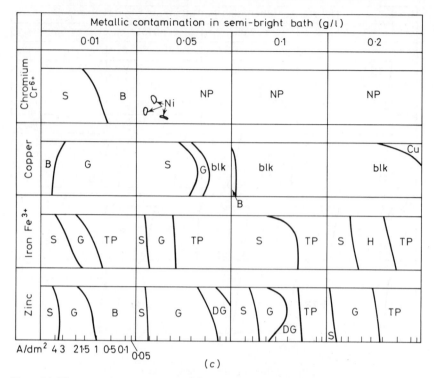

Figure 7.1. Diagrammatic representation of the appearance of Hull cell panels plated from solutions containing metallic contamination. (a) Watts nickel, (b) commercial bright nickel and (c) Watts nickel plus 0·1 g/l coumarin (semi-bright)

LEGEND

B Bright	H Hazy-bright	S Semi-bright	D Dull
G Grey	DG Dark grey	blk Black	TP Thin plate
NP No plate	Cu Copper	* Bright with skipping	

(after Dennis and Fuggle[6])

made worse by areas at the back of the cathode that are remote from the anode becoming anodic themselves, due to induced bipolarity, and consequently accelerated dissolution occurs. Metallic contamination is also often caused by dissolution of articles falling into the vat off jigs or wires and then being left to lie there, instead of being removed within a short time. This corrosive attack on the metals being coated and also the effect of spray and condensed water on electrode supporting rods are nowadays far more frequently the cause of trouble than the use of impure anodes or plating salts.

Heavy metal ions are the most potent source of trouble as far as inorganic contaminants are concerned. The most common metals that are likely to be encountered in practice are those present either in the other plating baths used prior to the nickel, which may be introduced via 'drag-out', or those present in the components being plated. In the category of metals likely to be carried over from solutions in the same process sequence, copper is the most frequently encountered. However, chromium, although normally used as a subsequent treatment, may be introduced by splashing or more probably

Table 7.1 METALLIC IMPURITIES IN WATTS NICKEL BATHS, THEIR EFFECTS AND REMOVAL

Metal	Approximate maximum acceptable concentration (g/l)		Visible effect	Recommended removal techniques
	Dull baths	Bright baths		
Aluminium	Unknown (not critical)		Possibly rough deposits	Precipitate at high pH
Cadmium*	0·5	0·05	Brightens dull, but causes darkness at low current density	Plate out at low current density
Calcium	0·5	0·5	Rough deposits	Add fluoride ion
Chromium (VI)	0·02–0·03	0·01–0·03	No deposit at low current density	Reduce to Cr(III) and precipitate at high pH
Cobalt†	2	2–40	Brightens dull, but causes bloom at low current density in bright deposits	Plate out at low current density
Copper	0·05–0·1	0·02–0·05	Dark deposits at low current density	Plate out at low current density
Iron (II)	0·2–0·5	0·1–0·5	Bloom at either high or low current density. Possibly pitting	Oxidise to Fe(III) and precipitate at high pH
Lead	0·01–0·03	0·0005–0·02	Dark deposits at low current density	Plate out at low current density
Manganese*	Unknown, but >0·2 <1		Little if any	Not known
Tin	Unknown	0·05–0·2	Dark deposits at low current density	Not known
Zinc	0·05–0·2	0·03–0·1	Dark deposits at low current density	Plate out at low current density

* Approximate values only are available for these metals.
† For dull deposits, a much higher maximum is acceptable, if the brightening effect of cobalt is not objectionable. The effect that cobalt has on bright baths is largely determined by the type of organic brighteners that are present.

by spray. Numerous metals may be nickel plated. but usually only copper. zinc. iron, lead, tin and aluminium or their alloys are of commercial interest. Electrode supporting rods are almost invariably of copper or brass.

The effect of different metals on nickel plating solutions and their electrodeposits varies greatly according to their nature and that of the bath. but in general they have a detrimental effect on appearance. levelling and ductility[4,5]. The relative effect of different metals depends to a certain extent on whether or not organic additives are present in the Watts bath and on the type of organic compounds present. Figs 7.1(a) and (b) show schematically the effect of equivalent concentrations of certain heavy metal ions on the appearance of Watts nickel and a specific bright nickel electrodeposit. as indicated on Hull cell panels plated using a current of 1A[6]: trivalent iron

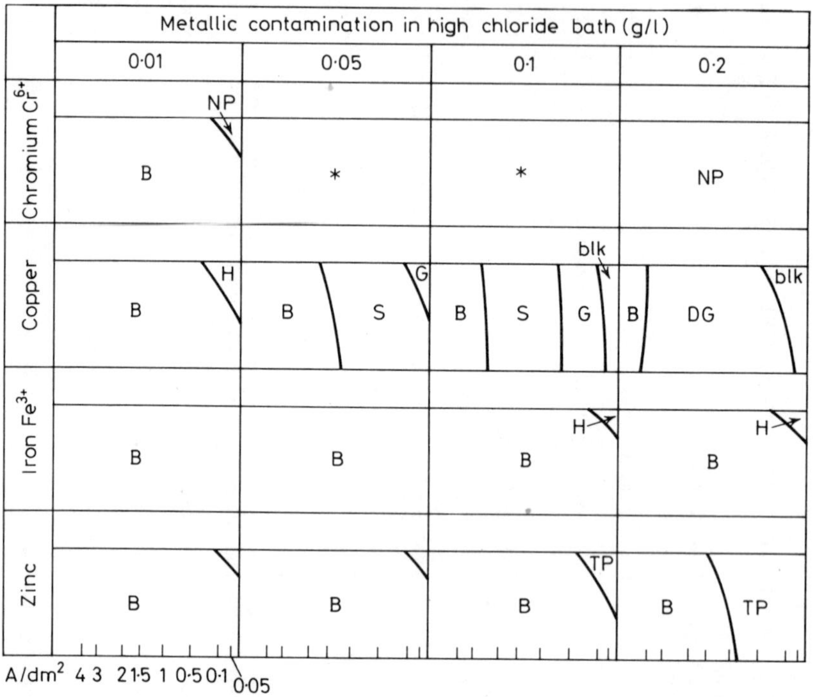

	Metallic contamination in high chloride bath (g/l)			
	0·01	0·05	0·1	0·2
Chromium Cr^{6+}	NP / B	*	*	NP
Copper	B / H / B	S	B / G / S / blk / G	B / DG / blk
Iron Fe^{3+}	B	B	B / H	B / H
Zinc	B	B	B / TP	B / TP

A/dm² 4 3 2 1·5 1 0·5 0·1 0·05

*Figure 7.2. Diagrammatic representation of the appearance of Hull cell panels plated from high chloride nickel solution plus brightener system used in Fig. 7.1(b) and containing metallic contamination (symbols as in Fig. 7.1). The panels marked with an * suffered very badly from skipping, having the majority of their surface unplated*

cannot attain the highest concentrations shown due to the relatively high pH of the nickel bath[7]. It is interesting to note that if other brightener systems are added to a Watts bath to formulate a bright nickel plating solution. the concentrations of certain metal contaminants which are permissible before their detrimental effects become noticeable can vary quite considerably by factors as great as 10 in some instances. e.g. some bright nickel solutions are very resistant to contamination by zinc[8]. The action of these metallic impurities on semi-bright nickel plating baths is also specific to the baths in question, and Fig. 7.1(c) shows the effect of the same cations on a coumarin-containing solution. Of the metals chosen, it will be noted that hexavalent chromium, copper and zinc are the most detrimental contaminants for both bright and semi-bright plating baths. Lead is also very harmful to both, although its solubility in high sulphate baths is limited. The introduction of zinc is almost inevitable if the bulk of the articles being plated consist of zinc alloy die-castings. However, it has been found that the use of a solution whose major anions are chloride rather than sulphate increases the tolerance to zinc of a bright nickel plating process. These high-chloride baths can often withstand up to ten times the content of zinc than can the corresponding baths based on the Watts solution, before the same degree of defects occur at low current density. Fig. 7.2 should be compared with Fig. 7.1(b) to observe the effect of the same four metallic contaminants when present in a high chloride or Watts bath, both containing the same brightener.

Manganese is sometimes present in a nickel solution, but usually only subsequently to a purification treatment in which potassium permanganate is employed. Aluminium is the only other metal that may be considered a common contaminant, and then only when that metal or its alloys are being plated. Table 7.1 gives some idea of the relative effects of certain common metallic impurities on the visual appearance of nickel plated at an average current density of 4 A/dm^2 from Watts baths, with and without brightening additions.

EFFECT OF METALLIC CONTAMINATION ON STRUCTURE AND PROPERTIES OF NICKEL DEPOSITS

SURFACE TOPOGRAPHY AND STRUCTURE

Only a few quantitative investigations[9] have been carried out to determine the effect of metallic impurity on mechanical properties of deposits, as distinct from appearance. This is no doubt due to the fact that some properties, e.g. ductility, are difficult to evaluate precisely. Recent work[10] has shown that in the case of a Watts nickel solution, a Watts nickel solution containing 0·1 g/l of coumarin and a typical commercial bright nickel solution, the appearance of the deposit deteriorates noticeably as a result of copper or zinc contamination before the ductility is significantly affected. Electron micrographs of the surface obtained using a scanning electron microscope reveal that the growth mechanism is modified by the presence of metallic impurity. Fig. 7.3 illustrates the surface of a Watts deposit plated at 4 A/dm^2 from a solution containing 0·0335 g/l copper. In addition to the large nodules, small ones can be observed forming all over the surface showing that the characteristic well-defined pyramid structure is not formed. At a higher concentration of copper (0·2 g/l at the same current density) the surface is completely covered with coral-type formations [Fig. 7.4(a)]. At low current density, fern-like dendritic growths occur [Fig. 7.4(b)]. To the naked eye, this latter deposit appears to be black and powdery. Bright nickels can be less sensitive to copper contamination than dull Watts nickel. With a copper content of 0·15 g/l and a current density of 4 A/dm^2, the deposit investigated has a similar number of nodules to the dull deposit plated from a solution containing 0·0335 g/l copper. The semi-bright nickel solution is even more sensitive to copper contamination than the dull Watts nickel, and also the surface topography varies with current density (Fig. 7.5).

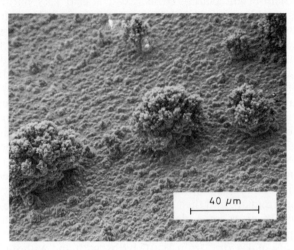

Figure 7.3. Watts nickel deposit plated at 4 A/dm^2, pH 4 and 55°C from a solution containing 0·0335 g/l copper; plating time 30 min (courtesy Arrowsmith, Dennis and Fuggle)

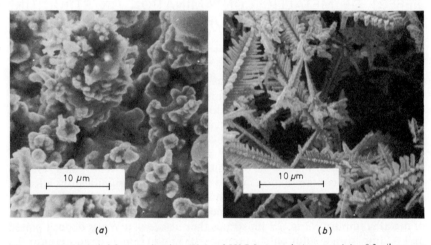

Figure 7.4. Watts nickel deposit plated at pH 4 and 55°C from a solution containing 0·2 g/l copper; plating time 15 min, (a) 4 A/dm^2 and (b) 1 A/dm^2 (after Dennis and Fuggle[6])

Zinc contamination can have a partial brightening effect on dull nickel. Scanning electron micrographs show that the surface topography is modified in such a manner that the pyramids are not as clearly defined as those present in deposits plated from uncontaminated solution; the peaks are rounded off as shown in Fig. 7.6(a). This effect is rather similar to that produced by adding coumarin to a Watts solution which also results in a semi-bright

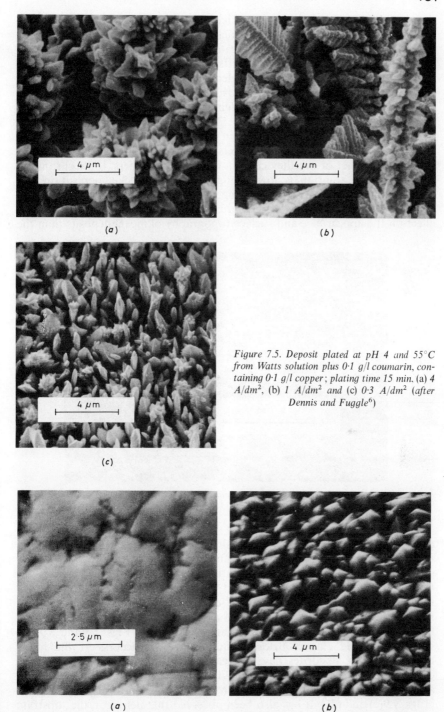

Figure 7.5. Deposit plated at pH 4 and 55°C from Watts solution plus 0·1 g/l coumarin, containing 0·1 g/l copper; plating time 15 min. (a) 4 A/dm^2, (b) 1 A/dm^2 and (c) 0·3 A/dm^2 (after Dennis and Fuggle[6])

Figure 7.6. Watts nickel deposit plated at 4 A/dm^2, pH 4 and 55°C from a solution containing zinc; plating time 15 min. (a) 0·1 g/l zinc and (b) 0·2 g/l zinc

deposit. At current densities at which a dull deposit is obtained in the presence of zinc, the pyramid structure is sometimes particularly well defined [Fig. 7.6(b)]. Zinc has a greater effect on the appearance of semi-bright nickel deposits than on Watts and bright deposits. The rounded pyramid growths characteristic of a deposit plated from an uncontaminated Watts bath containing 0·1 g/l coumarin become much rougher when zinc is present but there is no tendency for dendritic growth as in the case of copper contamination.

Chromium (VI) contamination, even at a low concentration, is particularly detrimental since it completely inhibits the deposition of nickel from Watts solutions whether or not they contain organic brighteners. If the chromium content of the solution is low enough to permit nickel deposition, the surface topography is usually unchanged from that of a deposit plated from an uncontaminated bath. Ferric iron contamination has a less detrimental effect on the appearance and surface topography of nickel deposits than the

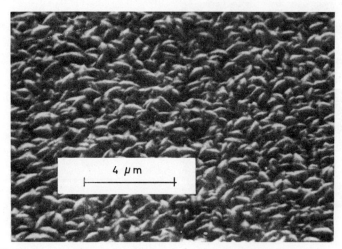

Figure 7.7. *Watts nickel deposit plated at 1 A/dm², pH 4 and 55°C from a solution containing 0·2 g/l trivalent iron; plating time 15 min (after Dennis and Fuggle[6])*

other three aforementioned metals, although it does result in modification of surface topography in some instances. For example, Fig. 7.7 shows a deposit plated from a Watts nickel bath containing 0·2 g/l Fe^{3+}.

In addition to the effect of current density, the time of plating also influences the surface topography of deposits plated from contaminated solutions. The results of plating from a Watts solution contaminated with 0·1 g/l of copper at 0·5 and 4 A/dm² are illustrated in Figs. 7.8 and 7.9. An initiation period elapses before dendritic growth occurs. At the lower current density, dendrites develop readily even when the coating is still quite thin [Fig. 7.8(b)], but at the higher current density renucleation appears to take place on pyramidal growths before dendritic growth occurs [Figs. 7.9(c) and (d)]. It has also been observed that defects in the substrate surface can lead to preferential growth of a Watts nickel deposit plated from a solution contaminated with copper [Fig. 7.8(a)]. This has also been demonstrated by deliberately introducing surface defects in the substrate by making microhardness indentations. Preferential growth occurs around the edge of the indentation as shown in Fig. 7.10. Deposition from an uncontaminated Watts bath onto a similar indented substrate does not result in

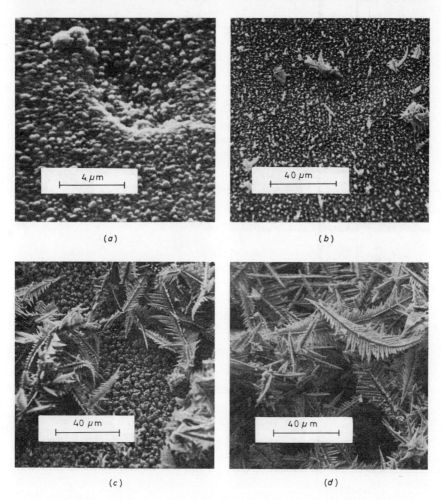

Figure 7.8. Deposits plated from Watts solution containing 0·1 g/l of copper at 0·5 A/dm^2 for varying times; (a) 8 min, (b) 12 min, (c) 24 min and (d) 2 h (after Dennis and Fuggle[6])

preferential growth. The above results indicate that a low quality surface finish is likely to be exaggerated in the deposit if the plating solution is contaminated, particularly if a low current density is employed. The effects of contamination on the appearance and surface topography of electrodeposits are greatest when the plating conditions result in a high concentration of contaminant metal in the deposit.

From the point of view of appearance, metals such as copper can only build up to a dangerous level in solution if the rate of contamination is high. However, it is evident that contamination should be avoided as far as possible to prevent detrimental effects on properties that would only be revealed after some period in service. If copper contamination is present,

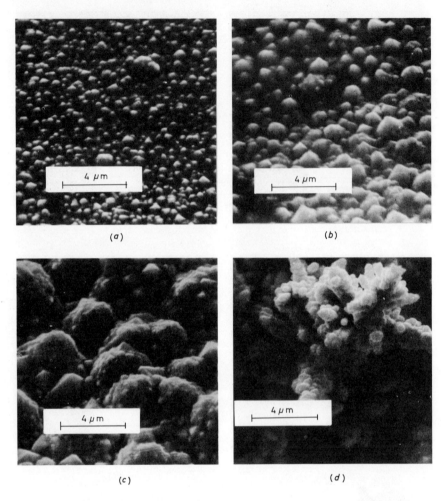

Figure 7.9. Deposits plated from Watts solution containing 0·1 g/l copper at 4 A/dm^2 for varying times; (a) 3 min, (b) 5 min, (c) 15 min and (d) 30 min (after Dennis and Fuggle[6])

the metal is continually plated out while the vat is in use. The rate of removal at 4 A/dm^2 is shown in Fig. 7.11. It is a simple matter from this graph to calculate the amount of copper that would be removed from a particular vat in, for example, a work-shift and hence the maximum quantity of contamination that could be continuously introduced without raising the copper content.

In the case of copper and zinc, and this is almost certainly true for many other metals, the ratio of the concentration of contaminant to nickel is

Figure 7.10. Watts deposit plated from a solution containing 0·1 g/l copper, at 0·5 A/dm², onto a substrate having micro hardness indentations in its surface. Preferential growth has occurred at the edges of the indentation

greater in the deposit than in the plating solution (Fig. 7.12). For example, in the case of copper in Watts nickel (curve B), at a concentration of 0·1 g/l copper, the ratio of Cu^{2+}/Ni^{2+} is 1·5/1000 while the ratio of Cu/Ni in the deposit is 12·5/1000. There is therefore a high probability that the concentra-

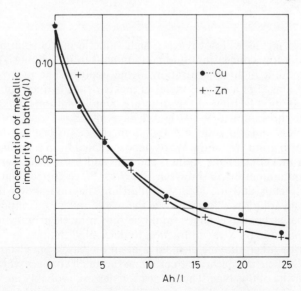

Figure 7.11. Rate of plating-out of metallic impurity from Watts solution at 4 A/dm², pH 4 and 55°C (after Dennis and Fuggle[10])

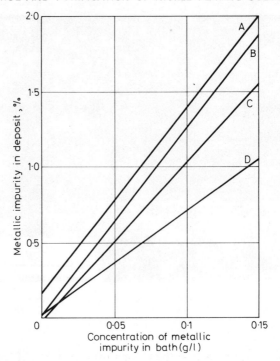

Figure 7.12. Relationship between the concentration of metallic contamination in solution and deposit.

LEGEND

A Watts plus coumarin plus copper   C Watts plus zinc
B Watts plus copper   D Watts plus coumarin plus zinc

(*after Dennis and Fuggle[10]*)

tion of most metals will not reach a high value in nickel solutions; on the other hand a low concentration of contaminant in the plating solution can result in quite a high concentration in the deposit. The incorporation of small percentages of a second metal in nickel deposits does not result in a great deal of modification of the structure. Small additions of copper, zinc, cobalt and chromium to Watts nickel have all been shown to result in a somewhat less uniform grain size and an increased dislocation density. The fairly uniform grain size of the Watts deposit plated from a purified solution containing nickel sulphate, sodium chloride and boric acid is modified by metallic contamination as shown in Fig. 7.13. Since copper and nickel are both f.c.c. metals and the binary equilibrium diagram shows that a continuous solid solution is formed over the whole range of composition, it is unlikely that structural changes would be shown by electron microscopy. It would be possible by X-ray diffraction to determine changes in d spacing as the copper or other alloying element content increases, but electron diffraction is not sufficiently accurate to evaluate the small changes which occur in electrodeposits of this type. The dislocation density probably increases since distortion of the lattice is likely to occur to accommodate the atoms of the alloying element. Since nickel and copper should alloy as a solid solution in accordance with the equilibrium diagram, it is not surprising that copper does not cause drastic reductions in ductility. However, electrodeposited alloys can have a completely different structure to ones of the same com-

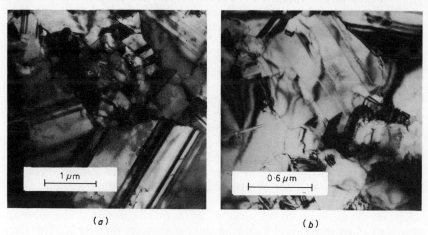

Figure 7.13. Electron micrographs of deposits plated from Watts bath containing metallic contamination; plating conditions: 4 A/dm², pH 4 and 55°C. (a) Watts solution plus 0·0067 g/l copper and (b) Watts solution plus 0·1 g/l zinc (courtesy Dennis and Fuggle[10])

Figure 7.14. Electron micrograph of a deposit plated from a semi-bright solution containing 0·0067 g/l copper; plating conditions: 4 A/dm², pH 4 and 55°C (courtesy Dennis and Fuggle[10])

position produced by conventional metallurgical processes. In the case of semi-bright nickel deposits contaminated with copper and zinc (Fig. 7.14), the structure is more clearly defined than in the absence of metallic impurity. This is probably due to a slight atomic rearrangement under the influence of stress in the deposit and it results in annihilation of dislocations of opposite sign. Usually the grain size of the contaminated deposit is less than that of the uncontaminated. No change can be detected in the structure of bright nickels on incorporation of a second metal. The grain size is extremely small and any changes in structure that occur are on such a small scale that they cannot be detected by thin foil electron microscopy.

MECHANICAL PROPERTIES

On many occasions. metallic impurities have been blamed for the loss of ductility of nickel coatings. but very little evidence is available to substantiate this. The effects of copper and zinc have been examined in some detail. and the loss in ductility has been shown to be quite moderate. certainly not catastrophic. The appearance would be quite unacceptable before the decrease in ductility is likely to reduce the value of the coating. This has been shown to be true for Watts solutions with or without organic additions[10].

The co-deposition with the nickel of other metals present in the solution produces alloys which have higher hardnesses than the pure metal. This may be beneficial for certain applications where an electrodeposit having high strength is required. and cobalt is particularly well documented in this respect[11-13]. Cobalt in the proportions at which it is usually present in commercial nickel salts has no significant effect on Watts deposits[14]. but does affect the proporties of deposits from sulphamate baths[15]. In some cases. the effect of larger quantities of cobalt in producing a stronger and hence a harder and less ductile deposit may be undesirable either in nickel plated from Watts or sulphamate baths[16].

The internal stresses of electrodeposits might be expected to be affected by alloying metals. even in low concentrations. and at least one metallic impurity, viz. iron has been shown to be deleterious by increasing tensile stress[17].

Other inorganic materials can be introduced accidentally or deliberately. Sodium and potassium ions have been claimed to embrittle deposits from Watts nickel solution. but more recent experience indicates that even high concentrations have no significant effect on the ductility of dull or bright nickel deposits[14]. Ammonium ions certainly do have a 'hardening' effect on Watts nickel and bright nickel deposits, although the initially greater hardness of the latter makes it more difficult to detect. However, small concentrations of around 1 g/l of NH_4^+ can have an auxiliary brightening effect without having a significant effect on the mechanical properties of the electrodeposits when bright plating is being performed. It is most unlikely that large quantities of these three cations could be introduced. except by desire or mischance. However, the water used for initial solution of the constituents of a bath can contain calcium ions and silica. amongst other impurities. and these unavoidably accumulate. if the same water is used for maintaining the level of that bath. Both of these can cause roughness if they become precipitated; in the case of Ca^{2+}, $CaSO_4$ (solubility 2·2 g/l at 50°C) readily precipitates if hard water is used for 'topping up' sulphate-containing solution. Another defect that results from the ensuing crystallisation of calcium sulphate is pitting of the deposit due to blocking of the holes of the air-agitation coils.

CORROSION RESISTANCE

Co-deposition of other metals with nickel causes a change in potential, but as shown by Ewing[9] not all metals result in a decrease in corrosion resistance. For example, he has demonstrated that the incorporation of copper in the plated coating decreases its resistance to the salt spray corrosion test, whereas zinc has the opposite effect. Changes in nobility can affect the performance of the coating in service, particularly if the contaminated deposit is one layer of a multi-layer nickel plus chromium coating. As described in Chapters 10 and 11, the most satisfactory results can be achieved with many of these systems only when the potential of each layer is controlled precisely.

ANALYSIS FOR INORGANIC IMPURITIES

The estimation of inorganic compounds in nickel plating baths often presents difficulties[4] and frequently their presence is diagnosed by the symptoms they cause rather than by direct identification. Fortunately, the blanket methods favoured for removal of metallic impurities are sufficiently embracing to remove most offending metals, whatever they may be. However, recognition of the specific deleterious element or elements might enable more positive steps to be taken to prevent a recurrence of the contamination. While the use of a Hull cell will, with experience, enable the presence of some metals to be positively identified[18], this is not always so, and the presence of more than one metallic impurity complicates their detection. Still more difficulties are experienced in relating the appearance of panels plated in the standard Hull cell to those that are obtained from nickel plating barrels, and so for the control of all baths used for barrel plating, a specialised cell has been devised[19] to simulate the conditions existing in this process. In this cell a V-shaped cathode rotates in the solutions under test, being separated from the anode by a perforated-plastics screen. In the case of bright nickel solutions which are devised so as to mitigate the effect of unwanted metals on their appearance, there may be a deterioration in some other properties before the presence of the impurity produces a visible effect, even at low current densities. For example, the Rousselots[20] have demonstrated that the presence of contaminating metals may affect the chromability of the nickel plate. Also, the co-deposition of certain metals, in particular copper[9], may adversely affect the corrosion resistance of nickel coatings. Thus, it would often be an advantage to be able to determine rapidly the concentration of metals in a nickel plating solution, e.g. so as to be able to assess the benefit of any purification treatment performed.

Many investigations have been conducted with this aim in mind. Most of them have been based on colorimetric techniques which are suitable for the small quantities which must be estimated, while a few have used polarography[21]. With both types of method some prior wet separation is usually required and because this can be tedious and require the use of very pure reagents, analysis for metallic impurities has neither been a rapid nor rewarding exercise. Serfass and his colleagues[22] have published the most comprehensive text on this topic. While the recent development of continuous liquid/liquid extraction apparatus shows promise for automating the colorimetric method, some disadvantages of this method still remain and hinder its wide-spread application. However, the colorimetric method is sometimes useful for metals whose salts absorb light of a very different wavelength to nickel salts, e.g. cobalt[23].

The invention of the Atomic Absorption Spectrometer has radically changed the analytical position. This instrument is so simple to use and so specific that almost any metallic element can be determined in nickel plating solutions[24], providing suitable lamps are available. Obviously, only lamps for the commonly encountered elements are essential for routine control analysis.

The only pretreatment required to the sample of nickel plating solution is simple dilution with water, depending on the anticipated content of the metal being estimated. Another reason for not using the undiluted solution is that the boric acid present blocks the burner, therefore in trace analysis the sample is diluted five times with water. Using this technique the approximate minimum concentrations of metals that can be determined in any Watts type of bath are 0·005 g/l for copper, zinc and manganese, 0·01 g/l for iron and chromium and 0·025 g/l for lead, tin and aluminium. Precise limits of detection vary, depending on the particular instrument used, the more sophisticated and costly models being the most sensitive.

REMOVAL OF METALLIC CONTAMINATION

The methods used for removal of offending metals may appear crude, but are reasonably successful[24-26]. It is obviously preferable, however, to prevent their concentrations rising to such a level as to cause defective deposits. It is often difficult to avoid introduction of contaminants even with careful operation, and so the routine application of removal treatments is frequently needed. Of these, low current density 'plating out', i.e. electrodeposition onto nickel-plated metal sheets or scrap, using much lower current densities than normally employed, is the most common and often the most effective. The most contaminant-rich nickel alloy obtained per coulomb[27] requires a low current density of 0·2–0·5 A/dm^2 and does not therefore involve excessive deposition of nickel. Purification can be a lengthy procedure if the concentration of the contaminating metal becomes excessive and is best performed as a routine operation during nights or week-ends or whenever the plant is idle for more than an hour or two. Medium current densities may have to be employed if gross contamination occurs and only a short time is available for chemical treatment. There must be a physical limit to the total size of the cathodes that can be inserted into any plating vat and still allow sufficient access of solution and anodes to maintain the efficiency of the process. Use of higher current densities will allow more current to be passed per unit volume and thus will reduce the total plating time required; since the proportion of nickel in the electrodeposit will be higher, the proportionate decrease in impurity content is not as great as the increase in deposition rate of alloy. Nevertheless, at the expense of more nickel usage, the 'plating-out' operation can be speeded up.

Iron is probably the commonest contaminant encountered in practice. Fortunately, its elimination is very simple. Raising the pH of the plating solution to at least 5·5 and preferably 5·8 is all that is required. Before doing so, it is often beneficial to oxidise the iron to the trivalent state by the addition of hydrogen peroxide, providing that this oxidation does not have detrimental effects on any organic additives that are present, for ferric hydroxide is much less soluble than ferrous hydroxide. In an air-agitated bath, the oxygen in the air passing through will have some oxidising effect, and this assists in preventing the accumulation of iron. The pH of the solution can be increased by careful addition of sodium hydroxide, nickel carbonate or freshly precipitated nickel hydroxide, and the resultant precipitate filtered off.

The well-known property of ferric hydroxide precipitates of occluding or adsorbing other metal hydroxides or organic compounds, which is so undesirable in quantitative analysis, can be of use in nickel plating baths to remove other impurities. Thus precipitation of iron can serve a dual function. This phenomenon has been utilised in a purification technique known as the *Liscombe* treatment, which involves the deliberate addition of ferrous sulphate. Although this is now regarded as obsolescent, it can sometimes be usefully employed as a non-selective method for 'cleaning' a nickel bath containing unidentified contaminants. The ferrous iron is then oxidised and precipitated and must be completely removed before the pH of the bath is lowered to its normal value.

High contents of heavy metal ions can be removed by precipitation as their sulphides. Any excess of sulphide remaining in solution after filtration is oxidised by hydrogen peroxide. It may be necessary in the case of bright nickel baths to perform a subsequent carbon treatment, followed by replacement of the organic additives.

The complexing or sequestering of various metal ions has been suggested as an alternative to their removal. For example, fluoborates and citrates have been used for complexing iron. Organic sequesterants have also been suggested for that metal and others, such as zinc. These compounds must obviously be selective so that they do not preferentially combine with the much higher concentration of nickel ions. There is no doubt that some of these do complex the undesirable ions and prevent them exhibiting their detrimental effects. However, the reaction is stoichiometric and consequently as the impurity content rises so must more of the complexant be added. High concentrations of certain complexants may themselves detract from the qualities of the electrodeposits. Because of this and because of the fear that if the complex were to break down the bath would be swamped with metallic impurities, little use has been made of these compounds in commercial practice.

As stated previously, chromium, particularly in its hexavalent form, is one of the most harmful metals that can enter a nickel plating bath. A specific means for removing it is recommended; this is based on the Liscombe method and uses ferrous sulphate to reduce the hexavalent chromium, which is then precipitated, together with ferric hydroxide, by adding freshly prepared nickel hydroxide.

Calcium when present in the form of its slightly soluble sulphate can be rendered innocuous by converting it to its very insoluble fluoride by addition of hydrofluoric acid or its alkali metal salts followed by filtration.

TYPES OF ORGANIC CONTAMINATION

It is questionable whether it is extraneous inorganic or organic compounds that have the most detrimental effect on the appearance or other properties of deposits obtained from nickel plating solutions. To ignore either is to court disaster. The external organic contamination can take many forms, but assuming that the plastics materials of construction of the plating vats and jig coatings are chosen correctly, it is most likely to take the form of grease-containing polishing compositions, lubricants or other soils carried into the plating bath on imperfectly cleaned work. Airborne dust can also enter and may include polishing materials, if the polishing shop is not physically separated from the plating plant.

Probably the most potent source of unwanted organic chemicals is intrinsic in the solution itself, if this is of the bright plating type. Most organic addition agents are not stable as is demonstrated by the frequent replacement additions that are required, particularly of the more active brighteners and levellers. Unless these addition agents are volatile, their disappearance can only result from their being incorporated in the deposit or changed ('broken-down') into other molecules. Consequently, their 'breakdown products' accumulate in the solution unless these are, in turn, volatile or insoluble. For example, 90% of coumarin contained in a Watts bath operated at pH 4 is cathodically reduced to melilotic acid and this degradation product is detrimental to the deposit. Sometimes, as in this particular case, they compete with the original compounds for adsorption on the growing electrodeposit and reduce their brightening or levelling effect. In other cases, the appearance of the nickel plate may be visibly affected, particularly at low current densities, where darkening may occur or, in severe instances, there may be total or partial absence of deposit, colloquially termed *skipping*. Pitting of the deposit may also occur and this fault is the main one caused by external contaminants of an oily or greasy nature.

ANALYSIS FOR ORGANIC IMPURITIES

Where the precursors of the extraneous organic matter are known, as in the case of addition agents, this knowledge assists in their identification and estimation. The position is also easier if only one organic additive is originally present, as in the case of the majority of semi-bright baths containing coumarin. Melilotic acid is the major product produced by the electrochemical reduction of this compound at the cathode and this can be estimated by measuring the absorption of the aged plating solution at a different wavelength to that used for coumarin. It is not always as simple to differentiate between two similar organic compounds having much the same structure, as mentioned above, even if they do absorb ultra-violet light[28]. Techniques which are of great assistance in characterisation of organic compounds, such as infra-red spectroscopy, are of little value where aqueous solutions are concerned. However, one method which has been found to be of help in specific cases is that of Vapour Phase Chromatography using a flame ionisation detector, since the latter is not rendered useless by the passage of water vapour. Thin-layer[28, 29] and paper chromatography have also been found useful tools in certain investigations.

It can be seen that the estimation of the breakdown products of known organic compounds is usually most difficult and these difficulties are enhanced if the organic matter is of a complex nature, such as fat mixtures, or of entirely unknown origin. For these reasons, chemical analysis for organic contamination is rarely attempted.

A crude method that has been utilised is to estimate the reducing power of the contaminated nickel plating solution as compared with that of a new one by titration of both with potassium permanganate. While this technique has little to recommend it for quantitative analysis, it has been found that certain organic contaminants are completely destroyed if sufficient potassium permanganate to produce a permanent purple colour is added to the affected bath[30].

REMOVAL OF ORGANIC CONTAMINATION

By far the most popular technique employed for abstraction of all types of organic contamination is the use of activated carbon[25, 31], i.e. the decolourising charcoal of the organic chemist. There are many varieties of activated carbon, but that formed from coconut shells is a grade found most suitable. However, it is no use expecting that even the best carbon will remove all organic matter from aqueous solutions, since these materials are selective in their action. Sometimes activated carbon is sufficiently selective to remove the breakdown products of addition agents without adsorbing a significant quantity of the additives themselves, but more often their structures are so similar that their adsorption coefficients are the same. A good example of this is given with nickel plating baths containing both coumarin and melilotic acid. The latter compound is readily removed by carbon, but so is coumarin. Therefore carbon treatment of the bath will result in the elimination of both compounds, providing sufficient carbon is employed. Even if a smaller quantity of carbon is added, the coumarin will not preferentially remain in the solution, but a partial removal of it and the melilotic acid will result. Theoretically, the nickel plating solution can be freed from organic compounds if sufficient carbon has been used and should then be in the same state as when first prepared, ready for the desired organic compounds to be added and able to give the same quality of electrodeposits as it did when new. Addition of hydrogen peroxide or potassium permanganate sometimes oxidises the organic matter into forms which are more easily removed by carbon, and it has been found that certain organic contaminants are almost completely destroyed if two separate additions of carbon are made, with an intermediate oxidation treatment.

In certain proprietary baths some of the brighteners will remain after carbon treatment. This is of little consequence if the content of their breakdown products or extraneous organic matter has been lowered: the unbalance of addition agents that may result can easily be rectified. If the additives are not adsorbed or are only very slightly adsorbed onto activated carbon, this property can be utilised by passing the plating solution continuously through this material, usually by putting a layer of it onto whatever filter medium is being used, or by introducing a carbon-containing cartridge into the filtration system. Provided that the carbon is renewed regularly at fairly frequent intervals, this continuous treatment can prevent the organic impurities—from internal or external sources—accumulating to a level at which harmful effects are produced. Thus the need for a complete purification of the whole of the solution is obviated, or at least the time between such treatments can be greatly extended.

For activated carbon to be effective, it must be in a fine form in order to possess a large surface area per unit mass, but not so fine that it will not be retained on a filter media. A size range of 70 to 100 μm is generally satisfactory. Care must be observed that the carbon used for removal of unwanted organic matter does not introduce detrimental inorganic material, since phosphates or heavy metals may be present in some grades.

Although carbon is by far the most popular and versatile material used for cleaning nickel plating solutions, other adsorbents have been found successful for specific organic compounds. For example, an amine which is ionised at the pH of the plating solution can be removed by ion exchange on activated earths. The Liscombe method is another technique useful on certain occasions.

It must be emphasised that plating out of an unknown organic impurity can be effective, although this operation may be protracted if low current densities are employed or if a high impurity content is present.

FILTRATION OF ELECTROPLATING BATHS

It is obvious that carbon and other substances used for purification must be removed from the solution before plating is recommenced, and although decantation can assist in this, some filtration of the solution will be required. This filtration of the plating bath is equally essential during normal operation. The incorporation in the electrodeposit of solid matter suspended in the solution will often result in these particles forming the nuclei of nodular growths. This is much more probable if the particles are conducting.

Unfortunately, the possibilities of solids entering an industrial plating bath are manifold. The external sources of airborne dust and heavier particles dropping from above are always a hazard, and, while certain precautions can be taken, most plating shops cannot be isolated from the dust created in many engineering factories or indeed from the grit still present in the atmospheres of far too many industrial towns. Also there is the possibility of abrasives, loose scale or other surface soils still clinging to the surface of the work when it reaches the nickel plating stage, due to the prior cleaning and/or pickling being unsatisfactory. In addition there may be danger of leakage of nickel grains through anode bags, for these must not be completely impervious to the plating solution, and the risk of a bag becoming accidentally perforated is always present.

For all these reasons, the continual passage of the plating solution through some material that will strain out all but the very smallest particles is most essential. Many types of filtration equipment are available, from the old type of plate filter to the modern types which can be cleaned by back-washing with water and so do not need to be dismantled and reassembled every time they become clogged. The actual filter media are manufactured from various natural and synthetic fibres, e.g. cloths of cotton, wool, Terylene or polypropylene. Filter papers or pads made of resin-impregnated pulp are also used, either independently or in combination with filter cloths. The degree of filtration is often improved by coating these fabrics with a layer, held on by means of the filter's own suction, of some insoluble powder, such as cellulose, asbestos, carbon or Kieselguhr, the last named material being suitable only for acid baths. If the carbon is the activated type, it serves a double purpose of filtration and purification, or if unactivated it is then more suitable for strongly alkaline solutions than for acid nickel baths. Other types of filter are based on the use of porous cartridges of various natures; resin-impregnated paper is an example of the disposable type and wound (overlaid) string exemplifies the type that can be cleaned. Further details of the construction and use of the different types of filters can be found in chemical engineering texts[32].

The rate at which the bath should be filtered depends partly on the input of contamination and partly on other factors such as the thickness of the nickel being plated and whether or not there are upward-facing surfaces on the articles being processed. For air-agitated solutions this rate should not be less than one volume of the tank per hour and preferably two or three times per hour. The rate of filtration can be less for still solutions but, even in these, particles may not settle immediately and, if they do, will probably be raised by the stirring needed to avoid stratification or to dissolve and disperse uniformly any additions that are made. Therefore the use of some filtration is most advisable with all types of nickel plating bath.

There is still some possibility of a persistent roughness which occurs on steel articles even when these have been satisfactorily cleaned and then plated in a well-filtered nickel solution. This can be caused by the steel becoming magnetised and thereby attracting and holding to it any ferromagnetic particles that have been previously encountered. These particles are held tenaciously and the best remedy is to demagnetise the steel before processing.

There is also a tendency for slivers which are raised by grinding or some other form of coarse abrasion of metal, particularly on steel, to be attached at one end but to stand proud and to remain so after normal acid dipping and alkaline cleaning[33]. More depth than usual of the metal surface must be dissolved in order to detach these slivers and this can be achieved by anodic dissolution, as in an acid etching or electropolishing process.

REFERENCES

1. LANGFORD, K. E. and PARKER, J. E., *The Analysis of Electroplating and Related Solutions*, 4th edn, Robert Draper, Teddington (1971)
2. FOULKE, D. G. and CRANE, F. D., *Electroplaters' Process Control Handbook*, Reinhold Publishing Corp., New York (1963)
3. ARMET, R. C., *The Modern Electroplating Laboratory Manual*, Robert Draper, Teddington (1965)
4. GREENALL, C. J. and WHITTINGTON, C. M., *Plating*, **53**, 217 (1966)
5. STAHL, G. and WILD, P. W., *Galvanotechnik*, **60**, 429 (1969)

6. DENNIS, J. K. and FUGGLE, J. J., *Trans. Inst. Metal Finishing*, **48**, 75 (1970)
7. EWING, D. T., BROUWER, A. A. and WERNER, J. K., *Plating*, **39**, 1339 (1952)
8. ALLEN, R., Editor of *Canning Handbook of Plating*, 21st edn, W. Canning & Co. Ltd., Birmingham, 377 (1970)
9. EWING, D. T., ROMINSKI, R., KING, W. and GORDON, W. D., *Effect of Impurities and Purification of Electroplating Solutions*, Research Reports Nos. 13 and 15 of American Electroplaters' Society (1949 and 1950)
10. DENNIS, J. K. and FUGGLE, J. J., *Trans. Inst. Metal Finishing*, **46**, 185 (1968)
11. ENDICOTT, D. W. and KNAPP, J. R., *Plating*, **53**, 43 (1966)
12. SAFRANEK, W. H., *Plating*, **53**, 1211 (1966)
13. MCFARLEN, W. T., *Plating*, **57**, 46 (1970)
14. ZENTNER, V., BRENNER, A. and JENNINGS, C. W., *Physical Properties of Electrodeposited Metals. I: Nickel*, Research Report No. 20 of American Electroplaters' Society (1952)
15. BELT, K. C., CROSSLEY, J. A. and KENDRICK, R. J., Proc. 7th International Metal Finishing Conference, Hannover, 222 (1968)
16. MARTI, J. L. and LANZA, G. P., *Plating*, **56**, 377 (1969)
17. CURKIN, L. H. and MOELLER, R. W., *Plating*, **41**, 1154 (1954)
18. NOHSE, W., Chapters 4 and 5 of *The Investigation of Electroplating and Related Solutions with the aid of the Hull cell*, Robert Draper, Teddington (1966)
19. CHAPDELAINE, E. G., *Plating*, **53**, 471 (1966)
20. ROUSSELOT, R. H. and G. E. Metal Finishing, **56** No. 12, 46 (1958); **57** No. 1, 58 (1959)
21. WILD, P. W., *Galvanotechnik*, **60**, 757 (1969)
22. SERFASS, E. J., LEVINE, W. S., PRANG, P. J., OYLER, J. E., DAVIES, R. M. and PERRY, M. H., *Determination of Impurities in Nickel Electroplating Solutions*, Research Reports Nos. 3 and 6 of American Electroplaters' Society (1949 and 1947)
23. SADAK, J. C. and SAUTTER, F. K., *Plating*, **56**, 1041 (1969)
24. WHITTINGTON, C. M. and WILLIS, J. B., *Plating*, **51**, 767 (1964)
25. POMERANZ, M., *Traitements De Surface*, **10** No. 85, 25 (1969)
26. FOULKE, D. G., *Metal Finishing*, **67** No. 6, 99 (1969)
27. CASE, B. C., *Proc. Am. Electroplaters' Soc.*, **34**, 228 (1947)
28. ASHURST, K. G., *Trans. Inst. Metal Finishing*, **40**, 74 (1963)
29. RUPPRECHT, W. E., *Analyst*, **94**, 126 (1969)
30. HOTHERSALL, A. W. and GARDAM, G. E., *J. Electrodepositor's Tech. Soc.*, **12**, 81 (1937)
31. HELBIG, W. A., *Proc. Am. Electroplaters' Soc.*, **29**, 68 (1941)
32. SALTONSTALL, R. B., Chapter 36 of *Electroplating Engineering Handbook*, 3rd edn, edited by GRAHAM, A. K., Reinhold Publishing Corp., New York (1971)
33. PINNER, W. L., *Proc. Am. Electroplaters' Soc.*, **40**, 83 (1953)

Chapter 8

Physical and Mechanical Properties of Electrodeposits and Methods of Determination

Electrodeposition provides a unique method of obtaining predetermined physical and mechanical properties from a particular metal, which in some instances cannot be achieved by any other metallurgical operation. A wide range of properties is possible for most electrodeposited metals and nickel is no exception. because of the special characteristics of the electrodeposition process obtainable through differing electrolyte solutions and consequently of the electrodeposits from them. Properties vary because distortions of the nickel lattice are induced by the presence of occluded matter. Electrodeposited nickel has been shown to have various elements incorporated in it. in particular. oxygen and hydrogen, and sometimes carbon. sulphur and nitrogen. Incorporated materials arise from the base electrolyte solution and from the organic compounds added to it. Structural modifications are also due to the nature of the electrocrystallisation process itself. i.e. the process of transfer of ions through the Helmholtz double layer and their incorporation in the growing cathode.

In Chapter 6. the physical and mechanical properties required of nickel coatings were outlined, and it is the purpose of this chapter to discuss the most important of these properties in more detail and, in particular, to describe the most important methods available for their evaluation. No attempt will be made to review all the methods suggested for the evaluation of the various properties.

DUCTILITY

Four types of ductility test can be carried out on electrodeposits. viz. bend. torsion. tensile and bulge. Furthermore, these can be performed either on stripped foils or on plated test specimens. Unfortunately. the results of one type of test cannot be correlated easily with those of another. The simpler forms of test require very little equipment; a bend test on a plated test specimen is carried out by bending over a mandrel. while a bend test on a foil can be carried out using a micrometer as a miniature vice. In the latter case, the foil is bent into a 'U' shape and inserted in the micrometer which is then screwed up until the foil cracks; in this way a semi-quantitative result is obtained. The thickness of the coating must always be taken into consideration in bend tests and, if a plated specimen is tested, the thickness of the substrate must also be known. Bend tests on plated components enjoy considerable popularity with platers on the shop floor since these are quick and easy to carry out. They suffer from the disadvantage that they can be classed only as *no* or *no-go* tests. If specimens are bent over a series of mandrels, the result can be quoted as greater than one value but less than another; obviously the smaller the increments in mandrel size the more accurately the ductility can be assessed. The percentage elongation E occurring at the surface of the coating after bending over a mandrel can be shown to be:

$$E = \frac{100\,t}{d + t} \qquad \ldots (8.1)$$

where t is the thickness of the substrate plus coating and d is the diameter of the mandrel. Failure is indicated by the cracking of the deposit and this is where experience is valuable if accurate and consistent results are to be

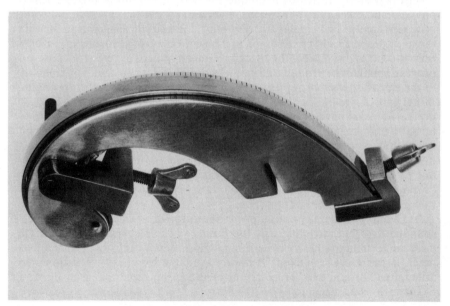

Figure 8.1. Spiral bend test apparatus for determination of ductility of electrodeposits. A test strip is shown clamped in position after bending

obtained. Only cracks which penetrate across the whole width of the specimen are considered to constitute failure in the case of bright nickel deposits. On the other hand, the more ductile semi-bright deposits rarely fail in this way and discontinuous cracking is usually accepted as the criterion for failure. This is still the method recommended in BS 1224:1970[1], but unlike the earlier versions of the standard the latest one advocates its use only for dull and semi-bright coatings. It is stated that these should have a ductility greater than 8% as determined by this method.

A modification of the simple bend test was introduced by Edwards[2]; this is known as the *Spiral Bend Test* and is suitable for determining the elongation of relatively brittle coatings. It makes use of a former having the profile of a logarithmic spiral, and the test piece is clamped at one end and bent around the former (Fig. 8.1). The periphery is divided into arbitrary units and the corresponding elongation can be obtained from the appropriate calibration chart. As with the simple mandrel test, the thickness of coating and substrate must be known and cracking occurs in a similar manner. The extent of cracking is influenced by the time that elapses after bending and to obtain consistent results the reading should always be noted either immediately or at a standard time after bending. The edges of test strips used for bend tests should be chamfered to remove the build-up of the deposit and also to eliminate any stress-raisers that might cause premature cracking.

Tensile tests provide a more accurate means of determining the ductility of electrodeposits, and in one form or another are used by most workers in the field at the present time. The earliest of these was the method put forward by Such[3] in which the ductility was expressed in arbitrary units. All modifications of this technique make use of a tensile testing machine, usually the Hounsfield tensometer, and some form of standard brass or steel test piece as substrate. The present authors[4] have used conventional flat strip-type specimens, the British Non-Ferrous Metals Research Association have used cylindrical specimens and Kendrick[5] has used wire, but these variations have no apparent influence on the results obtained. In a tensile test of this type, the thickness of the coating should not influence the value obtained. As in the bend test, the end point is reached when the deposit cracks, and similar problems concerned with the detection of the end point occur. In the case of bright deposits the failure is easy to observe but the extension is very small, usually about 1%. On the other hand, in the case of ductile deposits the failure point is difficult to detect, but the extension is large and may be as high as 25%. A hand lens or binocular microscope is of assistance in observing the onset of cracking which is taken as an indication of failure. However, cracks in brittle nickel almost immediately propagate across the width of the specimen. An extensometer is the most satisfactory means of measuring the small extensions, but a Vernier travelling microscope is adequate for extensions greater than 2%. The elongation is usually measured with the load off, but since soft fully-annealed substrates are used, very little elastic recovery occurs. Normally, the ductility of the nickel coating alone is assessed. Only in special circumstances is the ductility of a nickel plus chromium coating determined, since the ductility of the chromium plate is so low that it reduces the overall ductility of the composite coatings by the notch effect, and the chromium deposition process also induces hydrogen embrittlement.

An alternative type of tensile test piece was suggested by Cashmore and Fellows[6]. Thin steel of 125 μm thickness is used, the idea being that as soon as the nickel coating cracks, the load on the substrate is great enough to fracture it almost instantaneously. The steel strip is plated in a Perspex jig, and a small tensile specimen is stamped from this. The specimens require careful preparation to remove burrs from the edges. This test is satisfactory for fairly ductile deposits and has the advantage that no problem arises in locating the end point. It is not so satisfactory for brittle, bright deposits; failures frequently occur off the gauge length at the shoulders of the specimens.

True values for elongation of nickel deposits can be obtained only if the electrodeposit is tested in the absence of the substrate, as was done by Brenner and his colleagues[7]. However, extended plating times are necessary to produce electrodeposited nickel sheet sufficiently thick to allow flat tensile specimens to be stamped from it. The substrate can be removed from thick deposits by dissolution or by grinding, but in the case of the latter, precautions must be taken to avoid modification of properties. If the substrate is still present, it will influence the result to some extent, but this is not necessarily a disadvantage, since usually it is the behaviour of a plated component that is of interest. Hawkins[8] has shown that in the case of roll-bonded composites of copper and steel a linear relationship exists between their uniform elongations and the relative thicknesses of the two metals. It is probable that a similar relationship exists for composites consisting of steel and electrodeposited nickel, for example, where the ductility of the composite would be somewhat lower than that of the unplated substrate. Tensile testing of composite materials can also be utilised to determine the U.T.S. of one component if that of the other is known. However, the tensile strength of electrodeposits is likely to be of significance only in electroforming or engineering applications. Brook[9] determined the tensile strength of nickel deposits on copper substrates by assuming (a) that the total load carried by the copper substrate and the nickel deposit was distributed between the two in the ratio of their cross-sectional areas (this was assumed to be true because the values of Poisson's ratio for copper and nickel are similar), and (b) that the nickel and copper remained bonded during the test. The U.T.S. of fibre-reinforced composites is determined by the 'rule of mixtures' from the individual values of the matrix and fibres. This rule states that

$$\sigma_C = \sigma_M V_M + \sigma_F V_F \qquad \ldots (8.2)$$

where σ = stress, V = volume fraction and subscripts C, M and F refer to composite, matrix and fibre, respectively. Since, in this form, the rule of mixtures assumes that both fibres and matrix are subject to equal strains, it is known as the *equal strain hypothesis* and it has been suggested by Hawkins[8] that it can also be applied to layered composites.

Torsion tests on plated wire or other suitable specimens are probably of more use in evaluating the embrittling effect of hydrogen on the substrate than in determining the ductility of the coating. The extent of embrittlement of steel basis metal resulting from the preplating treatment and plating process in influenced by current densities, composition of the solutions and times of treatment. The concentration of hydrogen in electrodeposited nickel is much higher than the concentration of gaseous hydrogen in bulk nickel brought to equilibrium with gaseous hydrogen at 1 atm at 25°C. In torsion tests the brittleness can be assessed in terms of resistance to torsion by the relationship:

$$Y = \frac{100Q}{Q_0} \qquad \ldots (8.3)$$

PROPERTIES OF ELECTRODEPOSITS

where Y is the resistance to torsion (%), Q is the number of revolutions to fracture and Q_0 is the number of revolutions to fracture of a ductile reference standard.

The hydraulic bulge test is not often used to determine the ductility of nickel foils, but Read[10] obtained some rather unexpected results using this technique. Bright and dull nickel deposits apparently had approximately the same ductility when tested by this method. The Erichsen test can be used on plated panels, and for the purpose of the present classification is considered as a bulge test. It is not very sensitive and so is not often used for this purpose. However, the present authors have used a ball indentor to impart a standard amount of deformation to plated panels before carrying out corrosion tests (see Chapter 11).

COMPARISON OF BEND AND TENSILE TESTS

As indicated above, only two types of ductility test (bend and tensile) are of

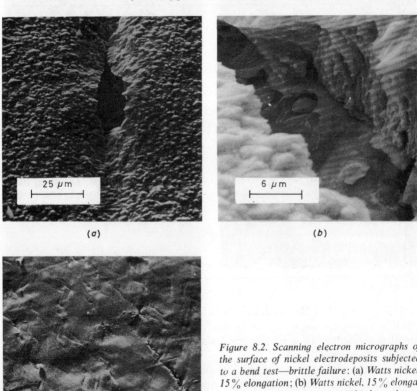

Figure 8.2. Scanning electron micrographs of the surface of nickel electrodeposits subjected to a bend test—brittle failure: (a) Watts nickel, 15% elongation; (b) Watts nickel, 15% elongation; (c) semi-bright nickel, 17% elongation

real significance as far as electrodeposited coatings are concerned. A brief examination of the mechanics of these test procedures reveals the reasons for different modes of failure and consequently different ductility values for the same electrodeposit. In a mandrel bend test biaxial stress occurs, and it

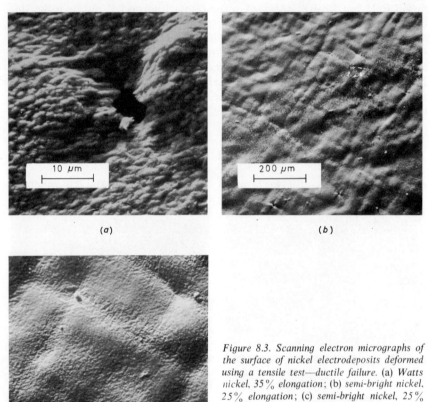

Figure 8.3. Scanning electron micrographs of the surface of nickel electrodeposits deformed using a tensile test—ductile failure. (a) Watts nickel, 35% elongation; (b) semi-bright nickel, 25% elongation; (c) semi-bright nickel, 25% elongation

is well known that this reduces ductility. The ratio of test strip width to thickness governs the extent of biaxiallity. However, when a strip of ductile metal is deformed in a tensile test, plane strain occurs (normal and shear stresses operative).

It is apparent from the characteristics of the aforementioned test procedures that the ductility of a particular deposit should be lower when estimated by a bend test than by a tensile test. Linsell[11] has shown this to be true for Watts and semi-bright nickel deposits. However, an interesting feature of these results is that biaxial stress has different effects on various materials. In the tensile test the Watts deposit had a greater ductility than the semi-bright, but in the bend test the order was reversed. These results indicate that there is no direct correlation between results obtained by bend and tensile tests. The results obtained by Read[10] using the bulge test also indicate that the performance when subjected to pure biaxial tension may be quite different from that when subjected to pure tension.

The results of the different mechanisms of deformation operative in bend and tensile tests are clearly indicated in Figs. 8.2 and 8.3. Distinct crosses and lines of necking occur after tensile testing, thus indicating ductile deformation, while the breaks which occur in the deposit during bending indicate a much more brittle failure. The surface condition of the substrate can also influence the ductility of the coating as illustrated in Fig. 8.4. In this case, the brass

Figure 8.4. Scanning electron micrograph of the surface of a Watts nickel deposit plated on to an abraded substrate (600 mesh grade emery). Brittle failure occurred in a tensile test at 5% elongation. The scratches were at right angles to the direction of applied load

substrate had an abraded surface, the scratch lines running in a direction at right angles to the direction of the tensile stress. A low ductility value was obtained and brittle failure occurred.

In view of the numerous problems associated with estimating the ductility of electrodeposited coatings it is preferable to select a test procedure which most closely simulates the conditions likely to be encountered in service. The most significant variables associated with ductility testing of electrodeposits can be summarised as follows:

(a) The stress system (uniaxial or biaxial) associated with the test procedure.
(b) The quality of surface finish.
(c) The means of determining the failure point.
(d) The composition and characteristics of the substrate.
(e) The thickness of the coating.

Structure of deformed nickel electrodeposits

The electron micrographs in Chapter 3 show that Watts deposits plated from purified solution have a fairly uniform grain size and that twinning occurs. On deformation in the tensile test, dislocation tangles form, particularly at grain boundaries and puckering of twin bands also occurs. The results of different amounts of deformation (Fig. 8.5) show that after a small

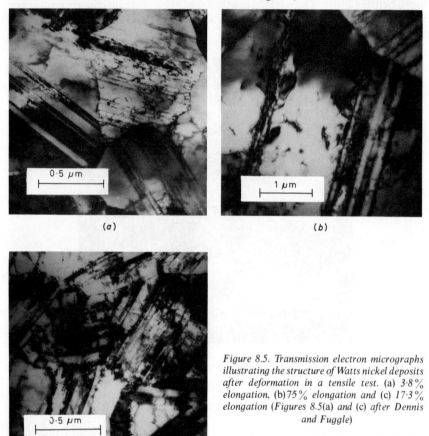

Figure 8.5. Transmission electron micrographs illustrating the structure of Watts nickel deposits after deformation in a tensile test. (a) 3·8% elongation, (b) 7·5% elongation and (c) 17·3% elongation (Figures 8.5(a) and (c) after Dennis and Fuggle)

amount of deformation the dislocation density increases to some extent in all regions of the grains. As the extent of deformation is increased, massive dislocation tangles form and from Fig. 8.5(c) [17·3% elongation] it is apparent that the original 'formation twins' have been bent and are tending to become parallel to each other in the direction of the applied load. Examination of the structures of deformed deposits plated from solutions containing organic additives has not proved to be very rewarding, since the unstrained structures either contain an abundance of dislocation tangles or the grain size is very small. Consequently, deformation phenomena are extremely difficult to observe.

ADHESION

The adhesion of electrodeposits to three main groups of basis materials are of interest. The simplest case and the one in which 'perfect' bond strength should be achieved is that in which the coating is deposited onto a cathode of the same metal and, under favourable conditions, the crystal lattice of the substrate is continued in the electrodeposit. This phenomenon of epitaxy, which can also occur when the substrate is different from the electrodeposit, has been discussed by numerous authors including Graham[12], Blum and Rawdon[13] and Finch, Wilman and Yang[14]. It should be observed that, as the majority of electrodeposits are finely crystalline, physical evidence of epitaxy is the exception rather than the rule and that epitaxy is not necessarily the only criterion for good adhesion. Hothersall[15], using metallographic techniques, and depositing from a solution that gave coarsely crystalline deposits and a substrate of well-defined grain structure, has shown that the grain boundaries of the substrate can be continued in the deposit. Finch et al[14] used electron diffraction to examine metals deposited at low current densities onto substrates of known crystal planes. They showed that the orientation of the substrate is continued up to a thickness of 30000 Å, but at greater thicknesses epitaxial growth no longer occurred.

This situation is unlikely to be encountered in many commercial applications as usually the coating and substrate are not the same metal. In this case, adhesion is represented by cohesive forces between metals. Hothersall[15] stated "that discharged ions take up their positions immediately adjacent to the basis metal lattice and within the field of molecular attraction". Interatomic forces involved are van der Waal's forces, covalent forces, metallic bonding forces and ionic and polar forces. However, the theory of bonding is discussed in detail elsewhere[16]. The structure of a dissimilar basis metal is more likely to be perpetuated in the coating at low current density and consequently good adhesion is then most likely to result.

The third group of substrates, viz. non-conductors, is of increasing importance today with the interest in the plating of ABS plastics (acrylonitrile–butadiene–styrene polymer) and a number of other plastics, which are described in Chapter 12. Non-conductors require the application of an initial conductive coating before an electrodeposited coating can be applied. Chemical reduction (electroless plating), or occasionally vacuum evaporation, is normally used to provide this conductive coating. With this particular system, three factors are involved in adhesion:

(a) Adhesion between the non-conductor and the thin conductive coating.
(b) Adhesion between the latter and the electrodeposit.
(c) Cohesion of the outermost layer of plastic to the underlying bulk of the moulding.

CAUSES OF POOR ADHESION

Surface contaminants and oxide films

The most obvious cause is inadequate cleaning (as discussed in Chapter 6).

grease, oxide and colloidal material being the most common contaminants. However, satisfactory adhesion may be achieved if the oxide film is not too thick, since it is dissolved or reduced at the cathode. Carbon or carbide may remain on the surface after etching (often colloquially termed *smut*) and this also causes low adhesion values.

Weak surface layers in the substrate

If the surface layer of the substrate is inherently weak or is embrittled as a result of the deposition process, failure will occur in the weak layer of the substrate and not at the bond. Although this may give a low value of adhesion it will be the best possible as far as a particular system is concerned unless further treatments, which result in the removal of that weak layer, can be employed. Hydrogen embrittlement is a common cause of failure in the surface layer of the substrate. Hothersall[15] reported figures showing that in the case of nickel plated on case-hardened steel, the adhesion value could be raised from about 85 N/mm^2 to almost 338 N/mm^2 by baking at 200°C for 30 min after plating to diffuse out hydrogen. Brittle intermetallics may be formed at the boundary in some instances.

Abrading and polishing of the substrate surface usually results in the formation of a distorted surface layer of lower strength than the bulk material, and an increase in adhesion is obtained if the surface is etched before electroplating. This has the advantage that it removes the work-hardened surface together with any oxide and soil, but has a deleterious effect in that a matt surface is produced which may be difficult to brighten even with a levelling decorative coating.

Initial layers of defective deposit

The formation of an initial defective layer of electrodeposit may also lead to poor adhesion. This can arise if cleaning solution is carried over on the surface of the component to be plated and either becomes incorporated in the deposit or detrimentally affects the cathode layer. A burnt deposit produced in the early stages has the same effect, and this can occur if the initial current density is not reduced when a large component, at a low temperature, is introduced into a warm plating solution so that locally, near the cathode surface, the temperature of the solution is reduced significantly.

Good adherence is necessary for metals such as nickel, where the tensile strength of the coating is fairly high, since stress is readily transmitted to the interface. An easily deformed coating is more likely to yield locally and so prevent the applied stress at the junction reaching a value high enough to cause detachment of the electrodeposit.

METHODS OF DETERMINATION OF ADHESION

Methods of testing adhesion have recently been reviewed critically by

Davies and Whittaker[17] and Plog[18]. Few of the accepted methods resolve the applied force used for evaluating adhesion into its individual components, and so absolute values cannot be obtained. Direct comparisons cannot therefore be made between results obtained using different tests and, in any of the quantitative tests employed, the force applicable in that particular test is used as a measure of adhesion.

As far as electrodeposited coatings are concerned, unless the adhesion is obviously low, the failure occurs almost invariably in the coating or the substrate; this is true even for plated plastics, failure occurring in the surface layer of the plastic. The measured value of adhesion can, therefore, be influenced by either the substrate or the coating. An ideal adhesion test has not yet been found, for its requirements are so demanding. Qualitative tests are simple and have some significance for the practical plater, but are of little value for precise specifications. The earlier quantitative tests specify a thick deposit, which is time consuming to produce; this is particularly disadvantageous if only a thin coating is used in service. Berdan[19] suggested a list of ideal features, but these are certainly not all satisfied by any test at present available. For example, adhesion can rarely be determined unless a special specimen is prepared and so Berdan's requirement that tests should be feasible irrespective of the cathode geometry cannot be satisfied. The test area depends on the type of test used and so at least a choice can be made with respect to this feature.

Qualitative tests

Numerous qualitative tests have been used; these have limitations depending on such factors as the geometry of the specimen, thickness of coating, initiative of the operator and the magnitude of the adhesion value. Operations which can be included in this category are bending, twisting, burnishing, buffing, abrading, chiselling, filing, cupping and thermal cycling. The bend test is popular since it is simple and at the most requires only the sawing through of the substrate, so that the specimen can be bent back in order to investigate whether the coating can be detached. Burnishing and buffing are useful for thin coatings for which the chiselling test would be unsatisfactory. The latter is of use for thicker coatings where it is possible to chisel off the coating at one point and then peel it back with some type of grip. Thermal cyling is particularly useful for complex shaped components which cannot be tested in other ways. One version of this is used for plated plastics, e.g. between a maximum temperature of 80°C and a minimum of −40°C. Such and Baldwin[20] have compared the behaviour of plated ABS panels in this type of test, with their peel-strength values.

Quantitative tests

Tensile and shear tests

Tensile and shear tests are the basis of many quantitative tests. The simplest of these require that some form of grip is attached to the coating. Solders, adhesives and even pressure-sensitive adhesive tapes have been used, but are unsatisfactory for high bond strengths, their limitations being the relatively low strength of the solder or adhesive and their ability to penetrate porous coatings. Brenner[21] electroformed a cobalt nodule onto the surface of the coating and used this as a means of gripping the deposit and applying the force.

The essential features of the method developed by Ollard[22] are illustrated

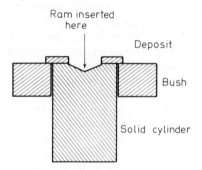

Figure 8.6. Essential features of the Ollard method for measurement of adhesion

in Fig. 8.6. This is suitable only for measuring the adhesion of very thick coatings, and these take a long time to prepare. The deposit is applied to one end of a solid cylinder and the cylinder machined down to leave the nickel projecting. A hole is drilled through the centre of the deposit, so that when the specimen is seated in the bush, a ram can be applied to push the substrate off the coating. It has been found that the dimensions of certain components have a bearing on the results, and so these are now usually standardised. For example, if the diameter of the central hole is increased relative to the outside diameter, higher results are obtained. Local stress concentrations can arise if the deposit is too thin or if the cylinder fits too loosely in the bush. High results are obtained if it fits too tightly and binds. Low values due to tearing are obtained if a stress concentration occurs at the outer edge of the junction of basis metal and deposit. If the coating is too

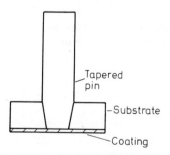

Figure 8.7. Tapered-pin method of determining the adhesion of electrodeposits to substrates

thin. the overhanging deposit does not provide an adequate means of support for the substrate to be pushed away. It is difficult to arrange for the push and pull on the coating to be consistently perpendicular and equally distributed over the interface area used in the calculation. Ideal conditions for determining bond strength by a tensile test have been put forward by Schlaupitz and Robertson[23], but their conditions are so critical that their technique is of interest only for academic work.

Modifications have been made periodically to the Ollard test; one interesting feature introduced by Gugunishvili[24] involves the use of tapered pins which fit accurately into tapered holes (Fig. 8.7). the end of the pin and the substrate surface being plated as one surface. The other ends of the pins are gripped in a spedial chuck and withdrawn using a tensile testing machine. the load required being recorded. Further improvements have been suggested by Knapp[25] and Rothschild[26].

Zmihorski[27] used a shear test to determine the adhesion of chromium to steel. A cylindrical steel rod was coated with a band of chromium of varying thickness and width. The coating was detached by drawing the rod through a hardened steel die which had a hole larger in diameter than that of the rod but less than that of the rod plus coating. A value can be determined for adhesion from certain known parameters.

Peel tests

Peel tests have a major advantage over tensile tests. in that a much thinner electrodeposited coating is required. From a practical point of view. the coating must be strong enough to permit continuous peeling of the deposit from the substrate. However. problems are likely to be encountered if the coating is brittle. when it will not permit continuous peel. Jacquet[28] introduced a method in which a grip is electroformed onto the plated component. but direct peel tests in which the coating itself is lifted and gripped. are much simpler when feasible. Contaminating treatments are often used to prevent adhesion at one end of a test strip, so that a tab for gripping can be formed. Precautions are necessary to ensure that the test surface itself is not contaminated and hence the test invalidated. Once part of the deposit has been lifted. the force required for continuous peeling can be applied and measured using a tensile machine. the result being recorded as load/unit width of the tongue of deposit detached. This type of test is very easy to carry out in the case of systems for which the adhesion value is relatively low. Such and Wyszynski[29] have used this technique for dull nickel on aluminium. Adhesion can be prevented at one end of the aluminium specimen by not applying the immersion zinc coating, which is an essential feature of the plating sequence for aluminium substrates. The latter process is necessary to overcome the problem of the natural oxide film which is always present on aluminium. Since the peel strengths of electrodeposits on aluminium are not often greater than 1800 g/mm. a spring balance equipped with a suitable chuck is usually adequate to measure the force required to detach the coating. A photograph showing the test procedure is shown in Fig. 8.8.

Peel tests are also used to determine the adhesion of electrodeposits to non-conductors. where the peel strength is usually a maximum of about

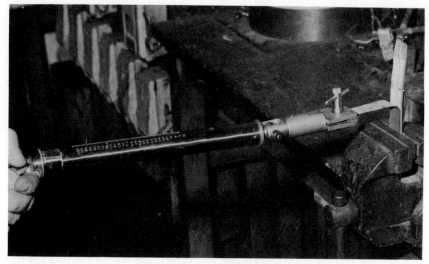

Figure 8.8. Jacquet-type peel test used to determine the adhesion of electrodeposited nickel to aluminium (after Such and Wyszynski[29])

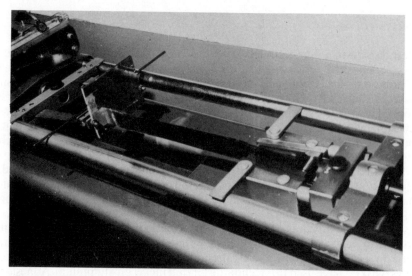

Figure 8.9. Jacquet-type peel test used to determine the adhesion of electrodeposits to ABS

500 g/mm. This can be carried out on the Hounsfield tensometer using an attachment of the type shown in Fig. 8.9. This is designed so that the force is always applied at right angles to the plated surface. Since the adhesion is low, a sensitive beam of the tensometer must be used. The detached end of the deposit is fixed to the chuck using an adhesive p.v.c. tape which is capable of carrying the load. The test strips peeled off are either 12·5 or 25 mm wide and the rate of peeling should be standardised. This technique has been used

to determine the adhesion of electrodeposits to ABS and other plastics and to evaluate factors which influence the bond strength.

Ultrasonic techniques

Ultrasonic methods of measuring adhesion have been employed in other fields. but have not been used to any great extent in the plating industry. Two types of test or a combination of them are used. Qualitative estimates of adhesion can be obtained using high frequency energy to detect flaws at the coating/substrate interface. Low frequency energy is employed to induce stresses at the coating/substrate interface sufficient to detach the coating.

Ultracentrifuge techniques

Ultracentrifuge techniques are considered by several authors to be the most satisfactory for detaching coatings from substrates. However. the equipment required is complex and expensive. and special test pieces are necessary. The specimen in the form of a rotor is rapidly revolved to develop a high centrifugal force. and if this is sufficiently high. the coating is detached in one piece; there are no problems in starting the test as is the case with peel tests. and the forces applied to the coating are at right angles to it. provided that hoop stresses are eliminated. The adhesion force A can be obtained from the expression

$$A = 4 \times 10^{-6} \times \pi^2 N^2 R T D \qquad \ldots (8.4)$$

provided that $T \ll R$. In equation 8.4 N represents the number of revolutions per second. R the radius of the rotor. D the density of the coating and T its thickness. If linear dimensions are in metres and D in kg/m^3 then the adhesion force is in N/mm^2.

The design of the equipment and the properties of the materials used for its construction impose a limit on the maximum speed of rotation possible. In the more recent designs the specimen in the form of a disc is attached to the rotor. Davies and Whittaker[17] quote the maximum speed for a 5 mm disc as being 30000 rev/s. For nickel that is 125 μm thick. the maximum adhesion value measureable is approximately 90 N/mm^2. Beams[30] has determined values for chromium on steel as high as 112 N/mm^2. Dancy and Zavarella[31] have measured even higher values. up to 500 N/mm^2 for chromium on steel in the quenched and tempered condition. Compact equipment suitable for this type of testing has been proposed already. and it is likely that eventually the method may become more popular and be incorporated in standards and specifications.

STRESS

Almost invariably. electrodeposited metals are in a stressed condition. This internal stress. which can arise from several sources. may be either tensile (tendency for coating to contract) or compressive (tendency for coating to expand). It can be divided into two categories. (a) that resulting from distortion due to lattice misfit at the interface between the coating and substrate and (b) the intrinsic internal stress which arises as a result of particular plating conditions and bath composition. It is possible to demonstrate intrinsic internal stress by producing a metal foil by deposition onto a substrate to which the coating does not adhere (e.g. Watts nickel deposited onto stainless steel). If the coating is carefully detached from the substrate. it curls in the appropriate direction depending on whether the stress is tensile or compressive. Highly stressed deposits cannot be obtained in the form of a foil in this manner as the deposit peels off continuously during deposition. Stress is affected by the presence in the solution of organic compounds in addition to the influence of the plating variables, and so such compounds are incorporated in the formulation of commercial plating solutions to act as 'stress relievers'. Their concentration is usually chosen so that the resulting deposit has a fairly low tensile or compressive stress. Compressive stress is normally less detrimental than tensile stress. since with the former the deposit does not tend to lift from the substrate. The effects of organic additions on stress have been discussed in Chapter 6 and it is the purpose of this section to describe the methods developed for estimating stress and to review the theories suggested to account for its occurrence.

METHODS OF MEASURING INTERNAL STRESS

The earliest attempt to measure internal stress was carried out by Mills[32] in 1877. This involved the electrodeposition of the metal onto a silvered thermometer bulb. Movement of mercury in the capillary indicated the occurrence of stress in the coating. The height of the mercury column increased when the stress was tensile and decreased when the stress was compressive; from this information it was possible to obtain an indication of the stress level in the coating. This method is now only of historical interest. but it serves to emphasise that the phenomenon and methods of evaluation have been known for many years.

Several of the methods used at the present time are based on the 'flexible strip' principle. If a thin strip of metal is plated on one side only. concave curvature will result if the coating has a tensile stress. and conversely. convex curvature if the coating has a compressive stress. In the earliest form of the test. stress was calculated from the resulting curvature. but alternatively it can be calculated if the value of the restoring force (the force required to prevent the strip from bending) is known. A number of variations of the strip principle. which was first used by Stoney[33] in 1909. have been employed. He used a thin steel ruler as the substrate and permitted the strip to bend during plating. Other investigators have prevented bending during plating. only allowing this to occur on completion of deposition. Kohlschuetter[34] attached a pointer to the bottom of the strip so that the progress of the bending could be observed during plating. The equipment was fragile and the weight of the pointer reduced its sensitivity. In the absence of a pointer some form of gauge or vernier microscope is required to measure the curvature of the strip. The techniques so far described are not suitable for routine testing, but could be employed for research work.

The technique of depositing predetermined thicknesses of nickel on thin strips for control of commercial plating baths was developed by Sodersberg and Graham[35], Phillips and Clifton[36] and Such[3]. Certain basic principles must be considered when using flexible strip methods to determine stress. The extent of bending must not be so great that the elastic limit of the substrate and/or deposit is exceeded, and Young's modulus and Poisson's ratio must be known for the materials concerned. The elastic modulus of a nickel electrodeposit can vary widely, depending on the solution and plating conditions; consequently, it is quite different from that of bulk metal. For continuous measure of stress as deposition proceeds, the strip must either be allowed to bend so that curvature can be measured continuously, or it must be possible to determine continuously the force required to prevent bending.

Spiral contractometer

An instrument for commercial use should be compact, robust, self-contained and rapid in operation, so that numerous routine tests can be carried out. Brenner and Senderoff[37] of the National Bureau of Standards designed the 'spiral contractor' which, although based on the flexible strip principle, was intended to overcome some of the experimental problems (Fig. 8.10). The test-piece is in the form of a strip of stainless steel wound into a helix. Tensile stresses in the coating subsequently deposited onto the helix cause it to unwind, and conversely, compressive stresses cause it to wind up more tightly, and the change in the radius of curvature of the helix is used as a means of calculating the stress. The change in this dimension is measured by determining the angular displacement of one end of the helix while the other is fixed. One end of the helix is fixed rigidly to a calibrated dial while the free end is attached to a coaxial torque rod which in turn is joined to a pointer which traverses the calibrated dial. The original design and dimensions of the instrument were described by Brenner and Senderoff[37] in 1949. The sensitivity of this instrument is to a large extent governed by the quality of the gear system used. Modifications of this apparatus have been constructed in which the gear system is replaced by jewel bearings and the dial by an optical lever system. The helices must be manufactured to a precise specification, the thickness being dependent upon the magnitude of stress to be measured.

A standard procedure should be adopted for preparation of the helices. Many deposits do not adhere adequately to stainless steel and hence the stress results would be invalid unless an intermediate coating were applied before commencing the stress determination. The helices are first plated with nickel from a Wood's nickel solution (this solution contains a low concentration of nickel ions and is operated at low pH) and then with copper from a cyanide copper solution. Almost all electrodeposits can be plated onto this surface, but it may be necessary to deposit onto other undercoats if the undercoat surface has a significant effect on the resulting stress. This has been shown to be so in the case of micro-cracked chromium[38,39]. Preparation of the helices is somewhat tedious and time consuming because in addition to the plating operations their interior surfaces have to be insulated with lacquer so that deposition is confined to the outer surface. The average thickness of the coating is calculated by weighing the prepared helix before and after the test. If the rate of deposition is constant the stress level can

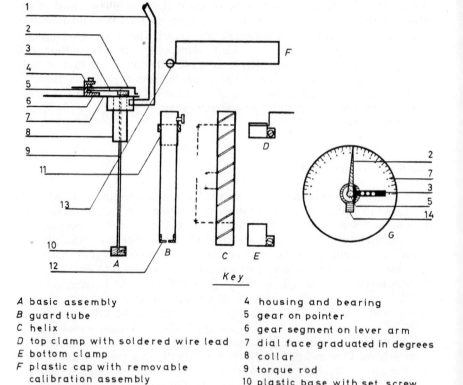

Figure 8.10. Spiral contractometer used to determine stress in electrodeposits (after Brenner and Senderoff[37])

subsequently be calculated for any thickness of coating. Since the specimen is in the form of a helix, it is possible to use a much longer plated length of metal strip, and the method is therefore more sensitive than the flexible strip method devised by Stoney. It also has the advantage that it is unaffected by vibration or by agitation of the solution.

Hoar and Arrowsmith method[40]

This is a method in which the strip is prevented from bending by the application of an external force so that relaxation of stress is not permitted. The force is applied electromagnetically. A soft iron armature is attached to the top of the specimen, and solenoid coils are mounted in suitable positions on the plating cell. The current is controlled continuously to prevent bending of the strip. A calibration curve for the coils is prepared by mounting the strip in the horizontal plane and attaching fractional gramme weights to the armature. A small galvanometer mirror is attached to the top of the strip so that an optical lever can be used to enable the strip to be retained in its initial position (null deflection method). The mean stress S can be calculated from the expression:

$$S = \frac{4M^3 G}{10^6 \times BTL(3M - L)x} \qquad \ldots (8.5)$$

where G is the restoring force, M is the length of the strip from support to armature, B is the width of the strip, T is the thickness of the strip, L is the plated length and x is the coating thickness. If G is expressed in newtons and all dimensions in metres, the stress is determined as N/mm^2. Gabe and

Figure 8.11. Equipment designed by Norris (based on the Hoar and Arrowsmith principle) for the determination of stress in electrodeposits (after Norris[43])

West[41] have discussed modifications to equation 8.5 in order to obtain greater accuracy. One of the most important features of specimen design is that the unplated length from the armature to the top of the deposit $(M - L)$ should be as short as possible. This is governed by the design of the plating cell and ancillary equipment. If the unplated length is too long the restoring force results in the specimen being distorted into an S shape. The test pieces employed in this method are usually thin strips of mild steel, of the order of 100 μm thick. These are much easier and quicker to prepare than the stainless steel helices used with the spiral contractometer; they are also expendable and so do not have to be stripped to recover a fairly expensive test piece. One disadvantage of this method is that the plating solution cannot be agitated as this would cause movement of the test strip, c.f. the spiral contractometer. It is essential to deposit a coating of uniform thickness to obtain accurate results. The current distribution can be improved in the Hoar and Arrowsmith method by positioning robbers on either side of the test strip, the circuit being designed so that only the current supplied to the test strip is measured[42].

Norris[43] has designed an instrument based on this principle which permits automatic control of the coil current (Fig. 8.11). The plating cell and associated components have been designed with the purpose of providing a routine instrument which is accurate and easy to operate. The main features are the partly submerged coils which enable the unplated length to be reduced to a minimum, the incorporation of 'robbers' to improve current distribution on the test piece and the use of an electrical circuit designed to provide d.c. outputs to meet all requirements from a single input. The instrument can be used manually, in which case an optical system is used which is sufficiently effective to enable tests to be carried out in subdued, natural or artificial light.

Stressometer

The third instrument which has been developed for research and commercial measurement of stress was introduced by Kushner[44] and is known as the *stressometer*. It also operates on the principle of a flexible specimen which is plated on one side, but in this case is in the form of a disc instead of a strip (Fig. 8.12). It is based on the symmetrical flexure of a circular plate or disc about its vertical axis. The disc-shaped test piece is clamped over the open end of a chamber which contains a suitable metering fluid. A capillary tube is connected to the chamber so that changes in volume of the chamber, due to curvature of the disc, can be detected. A clamping ring is used to mount the disc, and the cathode contact is usually made to the unplated side of the disc. The anode is also in the form of a disc so that uniform metal distribution can be achieved on the cathode. If the stress in the deposit under investigation is tensile, this results in concave flexing of the disc and liquid is expelled from the chamber so that the liquid level in the capillary rises; the opposite occurs if the stress is compressive.

The disc-shaped test piece is subjected to a uniform force over its surface and under these conditions the deflection of the centre of the disc is proportional to the uniform force causing bending, provided that the elastic limit is not exceeded and that the disc thickness is small compared with its diameter. The magnitude of the deflection is also dependent on the conditions at the edge of the disc, i.e. the method of mounting, but this is allowed for in calibration. By mathematical analysis of the conditions, Kushner has shown that the deflection at the centre of the disc will vary linearly and directly with the stress in the deposit. The volume displaced by the deflected disc can be

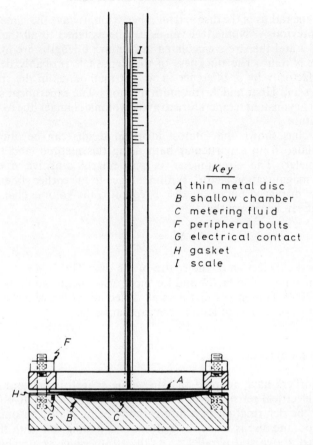

Figure 8.12. *Stressometer instrument used to determine stress in electrodeposits (after Kushner[44])*

equated to the volume change in the capillary and hence the change in level of the metering fluid is a linear measure of the deflection at the centre of the disc.

Kushner obtained the expression:

$$S = \frac{3r^2 \Delta L}{4KTH} \qquad \ldots (8.6)$$

where S is the stress in the deposit, r is the radius of the capillary. ΔL is the change in liquid level in the capillary, T is the thickness of the deposit. H is the thickness of the disc and K is the flexing constant. K can be determined by evaluating the change in liquid level in the capillary ($\Delta L'$), on application of a known uniform force Q. This can be carried out easily by making use of the plating solution itself. The instrument is submerged in the plating solution to a known depth and the deflection caused by a known hydrostatic pressure determined. Kushner expressed what he called the 'working equation' of the instrument as:

$$S = \frac{QA^2 \times \Delta L}{\Delta L' \times 4HT} \qquad \ldots (8.7)$$

where A is the radius of the disc and all other symbols have the same meaning as stated previously. Numerical values can be assigned to all the terms in equation 8.7 and the stress calculated in N/mm^2, if lengths are in mm and force Q in N/mm^2. The thickness of the deposit is probably determined most satisfactorily by a difference in weight method, as in the spiral contractometer and Hoar and Arrowsmith methods. The experiment should be carried out at constant temperature to avoid volume changes due to variation of temperature.

Kushner has shown that almost identical results can be obtained for deposits plated from a particular bath using this method, and the spiral contractometer. The stressometer is particularly sensitive due to the 'hydraulic magnification factor'. With reference to the earlier statement that the volume displaced under the disc is equal to the volume change in the capillary, it can be shown that:

$$\Delta L = \frac{YA^2}{3r^2} \qquad \ldots (8.8)$$

where Y is the deflection of the centre of the disc. The factor $A^2/3r^2$ is the hydraulic magnification factor and for most instruments would be between 1000 and 2000. This shows that a small deflection of the centre of the disc results in a large change of level in the capillary tube.

Strain-gauge method

Russian workers have adapted the stressometer instrument and used the change in electrical resistance of a strain gauge as an alternative means of measuring the deformation of a disc cathode being plated on one side. In their method, the disc is made the bottom of a plating cell, with the anode being placed above and parallel to it; the strain gauge is attached to the underside (unplated side) of the disc. Apart from the usual electrical precautions needed for measuring the small changes in resistance that occur, other difficulties may arise. Experience of the authors in the use of a similar cell unit, using the hydraulic magnification of the Kushner stressometer to indicate the deflections of the disc cathode, has shown that if the volume of the cell is too small, excessive changes in solution composition can occur during the time required for one plating test. In particular, the acidity and brightener contents may be unduly affected, with consequent changes in the mechanical properties of the nickel electrodeposit, unless acid and other additives are continually introduced at a satisfactory predetermined rate.

Further details of the Russian method, together with the rather complex formulae needed for calculation of the stresses induced, are given by Vagramyan and Solov'eva[45].

Flexible strip method applied to automatic control of stress

Schmidt[46] has described the use of a cathode consisting of individual flexible strips which are electrically connected, so that if they are placed at an angle to the anode as in a Hull cell, they will indicate the stress in metal deposited at several different current densities. The varying thickness of electrodeposit on each strip must, of course, be taken into account when calculating the stress. This device can be modified so that the displacement of the strips is detected by some means, such as variation in inductance of solenoids positioned near the strips. This change in inductance can then be employed as a method of controlling the stresses in the metal being deposited from a commercial plating bath, by using it as a means to adjust automatically the current density to a value which gives the desired stress.

Micro-stresses

The methods so far described are capable of measuring only macro internal stresses. These are the resultant stresses, the direction of which is constant throughout a large volume of the electrodeposit. Stress can also be detected on a micro scale within coherent lattice regions and within submicroscopic lattice regions. The former tends to cancel out over regions of macroscopic dimensions, but the direction changes from grain to grain. The stress which occurs within submicroscopic lattice regions is usually caused by the presence of impurity atoms in the structure. Micro-stresses can be evaluated by using an X-ray line broadening technique. However, the results are difficult to interpret since line broadening is also caused by small grain size, and many electrodeposits, especially the bright deposits, have an extremely fine grain size. Various authors[47,48] have devised equations for evaluating the effects of small grain size, but current work[49] is emphasising the problems encountered when attempting to determine stress in nickel electrodeposits by X-ray techniques. Watts nickel deposits often have a non-uniform and, in comparison to bulk metals, a fairly small grain size [see Fig. 3.5(d)]. As the current density is increased, the grain size becomes less uniform and a few very large grains are surrounded by numerous extremely fine grains. The average figure for the grain size has little meaning in such cases and the non-uniformity of grain size consequently decreases the accuracy of the calculation of internal stress.

Usually it is the macro-stress which is of importance to the electroplater, and fortunately this is the more simple to evaluate. Also, continuous measurements can be carried out as plating proceeds; this is not possible with X-ray methods which are not often used except for more academic research purposes.

THEORIES PROPOSED TO ACCOUNT FOR INTERNAL STRESS IN ELECTRODEPOSITS

Stresses in electrodeposits may result from lattice misfit between the deposit and substrate or from intrinsic stress. It is far more difficult to account for intrinsic stress, and no universally accepted theory has been proposed; it is more likely that stress is due to a combination of certain suggestions put forward in the theories discussed below. The causes of stress will almost certainly vary from one bath to another.

Lattice misfit

Lattice misfit is likely to occur when metal is electrodeposited onto a dissimilar substrate. The extent to which misfit is perpetuated in the coating is dependent on the type and surface characteristics of the substrate and the deposition conditions. The deposition of nickel on copper (both f.c.c. metals) serves to illustrate the way in which this type of stress can arise. The lattice parameter of nickel is 3·517 Å and that of copper 3·608 Å. If nickel is deposited on copper, in order to conform to the larger copper lattice, it must be in tension at the metal interface. The nickel lattice attempts to contract, and hence if no other internal stresses occur, a tensile stress would be recorded by the methods described earlier. Conversely, if copper is deposited on nickel a compressive stress would arise if there is no intervention by other phenomena.

The following theories have been submitted to account for the intrinsic stress.

Co-deposited hydrogen theory

This is one of the older theories, but is limited in that it is applicable only in cases where hydrogen can be incorporated in the structure and it can account only for tensile stress. It is particularly suited to explaining the cause of the high tensile stresses attained in chromium deposits. Copious amounts of hydrogen are liberated by the inefficient cathode reaction, and hydrogen is readily incorporated in the chromium lattice. This theory is based on the principle that hydrogen is co-deposited with the metal causing the lattice, initially, to be in an expanded state. As deposition proceeds the hydrogen diffuses away and the lattice shrinks so that a tensile stress is developed. One theory for the mechanism of chromium deposition suggests that the metal is deposited as a hydride which later decomposes. This would similarly lead to a contraction of the lattice.

It can be demonstrated easily that the nickel lattice is expanded by the absorption of hydrogen by making a thin strip of steel plated with nickel (stopped off on one side) cathodic in dilute sulphuric acid or a cyanide solution. Hydrogen enters the surface of the nickel, and the strip curves in a convex manner, just as it would if a coating were deposited which had a compressive stress. After switching off the current, the strip returns to its original shape after a sufficient period of time has elapsed. Bright nickel deposits are far more sensitive to this effect than dull nickel. Since macro-stress measurements are carried out on conventional substrates, the stress value obtained for chromium is the resultant of the true internal stress and the component of stress associated with distortion of the substrate lattice by absorbed hydrogen. In the initial stages of deposition, the internal stress in chromium can apparently reach a very high compressive value; this is because the absorbed hydrogen effect is of greater significance than the true internal stress in the chromium which is almost certainly tensile. As the thickness of the chromium is increased, the resultant value usually becomes tensile as the true stress becomes more significant. When the tensile strength of the chromium deposit is exceeded, the chromium cracks and stress relief takes place. In some cases the stress can even become compressive; this can be explained since chromium has good micro throwing power which results in deposition in the cracks. The cracks heal over as the deposit grows from the edge of the cracks. This wedge effect is said to cause expansion of the deposit and the sign of the resultant stress depends on the magnitude of the various phenomena discussed.

Dislocation theory

Hoar and Arrowsmith[50] proposed a theory based on the formation of vacancies in the deposit to account for internal stress. Surface energy gives rise to a surface tension, which is known to occur even in the case of solids at room temperature. This tension is likely to occur only in the surface layer of the deposit to a depth of about 2 Å. As the deposit is built up, any particular layer of atoms must lose its tension to the subsequent surface layer, if the bulk of the metal is to remain in equilibrium with the surface. If deposition takes place so rapidly that equilibrium cannot be attained, part of the tension will be 'frozen in'. It has been shown[51] that the surface layer of a solid must contain more vacant lattice sites than the bulk with which it is in equilibrium, and this may be the cause of surface tension. At the relatively low temperatures at which electrodeposition is carried out, excess vacancies can be 'frozen in' to the lattice at high deposition rates. A 'frozen in' vacancy may result in a line of vacancies at right angles to the cathode surface, since deposition is less likely to occur at a 'hole' in the lattice. This phenomenon can lead to the formation of edge dislocations oriented so that the side of the dislocation nearer the growing surface is the side short of atoms. A tensile stress will then occur in the deposit.

The above interpretation accounts only for tensile stress, but it can be extended to show how compressive stresses can arise. Organic molecules which lead to a reduction of stress in electrodeposits may be absorbed at lattice vacancies and so prevent the development of edge dislocations of the aforementioned type. However, stress reducers are usually large molecules and are known to be incorporated in the deposit. Their presence could result in edge dislocations of the opposite sign (side of dislocation nearer the surface overcrowded). If a sufficiently high concentration of this type of dislocation is formed, the stress may become compressive. The presence of compressive stress in deposits plated from solutions free from organic additions (e.g. zinc and cadmium) can be explained in a similar manner. If zinc hydroxide is co-deposited with the metal as extremely small entities, it is feasible that edge dislocations could be formed by a mechanism similar to that suggested in the presence of large organic molecules.

Excess energy theory

Glasstone[52], and Graham and Soderberg[35] have proposed a theory based on the fact that metal atoms will be at a much higher temperature at the instant of deposition than the bulk of the deposit. On cooling this would lead to contraction, i.e. tensile stress. In order that deposition may proceed.

a hydrated metal ion in solution must attain a certain energy level. The amount of activation polarisation concerned depends on the particular system, but the excess energy available after overcoming the energy barrier results in the formation of an expanded lattice which contracts at a later stage to give a tensile stress.

Wilman and Mubach[53] on the basis of electron diffraction studies of electrodeposits have also put forward a theory which suggests that the growing surface layer attains an instantaneous temperature of several hundred degrees, so that the lattice again forms in an expanded state to contract at a later stage with the subsequent development of tensile stress.

Kushner theory[54]

This is based on the proposition that the shell of water molecules surrounding a metal ion in solution is not immediately removed on discharge of the metal atom at the cathode. Diffusion away of the water molecules would result in a tensile stress. In the case of metals which have a compressive stress, this is accounted for by assuming that the water reacts with metal atoms to form oxides or hydroxides which cause an expansion of the lattice and hence compressive stress. Similarly, in the case of addition agents which cause compressive stress, it is necessary to assume the formation of complex ions of the addition agent with the deposit. This effect is considered to be analogous to inserting wedges in a spring.

Gabe and West theory[39]

This is concerned with the stress development and crack formation in chromium deposits. In this theory, it is assumed that the cathode film becomes entrapped between 'growth cells' of the deposit and so ultimately produces continuous or semi-continuous oxide membranes. Growth occurs in a series of convex 'cells', some of which appear to originate from nodules nucleated, in some instances, at surface cracks. An oxide layer at an internal surface provides sites for nucleating hydrogen from dissolved atomic hydrogen. The hydrogen pressure can build up to such an intensity that a brittle crack can be initiated.

Validity of theories

The co-deposited hydrogen theory is simple to understand but is limited in application; it can account for the stress in chromium in many instances, particularly when the effect of hydrogen on the substrate is also taken into consideration. The Gabe and West theory offers a feasible explanation of the behaviour of chromium. Both these theories have limitations for other metals. Dislocation theories can be applied to most metals and can explain the effect of the incorporation of organic and inorganic material in the lattice. However, the density of dislocations per unit area would have to be extremely high to produce the high stress levels attained in some metals.

The theories depending on the assumption that expanded lattices are formed due to high temperatures at the instant of discharge of metal ions and the hydrated ion theory involve many assumptions and, as yet, are not supported by much experimental evidence; in fact, on thermodynamic grounds, it is difficult to see how the depositing metal can attain a temperature of 400–500°C.

HARDNESS AND WEAR RESISTANCE

The hardness of electrodeposited metals can be varied both by a change in plating conditions and by the presence of organic additives in the plating bath, although the latter is not relevant to chromium deposition since few organic compounds are stable in such a strong oxidising solution. Nickel and chromium electrodeposits are usually relatively hard and it is not uncommon for their hardness to exceed that of the metal in the cold-worked condition. This is due to the occurrence of one or more of the following phenomena:

(*a*) Internal stress.
(*b*) Small grain size.
(*c*) Incorporation in the lattice of fine particles, such as oxides.
(*d*) Preferred orientation.

METHODS OF HARDNESS MEASUREMENT

Hardness is usually determined by an indentation technique, but scratch tests and various types of abrasion or rubbing techniques can also be employed. Each category of test provides different results and so do different tests within a particular category. For example, in the case of a particular piece of metal, the value obtained using a Brinell ball indenter is different from that obtained using a Vickers or Knoop pyramid indenter. Consequently, hardness is not a fundamental property of metals since its determination is dependent on other properties.

The essential feature of indentation techniques is that the size of the impression, made by applying a known load for a definite time, is measured using a microscope fitted with a calibrated eye piece. The hardness of electrodeposits is usually determined using a conventional microhardness technique employing a Vickers or Knoop pyramid indenter rather than by a macro technique. If the test is carried out on the surface of the deposit, the coating must be sufficiently thick to avoid the 'anvil' effect; otherwise, the substrate influences the hardness value obtained for the coating. The Knoop indenter is particularly suited for use on the surface of an electrodeposit since it produces a shallow depression. One diagonal is much longer than the other, and it is known that most of the elastic recovery takes place in the direction of the shorter diagonal. It is easy to measure the longer diagonal accurately and it indicates the true size of the 'unrecovered' indentation. Frequently, microhardness determinations are carried out on the metallographically-polished cross section of the deposit, but the obvious limitation in this instance is that the coating must be thick enough to accommodate the indentation. However, this procedure ensures that the impression is made in a smooth flat surface, whereas a certain amount of roughness may have to be tolerated if the indentation is made on the surface of the electrodeposit. Hardness can be calculated from the expressions given below, but for routine work the appropriate value can be obtained from tables when the dimensions of the impressions are known.

(a) The Brinell number (HB) equals load divided by the spherical area of the impression:

$$HB = \frac{P}{(\pi D/2)\{D - \sqrt{(D^2 - d^2)}\}} \quad \ldots(8.9)$$

where D is the diameter of the ball (mm), d is the diameter of the impression (mm) and P is the load applied (kgf).

(b) The Vickers hardness number HV, (formerly V.P.N.) equals load divided by the contact area of the impression:

$$HV = \frac{1 \cdot 845 P}{d^2} \quad \ldots(8.10)$$

where d is the length of diagonal of the impression (mm) and P is the load applied (kgf).

The necessity for a hard deposit is usually associated with wear-resistant properties, but hard deposits do not always provide good resistance to wear. Scratch and abrasion tests can provide information of greater value as far as this phenomenon is concerned. Weiner[55] has indicated that if the hardness value obtained in a scratch test is low, yet high in an indentation test, the wear resistance would still be poor. Values obtained in scratch tests are usually quoted as the load on the stylus necessary to produce a standard scratch width.

Numerous techniques have been suggested as means of evaluating wear resistance, but none are entirely satisfactory or as well known as the abrasion test used to determine the abrasion resistance of anodic films on aluminium[56]. Rotating abrasive discs and tungsten carbide wheels have been used to abrade coatings, the subsequent damage being assessed using a low-power optical microscope. Alternatively, small parts can be tumbled in a drum with a specific abrasive and the wear resistance assessed in terms of weight loss/unit time.

Chromium is particularly useful as a coating on softer and cheaper metals, to provide satisfactory wear-resistant properties for a component. As described in Chapter 5, this metal also has a low coefficient of friction which reduces wear caused by sliding friction, and this property is utilised in chromium-plated piston rings and cylinders.

ASSESSMENT OF SURFACE QUALITY—BRIGHTNESS, REFLECTIVITY AND SURFACE ROUGHNESS

The problem of assessing the aesthetic appeal of a metal surface has been discussed in Chapter 6. Considerable difficulties are encountered if attempts

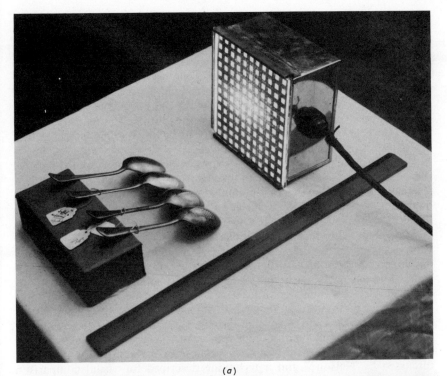

(a)

(b)

Figure 8.13. (a) Gardam grid in use to compare the quality of surface finish on spoons, and in (b) the difference in image clarity can be seen on the bowls of the spoons (courtesy The Worshipful Company of Goldsmiths)

are made to establish rigid specifications for the grade of finish which is acceptable. In some instances, a surface of high reflectivity may be desirable, but an etched or satin finish can also appear pleasing to the eye (e.g. etched aluminium, scratch-brushed stainless steel or 'satin' nickel). However, the assessment of extremely lustrous surfaces is of most interest in metal finishing practice. The essential feature of any instrument for comparison of surface finish is that it should place the surfaces in the same sequence as they would be placed by visual inspection, since this is the ultimate standard for acceptance. However, a physical test is unlikely to be in exact agreement with a visual test as personal factors are involved.

A simple but effective instrument known as the *Gardam grid*[57] (Fig. 8.13) can be used to assess the clarity of image formed on a bright plated surface. It consists of a box having an illuminated grid on one side. A measure of the quality and reflectivity of a surface can be obtained by measuring the distance, from grid to panel, at which the grid lines just cannot be resolved in the image. This technique enables a quantitative (although arbitrary) value to be given to the surface finish of a bright metal article. Ollard[58] developed an instrument which records only diffusely reflected light; it is claimed to be particularly suited to detecting small variations in highly polished surfaces. Small contour irregularities can be neglected on most articles since their effect on image distortion is masked by overall curvature, especially convex. Defects causing haze and bloom are mainly responsible for a surface having inferior appearance and hence lack of appeal. The problems associated with visual inspection of surface finish have been discussed by Nelson[59] and a technique described for visual comparison with a standard under controlled conditions.

GUILD REFLECTOMETER

Reflectivity can be determined using a Guild reflectometer or an instrument based on similar principles. This is the technique authorised by the British Standards Institution for the evaluation of the reflectivity of anodised aluminium[56]. One disadvantage of the Guild reflectometer is that the amount of diffuse reflection is influenced by directional scratches. A photograph and diagram of a modified Guild reflectometer are shown in Fig. 8.14. The advantage of this modified version over the original design is that the specimen can be placed on top of the instrument instead of vice versa, and this permits greater versatility. Total and specular reflectivity can be determined and so the percentage specular reflectivity can be calculated. A surface may have a high diffuse reflectivity but a low specular reflectivity. For example, a magnesium oxide coating, which has a brilliant white matt surface, results in approximately 100% diffuse and zero specular reflectivity, while a smooth, electropolished aluminium surface or a levelled, fully bright nickel electrodeposit has a high specular and a low diffuse reflectivity.

The inside of the reflectometer sphere is coated with a uniform layer of magnesium oxide, which is assumed to be a perfect diffuser, and hence the sphere a perfect integrator. Light dispersed into the sphere, from the surface under examination, falls on the photo-electric cell at the base, the current

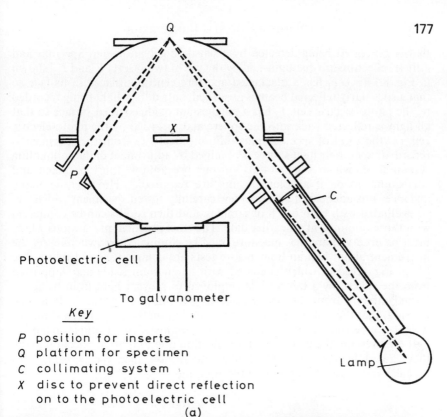

Key

P position for inserts
Q platform for specimen
C collimating system
X disc to prevent direct reflection on to the photoelectric cell

(a)

(b)

Figure 8.14. (a) *Reflectometer, and* (b) *photograph of modified Guild reflectometer. The two inserts can be seen in front of the apparatus. A magnesium oxide surface is shown between the inserts and a black box on top of the reflectometer; these are used for standardisation. A silica gel container is shown connected to the equipment; this is to prevent absorption of moisture when the equipment is not in use*

that is generated being detected by a sensitive galvanometer. Specular and diffuse reflection are distinguished by the use of two inserts I_1 and I_2 [shown in Fig. 8.14(b)]. I_1 has a blackened aperture centrally placed in its face so that a specularly reflected beam is absorbed, only diffuse light being recorded by the photo-electric cell. I_2 has a magnesium oxide covered surface so that all light is reflected back into the sphere and recorded by the photo-electric cell, i.e. the total of specular plus diffuse reflectivity from the specimen is recorded. A suitable light source is obtained by adjustment of the collimating system C shown in Fig. 8.14(a). Various precautions must be taken and corrections made if absolute results are required[56]. However, in many instances it is adequate to retain a few carefully stored specimens which can be evaluated with each batch of samples, and then results can be compared with those obtained at an earlier date. The chief practical precautions which must be observed are that moisture must be eliminated by switching on the instrument about half an hour before tests are commenced and that a disc [X in Fig. 8.14(a)], which is coated with magnesium oxide and supported from the sphere by a thin rod, is employed to prevent light from being reflected directly from the test surface to the photo-electric cell. It is also essential that the test surface should remain flat on the platform; a standard weight can be used for this purpose if the specimens are of simple shape.

If δ represents the galvanometer deflection due to diffuse reflection and τ that caused by the total of specular and diffuse, then the percentage of specular reflection R_S is given by the equation:

$$R_S = \frac{\tau - \delta}{\tau} \times 100 \qquad \ldots (8.11)$$

If τ_M represents the deflection when a perfect diffuser is placed on the platform, then the 'total reflectivity' R_T of a specimen expressed as a percentage is given by the equation:

$$R_T = \frac{\tau}{\tau_M} \times 100 \qquad \ldots (8.12)$$

The brighter the surface, the nearer the percentage $(\tau - \delta) \times 100/\tau$ approaches 100%, for the amount of diffused light is very low for an almost perfect mirror surface. Typical values for a nickel plus chromium coating are 62% for specular $(\tau - \delta)$ and 65% for total reflectivity (τ) giving a ratio of 95·4%.

Scott[60] has described the Guild reflectometer and other techniques for measurement of reflectivity. He states that total reflectivity is almost constant for any one metal, and thus the measurement of specular reflectivity can often provide sufficient information. Instruments for the measurement of specular reflectivity at fixed or variable angles of incidence are commercially available. They work on the principles described in the British DEF 1053 (Method No. 11)[61] and American ASTM D523-66T[62] specifications. Such instruments have more application for surfaces that are duller than the typical bright nickel electroplate. They are particularly suitable for satin surfaces where the reflectivity has frequently to be held within a specified range—not too bright or too dull. A typical range for the specular reflectivity of a satin nickel is between 20 and 30%. To measure this, an angle for incident light of 20° is usually employed in these *glossmeter*

PROPERTIES OF ELECTRODEPOSITS 179

instruments which have to be calibrated by the use of a suitable external standard. Scott has suggested a 45° prism, but polished black gloss and white ceramic tiles are also utilised.

SURFACE ROUGHNESS

Surface roughness, or as is more relevant to the present review, surface defects in an otherwise smooth surface, can be evaluated in a number of ways. The surface profile can be determined by using a surface analyser in which a stylus is traversed across the surface of the specimen and its vertical displacement magnified and either recorded graphically or integrated to provide a numerical value in terms of centre-line average[63] (C.L.A.) or root mean square (r.m.s.) values. This technique has been referred to in

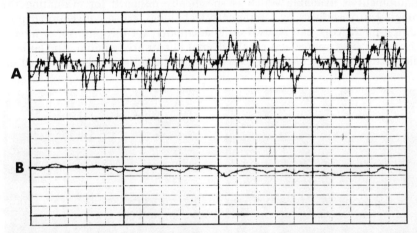

Figure 8.15. Talysurf traces showing levelling properties of a semi-bright nickel plating solution
LEGEND
A   *240 grit Aloxite finish on steel, C.L.A. value 0·24 μm*
B   *surface after plating with 30 μm of levelling nickel. C.L.A. value 0·09 μm; vertical scale × 10 000, horizontal scale × 100*

Chapter 2 as a means of determining the levelling power of a plating solution. This procedure has disadvantages in that soft surfaces can be damaged by the stylus[64,65] and in that one trace shows only the contours along a single line in the surface, which does not necessarily give an indication of the overall characteristics of the surface. However, the magnified profiles of the surfaces that are produced are useful indications of the degree of smoothing given by any nickel electrodeposit as illustrated in Fig. 8.15[66,67]. Optical methods such as interferometry and light profile microscopes are often more satisfactory for this purpose and so will be described below. Simple optical microscopy is only of very limited value and its uses have been discussed in other chapters.

Light profile microscopy

This technique is suitable for the investigation of surface contours which are too rough for examination by interference methods[68]. The salient feature of the method is that a slit image is formed on the test surface using oblique illumination[69]. To obtain high resolution and high magnification. a single lens is necessary and most of the effective lens aperture should be in operation. Oblique illumination causes severe chromatism. therefore monochromatic light must be used. Variations in surface profile are illustrated as lateral displacement of the straight-line image produced on the metal surface. This effect has been compared with that of a shadow cast across a set of steps by a vertical stick. when the Sun is low in the sky[70].

Interferometry

Equipment is available which is specifically designed for evaluating the roughness of surfaces by interference techniques. but a conventional projection microscope can be used quite adequately for this purpose. A monochromatic light source and a half-silvered mirror are required. The latter is placed on the stage of the microscope with the surface to be examined on

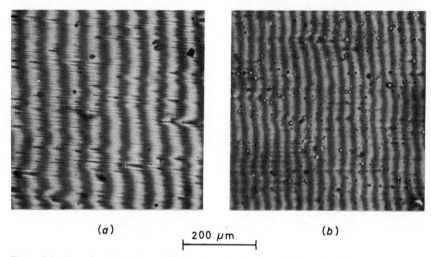

(a)   200 μm   (b)

Figure 8.16. Interference patterns obtained on an aluminium surface before and after electropolishing. (a) Rolled finish before electropolishing, and (b) after electropolishing for 6 min (considerable smoothing has occurred and a similar effect would be observed after plating on to a substrate using a levelling nickel solution)

top of it; it may be necessary to insert a thin piece of paper at one edge between the half-silvered mirror and specimen to form a wedge of small angle ϕ between them. Interference fringes can be produced when two surfaces are arranged in this way and illuminated by monochromatic light so that reflection can occur. The fringes can be photographed to permit assessment and for record purposes. If the thickness of the air film at the centre of a dark band is D, the thickness at the centre of the next dark band will be $D + (\lambda/2)$ where λ is the wavelength of the monochromatic light source. The thickness will increase by $\lambda/2$ for each dark band since the path difference must vary by λ. It can easily be shown that the separation of the fringes is a function of the angle between the two surfaces, i.e. ϕ. An estimate of surface irregularities is feasible since the distance between parallel dark bands is proportional to $\lambda/2$. Fig. 8.16 illustrates interference patterns obtained on an aluminium surface before and after electropolishing. It is apparent that considerable surface smoothing has occurred. Similar patterns showing improvement in surface quality would be obtained if a surface was examined using this technique before and after it was plated with a levelling electrodeposit.

PITTING

The pitting caused by hydrogen gas bubbles has been described previously in this book and elsewhere[71] and the ensuing porosity and its estimation by standard techniques have also been discussed. It was then stated that the tendency for these gas bubbles to cling to the cathode surface is lessened considerably by decreasing the surface tension of the electrolyte solution and by increasing the rate of movement past the cathode. It is obviously more important to be able to judge whether a particular bath is liable to produce gas-pitted electrodeposits than to ascertain subsequently how much pitting is present in the coatings obtained from it. Measurement of the surface tension of the nickel bath can be used as a means of checking that there is sufficient wetting agent present. The Du Nuoy ring detachment technique or the Stalagmometer dropping method can be used for this purpose. For air-agitated baths a surface tension of approximately 50 N/mm is the lowest that can be employed; where air agitation is not utilised, much lower surface tension values are necessary, usually within the range of 30 to 40 N/mm.

The estimation of surface tension reveals whether sufficient surfactants are present to reduce it to the required value, but does not provide any information as to their efficacy as 'anti-pits'. The fact that a certain organic compound will lower the surface tension of a nickel plating bath is not an adequate indication of its ability to prevent pitting. Some comparative 'pitting-test' is necessary to provide such information. However, pitting is usually prevented in production plating baths by either agitating the solution or moving the cathode, but agitation is a difficult variable to specify quantitatively and even more difficult to reproduce on a small scale. Rotating discs have been used to demonstrate the effect of varying speeds on the characteristics of electrodeposited coatings[72,73]. The same technique but with rotating anodes has been employed to illustrate such effects on electropolishing processes.

A similar type of test which gives results that can be directly related to production experience has been devised by Wyszynski[74]. In the test, use is made of the difference in linear velocity, depending on their distance from the centre of a rotating circle, of points on the radii of that circle. Since the adhesion of gas bubbles to a metal disc of uniform surface is constant for a given solution, their ease of removal from any point will depend on the angular velocity of that disc and be proportional to the distance of that point from its periphery. The rate of rotation selected must be sufficient to prevent the general attachment of hydrogen gas bubbles. except near the centre where the linear velocity is approaching zero. This frequently leads to a small circle in the centre being pitted. but. if a particular nickel plating solution is prone to cause pitting. then this pitted area becomes much larger and the radius of this circle can be taken as an empirical measure of the pitting tendencies of the bath.

Wyszynski chose a hemisphere of 0·6 dm d as a test piece. since this shape when used as a horizontal cathode with its open end upwards. does not trap bubbles on its under-side as does a disc. but allows them to rise and so become detached. This hemisphere or dome is suspended from the vertical shaft of an electric motor in a 1 litre beaker containing the solution under test. The disc nickel anode. electrically connected by a vertical and plastics-covered rod of that metal. is placed on the bottom of the beaker. The dome is revolved at an angular velocity of 6 rev/s giving a linear velocity of 1·1 m/s at its periphery. Nickel is normally electrodeposited at 4 A/dm^2 for 30 min. The dome is then removed from the solution and inspected for pits. This test enables the relative efficacy of wetting agents to be readily and conveniently assessed. Although the conditions chosen are arbitrary. it has been found to provide reproducible and realistic results, so that not only can comparisons of different surfactants be conducted. but the equipment can be utilised for the control of production plating baths.

REFERENCES

1. *Electroplated Coatings of Nickel and Chromium.* BS 1224: 1970
2. EDWARDS. J.. *Trans. Inst. Metal Finishing,* **35**, 101 (1958)
3. SUCH, T. E.. *Trans. Inst. Metal Finishing,* **31**, 190 (1954)
4. DENNIS, J. K. and SUCH, T. E.. *Trans. Inst. Metal Finishing,* **40**, 60 (1963)
5. KENDRICK. R. J.. *Plating,* **48**, 1099 (1961)
6. CASHMORE. S. D. and FELLOWS, R. V.. *Trans. Inst. Metal Finishing,* **39**. 70 (1962)
7. BRENNER, A.. ZENTNER. V. and JENNINGS, C. W.. *Plating,* **39**, 865 (1952)
8. HAWKINS, R.. to be published
9. BROOK, P. A.. *Plating,* **47**, 1269 (1950)
10. READ, H. J. and ALES, E. J.. *Plating,* **52**, 860 (1965)
11. LINSELL, R. G.. Project Report, Metallurgy Department of University of Aston in Birmingham (1969)
12. GRAHAM, A. K.. *Trans. Am. Electrochem. Soc.* **44**, 427 (1923)
13. BLUM, W. and RAWDON, H. S., *Trans. Am. Electrochem. Soc.,* **44**, 305 (1923)
14. FINCH, G. I., WILMAN, H. and YANG, L., *Electrode Processes.* Discussion of the Faraday Society. Butterworths, London, 144–158 (1947)
15. HOTHERSALL. A. W.. *Trans. Electrochem. Soc.,* **64**, 69 (1933); *Trans. Faraday Soc.,* **31**. 1242 (1935)
16. HUME-ROTHERY, W.. *Atomic Theory for Students of Metallurgy.* Institute of Metals (1960)
17. DAVIES, D. and WHITTAKER, J. A., *Metals and Materials,* **1** No. 2 (1967); *Met. Rev.,* **12** No. 112, 15
18. PLOG, H., *Galvanotechnik,* **61**, 155 (1970)
19. BERDAN, B., Chapter 15 in *Electroplaters' Process Control Handbook.* Edited by FOULKE, D. G. and CRANE, F. D., Reinhold, New York (1963)
20. SUCH, T. E. and BALDWIN, C., *Trans. J. Plastics Inst.,* **35**, 553 (1967)
21. BRENNER, A. and MORGAN, V. D., *Proc. Am. Electroplaters' Soc.,* **37**. 51 (1950)
22. OLLARD, E. A., *Trans. Faraday Soc.,* **21**, 81 (1925)
23. SCHLAUPITZ, H. C. and ROBERTSON, W. D., **39**, 750. 764. 862. 932 (1952)

24. GUGUNISHVILI, G., *Indust. Lab.* **24**, 333 (1958)
25. KNAPP, B. B., *Metal Finishing*, **47** No. 12, 42 (1949)
26. ROTHSCHILD, B. F., *Products Finishing*, **33** No. 9, 66 (1969)
27. ZMIHORSKI, E., *J. Electrodepositors' Tech. Soc.*, **23**, 203 (1947–1948)
28. JACQUET, P. A., *Trans. Electrochem. Soc.*, **66**, 393 (1934)
29. SUCH, T. E. and WYSZYNSKI, A., *Plating*, **52**, 1027 (1965)
30. BEAMS, J. W., *Proc. Am. Electroplaters' Soc.*, **43**, 211 (1956)
31. DANCY, W. H. and ZAVARELLA, A., *Plating*, **52**, 1009 (1965)
32. MILLS, E. J., *Proc. Roy. Soc.*, **26**, 504 (1877)
33. STONEY, G. G., *Proc. Roy. Soc.*, **82**, 172 (1909)
34. KOHLSCHUETTER, V., *Z. Electrochem.*, **24**, 300 (1918)
35. SODERBERG, K. G. and GRAHAM, A. K., *Proc. Am. Electroplaters' Soc.*, **34**, 74 (1947)
36. PHILLIPS, W. M. and CLIFTON, F. L., *Proc. Am. Electroplaters' Soc.*, **34**, 97 (1947)
37. BRENNER, A. and SENDEROFF, S., *Proc. Am. Electroplaters' Soc.*, **35**, 53 (1948)
38. DENNIS, J. K., *Trans. Inst. Metal Finishing*, **43**, 84 (1965)
39. GABE, D. R. and WEST, J. M., *Trans. Inst. Metal Finishing*, **40**, 197 (1963)
40. ARROWSMITH, D. J. and HOAR, T. P., *Trans. Inst. Metal Finishing*, **34**, 354 (1957)
41. GABE, D. R. and WEST, J. M., *Trans. Inst. Metal Finishing*, **40**, 6 (1963)
42. DENNIS, J. K., *Corrosion Technology*, **12**, 36 (1965)
43. NORRIS, P. J., *Electroplating and Metal Finishing*, **21**, 85 (1968)
44. KUSHNER, J. B., *Plating*, **41**, 1146 (1954)
45. VAGRAMYAN, A. T. and SOLOV'EVA, Z. A., *Technology of Electrodeposition*. Robert Draper Ltd., Teddington, 214 (1961)
46. SCHMIDT, F. J., *Plating*, **56**, 395 (1969)
47. LONGFORD, J. I., *J. App. Crystallography*, **1**, 48 and 131 (1968)
48. WILSON, A. J. C., *J. App. Crystallography*, **1**, 194 (1968)
49. WILLIAMS, E., Wolverhampton Polytechnic, private communication
50. ARROWSMITH, D. J. and HOAR, T. P., *Trans. Inst. Metal Finishing*, **36**, 1 (1958)
51. GURNEY, C., *Nature*, **160**, 166 (1947); *Proc. Phys. Soc.*, **A62**, 639 (1949)
52. GLASSTONE, S., *J. Chem. Soc.* **127**, 2892 (1926)
53. MUBACH, H. P. and WILMAN, H. W., *Proc. Phys. Soc.*, **B64**, 905 (1953)
54. KUSHNER, J. B., *Metal Finishing*, **56** No. 4, 46; No. 5, 82; No. 6, 56; No. 7, 54 (all 1958); *Metal Progress*, **81** No. 2, 88 (1962)
55. WEINER, R. and KLEIN, G., *Metalloberflache*, **B7**, 1 (1955)
56. *Anodic Oxidation Coatings for Aluminium*, BS 1615:1961
57. GARDAM, G. E., *J. Electrodepositors' Tech. Soc.*, **26**, 27 (1950)
58. OLLARD, E. A., *J. Electrodepositors' Tech. Soc.*, **24**, 1 (1949)
59. NELSON, J. H., *J. Electrodepositors' Tech. Soc.*, **21**, 113 (1946)
60. SCOTT, B. A., Chapter 6 in *Quality Control in Metal Finishing* (Ed. by ISSERLIS, G., Robert Draper Ltd., Teddington (1967)
61. *Standard Methods of Testing Paint, Varnish, Lacquer and Related Products*. British Ministry of Defence Specification DEF-1053, Method No. 11 (1968)
62. *Tentative Method of Test for Specular Gloss*, American Society for Testing and Materials. Specification ASTM D523-66T
63. *Centre-Line Average Method for the Assessment of Surface Texture*. BS 1134:1961
64. DENNIS, J. K. and FUGGLE, J. J., *Trans. Inst. Metal Finishing*, **47**, 177 (1969)
65. SMITH, J. F., *Trans. Inst. Metal Finishing*, **46**, 199 (1968)
66. SUCH, T. E., *Trans. Inst. Metal Finishing*, **32**, 26 (1955)
67. KATZ, W., *Metal Finishing Journal*, **16** No. 183, 68 (1970)
68. TOLANSKY, S., *Z. für Elektrochemie*, **56** No. 4, 263 (1952)
69. TOLANSKY, S., *Scientific American*, **191**, 54 (1965)
70. LORKING, K. F., *Trans. Australian Inst. Met.*, **5**, 109 (1952)
71. BLOOR, D. W., *Electroplating and Metal Finishing*, **20**, 343 (1967)
72. LEVICH, V. G., *Acta Physiochim. U.S.S.R.*, **17**, 257 (1942)
73. ROGERS, G. T. and TAYLOR, K. J., *Nature*, **200** No. 4911, 1062 (1963)
74. WYSZYNSKI, A. E., Research Laboratory, W. Canning and Co. Ltd., private communication

Chapter 9

Chromium Plating

As explained in Chapter 1, since the subject of this book is the electro-deposition of nickel, with or without a subsequent electroplated chromium layer, chromium plating will be discussed only to the extent necessary to make clear the interrelations between it and nickel plating. Consequently, all that is required is some description of the properties of chromium electrodeposits and of the variations in solution composition and plating conditions required to achieve certain desired properties. Therefore, the chromium plating process will not be dealt with in the detail that its importance otherwise merits, since it is already adequately covered in other publications, notably by Morriset et al.[1]. In particular, hard chromium deposition will be mentioned only briefly as it is comparatively rare for these thick deposits to be applied over nickel undercoats, and normally they are plated directly on to the basis metal. For further details of this important technology, the reader is referred to the books of Greenwood[2,3].

At present, chromium is normally deposited on a commercial scale from aqueous solutions of chromium trioxide, i.e. chromic anhydride, CrO_3 which is commonly called chromic acid, in spite of many attempts having been made to use a solution in which the metal is present in the trivalent form. However, chromium cannot be deposited from a bath containing only chromic acid, for another chemical which acts as a catalyst must also be present. Lack of knowledge concerning which anions performed as catalysts accounted for many of the early problems encountered during the development of chromium plating electrolyte solutions (see Chapter 1). Although many papers have been published that describe baths based on trivalent chromium, these have not yet been commercially successful for depositing coatings, although a 'trivalent bath' based on a mixture of water and an organic solvent has recently been developed and is claimed to be satisfactory.

Plating solutions used for chromium deposition from its hexavalent state are relatively simple in formulation, especially when compared with 'organic' bright nickel baths. Few chemicals, particularly organic compounds, are stable in the highly oxidising chromic acid solutions, especially during electrolysis, and so the possibility of modifying the structure and properties of chromium deposits by means of additions to the bath is very restricted. Variation of the type and concentration of catalyst or the use of more than one catalyst offers the only significant change of formulation of the basic chromic bath. The characteristics of the deposit are influenced by the concentration of the chromic acid, its ratio to sulphuric acid and the usual plating variables (current and temperature). The source of current is also most important in that its waveform or 'ripple' greatly affects the nature of the deposits, so much so that single-phase rectification is not adequate. The pH in such a strong acid liquid cannot be varied significantly in the bulk of the electrolyte solution.

THE MECHANISM OF DEPOSITION FROM CHROMIC ACID PLATING BATHS

Although the baths themselves are simple, the mechanism of chromium deposition from them is far more complicated than that for most metals, because, not only is chromium present in the form of complex anions, i.e. $Cr_2O_7^{2-}$, but it can exist in several valency states and its trivalent ion readily forms complexes. None of the theories put forward to account for the deposition mechanisms are entirely satisfactory, but some Russian workers[4,5] and more recently Ryan[6] have reviewed the modern concepts. In their text, Sully and Brandes[7] critically review the opposing theories, while Morrisset[8] and Weiner[9] have provided briefer summaries in their comprehensive books. It is intended in the present work to give only a synopsis of proposed theories and to include references to more detailed sources. Several of the early theories were based on the assumption that deposition took place from Cr(III) or even Cr(II), although chromium is introduced into the bath as Cr(VI). A number of electrochemical reactions have been suggested as possible means of initially forming Cr(III). Silverman[10] was a protagonist of the complex formation theory; he suggested that the addition of SO_4^{2-} to the bath results in the formation of complexes, deposition ultimately taking place from a complex cation. Kasper's[11] theory indicates that deposition takes place by direct reduction of Cr(VI), and later Ogburn and Brenner[12] showed, using a radio-tracer technique, that only chromium originally present in the hexavalent state was deposited at the cathode. Snavely[13] has suggested that the final reduction to metallic chromium takes place via an intermediate hydride, of short but finite life, which is formed in the cathode film. This hypothesis still has some protagonists, for example Knodler[14], and Raub and Muller[15].

Prior to Snavely, Rogers[16] had indicated the importance of the cathode film and its pH. Most of the recent investigations have been concerned with examination of the cathode film, particularly the effect of the catalyst anions (e.g. SO_4^{2-} and F^-) on its composition and formation. It is now apparent that deposition does not take place from ions within the bulk electrolyte solution but from entities (complexes) within the cathode film. It is likely that chromium ions enter the cathode film in the form of CrO_4^{2-} anions and that complex chromium-chromate cations are formed from which metallic chromium is subsequently deposited. Sulphate ions act as exchange ions in the formation of the cation complexes. Atomic hydrogen as well as electrolytic reduction is involved in the reduction of Cr(VI) to Cr(III). This reaction must occur since, as stated above, only chromium initially present in the hexavalent state is found in the deposit. Ryan[6] concludes that it is probable that the final reduction involves the reactions $Cr^{3+} \rightarrow Cr^{2+}$ and $Cr^{2+} \rightarrow Cr^0$. Maximum current efficiency is associated with a critical cathode film thickness and with the concentration of Cr(III). Variation of either temperature, current density or bath composition results in changes in the characteristics of the film and therefore causes changes in current efficiency and deposit characteristics. Many phenomena known to occur in commercial solutions can be accounted for by consideration of such experimental information. For example, if the operating temperature is increased beyond the range 25°–55°C, the current efficiency decreases because the thickness of the cathode film decreases below the critical value.

However, Weiner, who has also extensively investigated this topic, does not agree with all of Ryan's conclusions. In an English summary[17] of his work, Weiner puts some emphasis on the actual metal forming the cathode, in particular whether or not that metal is normally passivated by chromic acid. He states that the function of the catalyst, such as SO_4^{2-} or F^-, is to depassivate the cathode by chemical dissolution, rather than to affect the deposition of chromium by some electrochemical mechanism within the cathode film. He bases his suggestions on the differences in shape of cathode current density/potential curves for various metals in chromic acid solutions, with and without catalysts. Weiner believes his theory is confirmed by his finding that in the absence of catalyst no substrate metal could be found in the cathode film, whereas the substrate metal and its oxides were detected in the film when a catalyst was present.

BATHS BASED ON CHROMIC ACID

DECORATIVE CHROMIUM DEPOSITION

The solutions that are mostly used for decorative purposes are still of the type investigated by Sargent[18], that is chromic acid with SO_4^{2-} as catalyst. The actual concentration of the bath is not critical but the ratio of chromic acid/sulphuric acid is most important. This ratio usually falls within the range 80:1 to 120:1, with 100:1 being the popular value. The concentration of chromic acid is normally within the range 150–550 g/l. The concentrated solutions are more expensive to install than the more dilute type, and a higher drag-out rate is incurred in operation. However, the concentrated solutions are less sensitive to minor changes in concentration and metallic contaminants and are therefore easier to control. The throwing power of chromium baths is notoriously poor, but the baths containing 250–300 g/l of chromic acid are the best in this respect. Conductivity reaches a maximum at concentrations between 400 and 500 g/l and so the lowest applied e.m.f. is then required to give a specific current density; 4–10 V is the normal range required, the actual voltage depending on the current density used, the inter-electrode distance and total area being plated. The current efficiency of chromium baths is very low compared with baths used for plating most other metals. Baths used for depositing regular chromium have cathode efficiencies ranging between 6 and 16%, the more dilute baths having the higher efficiencies.

There is a close relation between the solution temperature and current density needed to give a satisfactory bright plate, i.e. the higher the current density the higher the solution temperature that is required. For each particular concentration of chromic acid there is an optimum temperature which gives the widest range of current density at which bright deposits are obtained, as shown in Fig. 9.1, taken from the work of Haring and Barrows[19]. It should be noted that they used steel as a cathode metal;

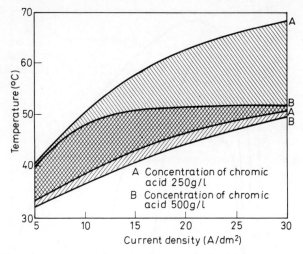

Figure 9.1. Effect of temperature and current density on the bright plating range of two chromic acid solutions, using a steel cathode (after Haring and Barrows[19])

many other workers have obtained somewhat different results using different substrates and also with varying ratios of chromic acid/sulphate ion, but the same principles apply.

Efficiency values over 12% are not readily obtainable from the chromic acid plus sulphuric acid solution, for they entail the use of high current densities and low temperatures, such that 'burning' of the chromium deposit

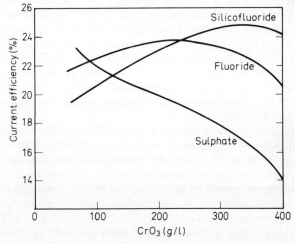

Figure 9.2. Current efficiency versus chromic acid concentration (after Bilfinger)

is likely to occur. However, the presence of other anions in place of all or part of the sulphate, in particular the fluoride or similar ions, such as fluosilicate or fluoborate, has been found to produce higher efficiencies (Figs. 9.2 and 9.3).

Fischer[20] extensively investigated the effect on cathode efficiency of these

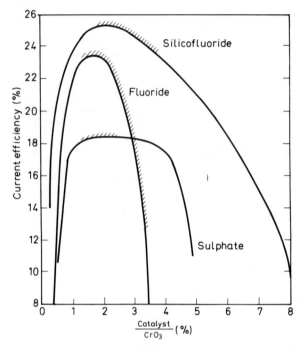

Figure 9.3. Current efficiency in 250 g/l chromic acid baths at 55°C, current density 50 A/dm²; shading indicates bright plating range (after Bilfinger)

fluoride-containing ions, and more recently Parthasaradhy[21] has presented a detailed survey of the numerous variants of these types of bath that have been described in the literature. In addition to the fluosilicate, other complex fluorides, such as fluoaluminate[22] have been tried. Again, as with solutions containing only sulphate as the catalyst, the highest efficiencies are obtained at the highest current densities but the current density range for bright plating is stated to be wider. Since these baths will give deposition at efficiencies up to 25%, without excessively high current densities, they are steadily becoming more popular.

With efficiencies of the order quoted, it is obvious that copious amounts of hydrogen must be liberated at the cathode simultaneously with metal deposition, and this leads to serious hydrogen embrittlement of the substrate, if it is susceptible to this phenomenon. To meet certain service requirements, it may be necessary to heat treat the plated component in order to remove the embrittlement. Low cathode efficiency is obviously wasteful as far as power consumption is concerned, but the time taken to plate a particular thickness is not so much greater than it would be for many other metals which are plated at much greater efficiency but lower current density since fairly high current densities are used for decorative chromium plating, 12 A/dm^2 being typical.

Chromium electrodeposited from chromic acid baths of the aforementioned type is hard, brittle and has a high tensile stress. When the thickness of the deposit exceeds a certain value, the total internal tensile stress is sufficient to cause the deposit to crack as the tensile strength of this chromium is exceeded, as described by Fry[23]. Deposits plated from the dilute solutions crack at a lower thickness than those plated from the more concentrated solutions. A coarse crack pattern is formed (macro-cracks) and this can be particularly detrimental to the performance of the composite nickel plus chromium coating in service. These cracks are often observed at high current density regions on components of complex shape. They induce localised attack of the nickel undercoat, which leads to rapid penetration to the substrate. This is an unfavourable system from the corrosion aspect since the anode area is small and the cathode area large, which is conducive to rapid localised attack on the former. If the chromium layer is less than a certain minimum thickness the deposit will be porous. Consequently, chromium deposits nearly always contain discontinuities, either on a macro or micro scale. The brittle nature of the deposit results in further cracking on slight deformation, either during assembly or in service. For these reasons, nickel is exposed at small defects in chromium coatings and becomes the anode of the Cr/Ni bimetallic couple when the deposit is exposed to a moist and corrosive environment; this bimetallic couple leads to rapid electrochemical dissolution of the nickel. The mechanism of attack is discussed in later chapters but modified nickel plus chromium coatings have been developed recently with the object of controlling this form of attack.

HARD CHROMIUM DEPOSITION

Chromium used for engineering purposes is usually termed *hard* chromium, although this is a somewhat misleading name (see Chapter 5); thick chromium would be far more appropriate. The baths used are very similar in formulation to those used for the deposition of thin decorative chromium except that the more concentrated solutions are not often employed, the usual composition range being 150–300 g/l of CrO_3. Solutions containing fluoride-type ions are frequently utilised because of their greater cathode efficiency. Higher temperatures (over 50°C) and higher current densities, up to 80 A/dm^2, are used than are employed for decorative chromium. High temperatures are necessary to prevent burning and the formation of rough deposits at the rapid rate of deposition given by these high current densities. In order to obtain the maximum hardness, it is necessary to consider both bath temperature and current density as indicated in Fig. 9.4. Table 9.1 shows the optimum temperatures of the plating solution and how these are related to certain ranges of cathodic current densities when plating from chromic acid baths catalysed with sulphuric acid (100:1 ratio) in order to obtain the hardest deposits; it is interesting to note that these deposits are also the brightest. The presence of fluoride-type ions enables faster deposition rates to be employed, but greater care may be required in some cases because of their greater etching effects on unplated areas. The properties required of hard

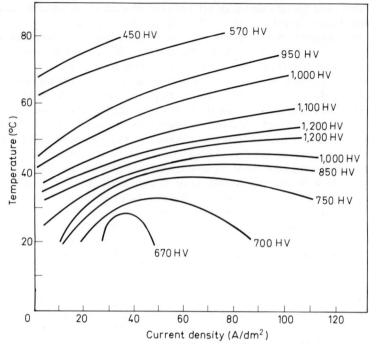

Figure 9.4. Hardness as a function of current density and temperature; curves of equal hardness (after Wahl and Gebauer)

Table 9.1 OPERATING CONDITIONS OF CHROMIUM PLATING BATHS TO GIVE THE HARDEST ELECTRODEPOSITS

Current density	Temperature (°C)	
(A/dm^2)	250 g/l chromic acid solution*	400 g/l chromic acid solution*
22	48–50	43–45
33	50–54	45–49
44	50–55	45–50
66	52–56	47–51
110	53–57	48–52

* Catalysed only by sulphuric acid with $CrO_3:H_2SO_4 = 100:1$ (Table reproduced from Ministry of Defence Specification DEF-160 entitled *Chromium Plating for Engineering Purposes*, H.M.S.O., London, 1967)

chromium coatings and the thicknesses suggested for different applications are described in the relevant British Standard[24]. Since very high current densities are used when depositing thick coatings, it is frequently necessary to have the vat fitted with facilities not only for heating but also for cooling, the latter being necessary due to the heating effect of the large current passed per unit volume.

SELF-REGULATING HIGH-SPEED BATHS (S.R.H.S.)

This type of solution is again based on chromic acid, with silicofluorides (or other complex fluorides) and sulphates being used as catalysts to give the higher cathode efficiency, but it differs from previously described baths in one significant aspect. Control of the catalyst anion concentrations is facilitated by using a mixture of salts which, by the common-ion effect, result in their solubilities being those necessary to maintain automatically the desired concentration of catalysts in the bath. Undissolved salts remain at the bottom of the tank and periodic stirring ensures that not only is the catalyst replenished in solution but that optimum concentration cannot be easily exceeded. A mixture of salts is required because no single salt exists which has the correct solubility. Strontium sulphate is the basis of most of the mixtures, but this has a solubility which gives a maximum of 2·1 g/l of SO_4^{2-} at 38°C, which is insufficient for chromic acid concentrations much greater than 210 g/l and so other sulphates or fluorine-containing salts must be added to provide the correct catalyst concentration for baths of higher concentration. In order to control the sulphate content, strontium chromate is often added, since the reserve of strontium ions this salt provides will combine with the majority of any additional sulphate ions that are accidentally introduced, as for example by carry-over, and remove them by precipitation as strontium sulphate. If the intrinsic solubility of potassium silicofluoride is greater than that required to furnish the correct content of silicofluoride ions, then the solubility of that salt can be suppressed by addition of another potassium salt, usually potassium dichromate. A typical mixture used for preparing an S.R.H.S. bath contains chromic acid, strontium sulphate, strontium chromate, potassium silicofluoride and potassium dichromate.

However, it is by no means essential in this type of solution to have the content of all catalyst ions controlled by the common-ion effect. Their suppressive effects on catalyst solubility depend on the concentration of the particular common ions chosen. According to Stareck[25], the maximum concentration of sulphate that can be held in solutions can be reduced to as low as 0·2 g/l, if as much as 12 g/l of Sr^{2+} is present. Again, the equilibrium solubility of potassium silicofluoride results in a concentration of 6 g/l of SiF_6^{2-}, but this can be lowered to 1·5 g/l, if 16 g/l of K^+ are present. If more than 6 g/l of SiF_6^{2-} are required, sodium silicofluoride will give a maximum of 11 g/l and, of course, the concentration of SiF_6^{2-} can then be regulated by the presence of sodium ions. A most important feature of these self-regulating solutions, which contain fluoride complexes such as fluosilicate or fluoborate as catalysts, is that the cathodic current efficiency obtainable at a particular current density increases as the temperature is increased due to the increased solubility of these compounds leading to a higher total catalyst content, while in solutions catalysed only by sulphate ions the cathodic current efficiency falls with increase in temperature.

CRACK-FREE CHROMIUM

Crack-free chromium is deposited from another variant of the chromic acid solutions, usually from those of fairly high concentration, since this favours lower stress and consequently fewer cracks for a given thickness of coating. However, this type of deposit is still very hard and brittle, and being thicker

Table 9.2 TYPICAL VARIANTS OF PLATING PROCESSES BASED ON CHROMIC ACID SOLUTIONS

Type of deposit	Composition (g/l)					Operating conditions		
	Chromic acid	Sulphate ion	Silicofluoride ion (SiF_6^{2-})	Strontium ion	Potassium ion	Temperature of solution (°C)	Current density (A/dm²)	Cathode efficiency (%)
DECORATIVE (i.e. conventional or regular)								
(i) Single catalyst	500	5	Nil	Nil	Nil	38	10	8
(ii) High efficiency	250	1	2	Nil	Nil	45	15	15
(iii) Self-regulating	250	1[1]	2[2]	4[3]	14[4]	45	15	15
CRACK-FREE								
(i) Single catalyst	500	3·5	Nil	Nil	Nil	50	25	8
(ii) High efficiency	450	2·5	1	Nil	Nil	45	20	12
MICRO-CRACKED	175	0·35	3	Nil	Nil	45	16	18
HARD								
(i) Single catalyst	250	2·5	Nil	Nil	Nil	55	50	12
(ii) High efficiency	300	1·5	4	Nil	Nil	55	50	22
(iii) Self-regulating	200	0·8[1]	3[2]	6[3]	6[4]	55	50	22

Notes

1. Added by means of strontium sulphate.
2. Added by means of potassium silicofluoride.
3. Added by means of strontium sulphate and strontium chromate and so not all present in solution (i.e. in ionised form).
4. Added by means of potassium silicofluoride and potassium dichromate and so not all present in solution (i.e. in ionised form).

Properties of chromium plating processes are affected by the presence of metallic impurities that commonly accumulate in the solution; for example chromium in the trivalent form and iron are frequently found in hard plating baths and, in addition, nickel, copper and zinc are often present in commercial decorative plating solutions. The table above therefore refers only to pure, freshly prepared baths.

than decorative chromium it is even more likely to crack during assembly or in service when present on components that are flexed or deformed, thus giving a macro-crack formation. For this reason the thick (1 μm) crack-free decorative coatings have not enjoyed the popularity that might have been anticipated. On rigid components, crack-free chromium is successful since the tenacious oxide on the chromium confers very great corrosion resistance to that metal, which protects the underlying nickel coating and hence the substrate[26, 27].

In baths used for the plating of crack-free chromium, the catalyst is present in lower concentration than usual; it can be sulphate only or a combination of sulphate and fluoride or silicofluoride. In any case, there will be a higher ratio than normal between the chromic acid and catalyst. Since these solutions are also usually operated at fairly high temperatures, they are sometimes termed the H.T.H.R. (high temperature high ratio) type. Crack-free chromium appears much bluer in colour than conventional, decorative chromium. A typical bath which gives this type of deposit is included in Table 9.2.

MICRO-CRACKED CHROMIUM

Dilute solutions containing mixed sulphate/fluoride or silicofluoride catalysts have been formulated to enable chromium to be deposited such that at a thickness of approximately 1 μm the deposit is uniformly micro-cracked, that is the plate has a continuous and uniform network of cracks. Subsequently, when this micro-cracked chromium is used as the top layer in a decorative nickel plus chromium coating, the nickel can be attacked at many of the cracks, instead of at a few isolated points as in the case of a macro-cracked top layer. This does not prevent attack of the nickel, but corrosion is spread out over a large area with a consequent decrease in the anodic current density and the nickel coating is not penetrated rapidly.

Many factors influence the crack pattern, including concentration of chromium trioxide, concentration and type of catalyst, concentration of Cr(III) and other metallic impurities, temperature, current density and type and condition of the underlying metal. The ideal crack density is between 25 and 80 cracks per millimetre, and so it is essential to control the plating conditions fairly precisely, particularly if satisfactory coverage with chromium is to be achieved. As all the necessary criteria were not at first understood, the first micro-cracked chromium coatings were obtained by deposition of two superimposed chromium layers, i.e. a duplex deposit[28, 29], the first layer having good coverage and the second conferring the micro-cracking. Since chromium was being deposited on chromium, it was essential that the second bath contained fluoride-type ions.

Chromic acid solutions containing small quantities of selenium have been used to produce micro-cracked chromium not only in a duplex system but also as a single layer from one bath[30]. These have not been commercially utilised to any great extent for, although they result in a uniform crack pattern, they can have a very blue appearance, particularly if the selenium content rises slightly. The solutions without selenium are far superior in this respect and some have now been developed to produce micro-cracked chromium having good coverage. Whilst these baths deposit chromium having a slightly hazy appearance if compared directly with conventional chromium, especially when examined at a glancing angle, the micro-cracked plate has a good bright appearance when viewed at 90°. Deposits at least 0·75 μm thick must be deposited in order to achieve a satisfactory crack density, and this thickness is obtained from most commercial solutions in about 8 min, since these usually contain fluoride ions, which give these baths quite high cathode current efficiencies. Solutions which provide micro-cracked deposits are operated at somewhat higher current densities and temperatures than conventional, regular baths. Their chromic acid contents lie between 150 and 250 g/l. Since their sulphate content is often critical, this type of solution may have its catalyst self-regulated as described in a previous section.

Variables affecting micro-cracking

Such and Partington[31] have shown the effect of varying concentrations of sulphuric acid on the onset of micro-cracking. This can be illustrated by reference to mean stress/thickness curves, but is shown far more clearly by the instantaneous stress/thickness curves in Fig. 9.5. As the thickness increases.

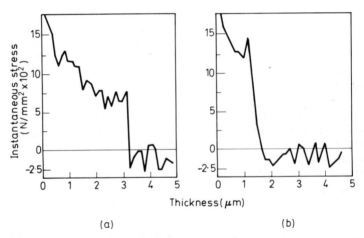

Figure 9.5. Influence of sulphuric acid concentration on the instantaneous stress variation in chromium deposited on steel at 45°C and 22 A/dm² from a micro-cracked chromium solution; (a) 0·25 g/l sulphuric acid and (b) 0·55 g/l sulphuric acid (after Such and Partington[31])

the stress accumulates until the tensile strength of chromium is exceeded and cracking occurs. This phenomenon is indicated in the graphs by the sudden fall in stress. It is apparent therefore that, for the particular solution investigated, cracking occurs at a lower thickness as the sulphuric acid content is increased. This is important since, to achieve good results in service, it is more essential to have the correct crack density than to have a certain minimum thickness of chromium. Similarly, Cr(III) influences the crack pattern; when the concentration exceeds 3 g/l, slightly greater thick-

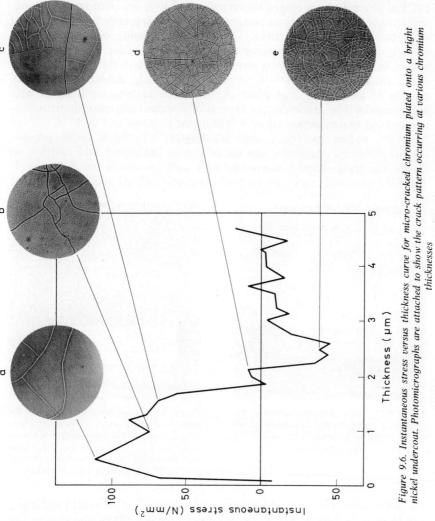

Figure 9.6. Instantaneous stress versus thickness curve for micro-cracked chromium plated onto a bright nickel undercoat. Photomicrographs are attached to show the crack pattern occurring at various chromium thicknesses

nesses of chromium have to be deposited to obtain the same crack density as that found in slightly thinner plate deposited from a solution free from Cr(III). Other metallic impurities such as zinc, copper and nickel have the same effect. Fig. 9.6 shows an instantaneous stress/thickness curve for micro-cracked chromium on bright nickel; attached to it are photographs representative of the crack patterns occurring at particular points, that is particular plate thicknesses on the curve. This illustration indicates how the crack pattern develops with increasing thickness of chromium. The appearance shown in Fig. 9.6 (d) has been termed zoning by one of the present authors in an earlier publication[32]. It consists of a clearly defined macro-crack pattern in which the large areas of the macro pattern are filled in with micro-cracks. The crack pattern typical of commercially acceptable micro-cracked chromium deposited on bright nickel is shown in Fig. 9.7. Cracking has been shown by Jones and Saiddington[33] to be a cyclic process. As the thickness increases, one set of cracks heals over and another set forms, the crack density increasing with each successive set of cracks. In Fig. 9.7 two sets of cracks are visible, one clearly defined and one quite faint,

Figure 9.7. Optical photomicrograph of the surface of micro-cracked chromium plated onto a bright nickel undercoat; two sets of cracks are visible

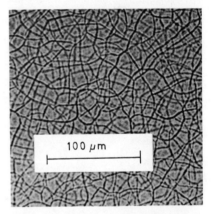

Figure 9.8. Optical photomicrograph of the surface of micro-cracked chromium plated onto a bright nickel undercoat at the same conditions as that shown in Fig. 9.7, except that the plating time was increased from 10 to 15 min; many cracks are visible

the latter being the ones which have healed over. Healed cracks are much less likely to be sites at which corrosion can take place than new cracks. For purposes of assessing crack density only the freshly formed ones should be counted. If all the cracks visible when using an optical microscope at a magnification of $\times 200$ were counted, the crack count obtained would be approximately twice the effective crack density. Fig. 9.8 shows a very high crack density and, since several sets of cracks have been formed, the zone boundaries are less clearly defined.

Crack density varies with current density as well as with thickness, and the correct frequency of cracks is only obtained over a specified current density range. It is difficult therefore to produce a uniform crack pattern at the extremes of current density on a shaped component. This is the main

CHROMIUM PLATING

disadvantage of micro-cracked chromium compared with micro-porous chromium since, to obtain satisfactory corrosion resistance in service, the frequency of discontinuities in the chromium must be uniform even at recesses and elevated regions, if these are present on significant surfaces. One means of overcoming this problem is to cause the current density to fluctuate over the surface of the component being plated. This can be done either in a periodic manner or by step-wise decreases of the current density from that initially applied[34]. In this way, any type of crack pattern specific to a particular current density is no longer confined to certain areas, so that the surface has a uniform appearance.

The underlying metal has a significant effect on the form of the crack pattern. Zoning on bright nickel has already been described. Matt surfaces

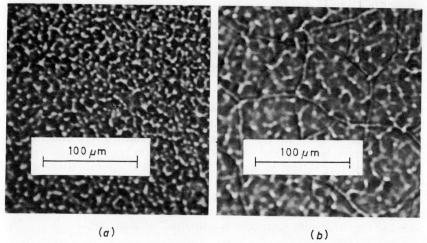

Figure 9.9. Optical photomicrographs of the surface of micro-cracked chromium plated onto unpolished semi-bright nickel. (a) No cracks after plating for 10 min and (b) cracking has commenced after plating for 15 min at the same conditions (after Dennis[32])

Figure 9.10. Optical photomicrograph of the surface of micro-cracked chromium plated onto copper polished in one direction only. Directional cracking has occurred after plating for 10 min (after Dennis[32])

have been found to retard the formation of micro-cracks in chromium. Fig. 9.9(a) shows the appearance of a chromium deposit on unpolished semi-bright nickel; it was plated at the same conditions and for the same length of time as that shown in Fig. 9.7. However, cracks begin to form if plating is continued for longer periods [Fig. 9.9(b)]. On certain polished substrates (dull Watts nickel, semi-bright nickel or steel) micro-cracking does occur but the cracks are much less clearly defined and zoning cannot be observed. Beacom, Hardesty and Doty[35] have shown that directionally orientated cracks occur in conventional chromium deposits on polished dull nickel. The present authors have not observed this effect in chromium plate deposited onto polished nickel from commercial solutions formulated to produce micro-cracked deposits, but did observe it when polished copper was the substrate (Fig. 9.10).

Micro-porous Chromium

Micro-porous chromium is included in this chapter, only because the plating solution used for this deposit is the conventional regular solution. The pores in the chromium layer result from the incorporation of small inert particles in the underlying nickel coating. This is usually applied as a thin layer subsequent to bright nickel plating. In service the micro-pores behave in the same manner as the cracks in micro-cracked chromium; the corrosive attack being spread over 100 to 300 pores per square millimetre. The value of this system as a corrosion-resistant coating is discussed in Chapter 10.

Attempts to incorporate inert particles in the chromium layer in order to produce pores have met with little commercial success, whether the material added to the chromium solution was graphite[36] or alumina[37].

Micro-cracked Nickel

Another technique has been devised to produce micro-cracked chromium from conventional baths, which would normally produce standard uncracked chromium at the thicknesses of around 0·25 μm normally employed for decorative deposits. This technique is based on the deposition of a highly tensile stressed and brittle layer of nickel subsequent to the bright nickel coating and immediately prior to the chromium plating operation. This

Table 9.3 COMPOSITION AND OPERATING CONDITIONS FOR BORNHAUSER TETRACHROMATE BATH*

Chromic acid	300 g/l
Sodium hydroxide	50 g/l
Sulphuric acid	0·6 g/l
Ethanol†	1 ml
Solution temperature	15°–22°C
Current density	20–100 A/dm^2

* Data abstracted from Dr. O. Bornhauser's German Patent No. 608 757 [38]
† To reduce some of the Cr(VI) to Cr(III) (approx. 4 g/l Cr^{3+}).

nickel layer can either be so stressed that it cracks spontaneously as deposited, or it can crack because of the extra strain induced in it by the high tensile stresses in the superimposed chromium. Further details of this process are also given in Chapter 10.

As for micro-porous chromium, the advantage claimed for this process is its ability to produce micro discontinuous chromium plate in low current density areas with chromium deposits little thicker than those required for standard decorative coatings.

TETRACHROMATE BATHS

This variety of chromium plating bath, invented by Dr. O. Bornhauser[38], has enjoyed some popularity in Germany and France since it was first marketed in the early 1930s. It has been described in English by Taylor[39], but has not found much application in the U.K. Sodium tetrachromate, $Na_2Cr_4O_{13}.H_2O$ is its principal constituent and the process is operated at lower temperatures and higher current densities than the conventional chromic acid baths. A typical formula for a Bornhauser solution is given in Table 9.3.

Chromium deposited from the Bornhauser bath is dull and grey and so must be polished to obtain a bright appearance but, being soft, it is relatively easy to polish. Insoluble anodes of lead are used. The cathode efficiency of around 30%, combined with a high current density, results in a rapid rate of deposition; this is fortunate since this process was developed for plating directly onto basis metals, particularly zinc alloy die-castings, without an undercoat of nickel. However, the corrosion protection afforded does not compare with that given by nickel plus chromium coatings.

Recent improvements to this process have been claimed in the literature[40] but the chromium has usually still been deposited in the dull condition. Roggendorf[41] has described tetrachromate solutions containing metals such as indium, vanadium and selenium, which give bright plate, although at lower cathode efficiencies and maximum current densities than obtainable from dull baths. These solutions are commercially available in three variations[42], which are capable of depositing regular, micro-cracked or crack-free chromium.

BLACK CHROMIUM

Attempts to produce chromium electrodeposits in black form have been tried for many years and there are many references in the literature to processes that have never achieved wide usage for various reasons. Ollard[43] was one of the earliest, but Graham[44] and some Indian workers[45] have also made important contributions. It was not until the later 1960s that two American processes[46, 47] were made sufficiently practicable to achieve commercial popularity. As with most previous solutions, these two were both based on chromic acid with all sulphate removed, one depending upon nitrate and fluoride ions as catalysts and the other on the presence of acetate ions.

The respective basic bath formulations are given in Table 9.4, but the additions to provide the catalysts are proprietary and so their nature and concentration can only be gleaned from the respective patents[48, 49].

It can be seen that these solutions operate at near ambient temperatures

Table 9.4 COMPOSITION AND OPERATING CONDITIONS OF COMMERCIAL BLACK CHROMIUM PLATING BATHS

	Bath A*	Bath B†
Chromic acid	470 g/l	300 g/l
Trivalent chromium [Cr(III)]	8 g/l	10 g/l
Solution temperature	16°–32°C	18°–27°C
Current density	30–50 A/dm^2	15–30 A/dm^2

* Data for Bath A are extracted from Branciarolli and Stutzman[46].
† Data for Bath B are extracted from Longland[47].

and at current densities not much more than those required for conventional chromium plating. Most common metals can be plated directly with black chromium or an undercoat of nickel, or decorative chromium can be used. It is essential to eliminate all sulphate ions by the continual presence of barium salts in order to obtain a true black colour, whose lustre depends mainly on that of the substrate. The deposit then contains a large proportion of oxide and this may explain its good corrosion resistance, which is, of course, enhanced by the post-plating waxing or oiling treatments that are deemed to be necessary to remove smut and provide a uniform colour. The main applications for this finish are for its aesthetic appeal, particularly in contrast to normal chromium plate. Colours other than black are said to be obtainable, in particular a method of producing a gold colour has been patented[50].

DEPOSITION FROM TRIVALENT BATHS

Much of the earlier work conducted on the electrodeposition of chromium was concerned with trivalent baths and these have also been the subject of many recent investigations. In spite of this, this type of bath has not yet been used industrially for the deposition of chromium coatings although it is employed for electro-winning of that metal[51]. Chromic chloride or sulphate have been the favourite salts used as the basis for possible successful electrolyte solutions, which often also include complexing agents such as ammonium salts, or glycollic acid. Baths based on fluoborate and sulphamate ions have also been investigated. The results achieved with trivalent chromium electrolyte solutions up to 1947 were reviewed by Parry et al.[52]. Yoshida and Yoshida[53], Machu and Eli-Ghandour[54] and Zell[55] have more recently described their trials of these types of bath.

The use of non-aqueous solvents, in particular organic liquids, has also been tried, and a critical review by Chisholm[56] has been published. None of these trivalent baths appeared likely to be a competitor to that based on chromic acid.

However, recent work by the British Non-ferrous Metals Research Association[57] has shown the advantages of plating chromium from Cr(III) baths based on a mixture of organic solvents and water. The current efficiency (40–50%) is much higher than can be achieved using Cr(VI) baths and, as deposition also takes place from a lower valency state, the rate of deposition of metal is about six times faster than from the Cr(VI) solution. The covering power and macro throwing power are also better for this type of solution. The chromium deposit is relatively ductile and softer than conventional plate, the hardness being about 400–600 HV. The deposit has a darker

Table 9.5 COMPOSITION AND OPERATING CONDITIONS OF B.N.F.M.R.A. TRIVALENT CHROMIUM BATH*

Chromic chloride, $CrCl_3 \cdot 6H_2O$	213 g/l
Sodium chloride	36 g/l
Ammonium chloride	26 g/l
Boric acid	2 g/l
Dimethylformamide	400 g/l
pH	1·1–1·3
Temperature of solution	20°–30°C
Current density	10–15 A/dm^2

* Data extracted from Ward et al.[59]

appearance than conventional chromium deposits and resembles stainless steel. Below a thickness of 0·75 μm the deposit is micro-porous, the density of pores being greater than in conventional micro-porous chromium coatings, but above a thickness of 0·75 μm it appears to be micro-cracked. The coating can be used in conjunction with nickel undercoats to provide good corrosion resistance in an outdoor environment. The corrosion mechanism is similar to that occurring for nickel plus micro-cracked or micro-porous chromium systems.

Many formulations are listed in the patent[58], but one shown in Table 9.5 is typical of these baths. The organic solvents used (dipolar aprotic solvents) have high dielectric constants which permit ionisation of the dissolved metal salts and they readily form complexes with metals. Dimethyl formamide is preferred for economic as well as technical reasons. Aqueous trivalent baths have previously been unsuccessful mainly due to their limited range of operation. Below about pH 1 hydrogen evolution is excessive and above pH 3 basic chromium compounds precipitate. Current efficiency for metal deposition is therefore low and the copious liberation of hydrogen causes the pH of the cathode film to change rapidly, thus making control difficult. The organic solvents are advantageous because the hydrogen ion concentration is low and there is less tendency to co-deposit hydrogen with the metal, since they contain a highly electronegative oxygen atom.

A solution having the formulation shown in Table 9.5 has one disadvantage in that chlorine is liberated at the anode; this is soluble in the dimethylformamide and soon results in inferior metal deposits. This can be prevented by plating in a cell fitted with a diaphragm (pore size 1 μm) between anode and cathode. If a molar solution of ammonium or sodium acetate is used in the anode compartment, ethane and carbon dioxide are evolved at a carbon anode. This arrangement necessitates the use of higher e.m.f.s. (10–25 V) than are required for the deposition of chromium from Cr(VI) baths.

At the time of writing, organic Cr(III) baths have been used on a pilot scale, but they appear to offer real possibilities of depositing chromium at a faster rate than from Cr(VI) baths and also of providing deposits having superior properties. The initial cost of the plating solution is much higher than that of Cr(VI) baths, but the high overall efficiency of deposition together with other minor advantages enables a particular thickness of chromium to be deposited at about the same cost as from a Cr(VI) bath. Accurate estimates of cost are impossible since, if the bath came into large scale operation, the cost of the necessary raw materials would almost certainly decrease. Its commercial exploitation has been hindered by the necessity for a porous diaphragm but experimental work is now in progress on modifications to the bath composition which will suppress the evolution of chlorine and hence render the diaphragm unnecessary.

STRESS IN CHROMIUM DEPOSITS

Chromium deposits usually have large internal tensile stresses and so naturally their causes and effects have been investigated by a number of workers[60-65]. However, since hydrogen can be absorbed by many metals and result in the distortion of the crystal lattice, caution must be observed

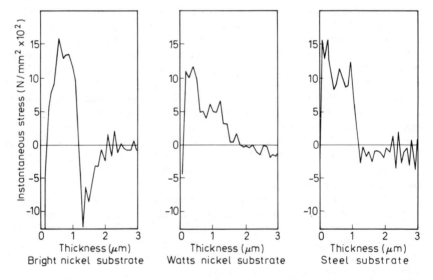

Figure 9.11. *Effect of substrate on the instantaneous stress in micro-cracked chromium (after Dennis[32])*

when determining stress in chromium deposits because the co-deposited hydrogen may simultaneously induce stress in the surface layers of the underlying metal. Fig. 9.11 illustrates how the substrate can effect the measured value. Hydrogen adsorption results in high compressive stresses in the surface layers of bright nickels, but dull nickel deposits and mild steel are much less sensitive to this effect and consequently cause only slight modification of the true internal stress in the chromium coating Fig. 9.12(a) shows that the mean stress in crack-free chromium is less than the mean stress in conventional regular chromium. This is the reason why thick coatings of the former can be deposited relatively free from cracks. The total stress does not exceed the tensile strength of the deposit until the coating is relatively thick. In the

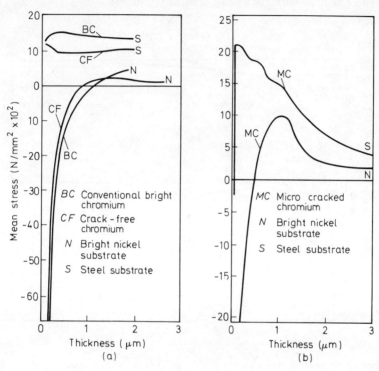

Figure 9.12. Mean stress versus thickness curves for (a) decorative, crack-free and (b) micro-cracked chromium deposited on bright nickel and steel (after Dennis[32])

case of micro-cracked chromium, a high tensile stress is developed very rapidly with increasing thickness and this causes the coating to crack and stress relieve itself [Fig. 9.12(b)].

Fig. 9.13 demonstrates the effect of hydrogen on the measured value of internal stress. This was compiled by fitting together three results on the one curve. Three stress measurements were carried out using the Hoar and Arrowsmith method, one being halted at point A, a second at point B and a third at point C. The strip was left in the warm plating solution after switching off the current, and the change in stress, without further deposition, was evaluated. The warm solution assisted in the removal of hydrogen by diffusion, and it can be seen that at all three points on the curve the stress became more tensile.

Some of the theories devised to explain these high stresses have been mentioned in Chapter 8, and Hume-Rothery and Wylie[66] have postulated that stress is related to texture orientation of the chromium with deposits having the most preferred orientation being the least stressed. Certainly, the internal stress in micro-cracked chromium deposits is influenced by the condition or texture of the surface of the underlying metal in addition to the effect of hydrogen absorption already discussed. On bright substrates, cracking occurs rapidly as indicated by a very rapid fall in instantaneous stress and to a lesser extent by a sharp fall in mean stress [Figs. 9.11(a) and 9.12(b)]. On matt surfaces a large sudden fall in stress does not occur [Fig.

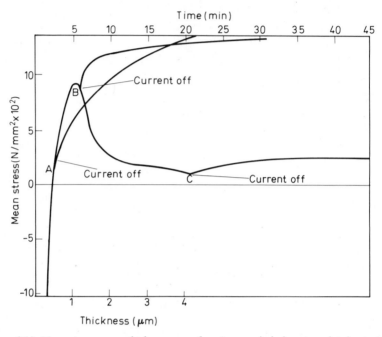

Figure 9.13. Mean stress versus thickness curve for micro-cracked chromium plated onto bright nickel showing the effect on stress of stopping the deposition at various intervals, but allowing the test piece to remain in the warm plating solution. The internal stress becomes more tensile as hydrogen diffuses away from the surface layers of the bright nickel undercoat showing that hydrogen absorption causes a compressive stress in the latter. Note: The curve was constructed by fitting together the results of three separate tests

9.11(b)] and it has been shown by examination of deposits after various plating times that this is in agreement with the fact that cracking is delayed on matt surfaces, and even if the deposit is increased in thickness, does not take place so readily as on smooth, bright surfaces.

Hardesty[67] has indicated that stress in the substrate can influence the form of crack pattern in conventional chromium. He attributes this to the tensile stresses induced in the surface layers by polishing. He estimated that a tensile stress of approximately 85 N/mm^2 exists in the surface layers of a polished nickel deposit. Dennis[32] did not detect any influence on the cracking characteristics of micro-cracked chromium due to differences in internal stress in bright nickel undercoats. The same zoned pattern occurred when bright nickels having internal stresses of 70 N/mm^2 tensile and 95 N/mm^2 compressive were used as undercoats. Initial compressive stresses, due to adsorption of hydrogen in the underlying metal, were recorded for the chromium deposit in both instances.

SPRAY CONTROL

Since the standard anodes used are insoluble and the cathode efficiencies far below 100%, copious volumes of oxygen and hydrogen gas are evolved during chromium plating. The bubbles of gas formed have great energy, and so when they burst they throw tiny droplets of chromic acid solution into the air as a fine spray. If this toxic spray or mist enters the respiratory system, it can cause great damage and so legislation has been introduced to ensure that precautions are taken to avoid this. The spray can be removed by lip extraction of the air immediately above the solution level into a fume hood by means of a powerful fan to keep the Cr(VI) content $< 10^{-4}$ g/m^3.

An alternative procedure is to reduce the surface tension of the solution and form a foam blanket on it by addition of certain surfactants. However, standard hydrocarbon wetting agents have rather short lives in this hot and highly oxidising solution, particularly during electrolysis. The only types of surfactants which are stable are the fluorinated organic compounds in which all or most of the hydrogen atoms have been replaced by fluorine[68]. Aliphatic long chain compounds having polar groups are best for this purpose, e.g. perfluorinated carboxylic or sulphonic acids or their salts having 6 to 12 carbon atoms[69,70]; the sodium salt of perfluoro-octyl sulphonic acid [$CF_3(CF_2)_6CF_2SO_3Na$] is typical of these. The effective concentration of these compounds obviously depends on their surface activity. The British Department of Employment and Productivity Chromium Plating Regulations state that the surface tension of the chromic acid baths shall not be more than 40 N/mm. This is usually achieved by use of concentrations of these chemicals varying between 0·2 and 0·5 g/l. Because of their low solubility in aqueous solutions, these perfluorinated compounds are usually added as a mixture with alkali metal carbonates or bicarbonates. The violent carbon dioxide evolution that then occurs in the acid solution disperses the solid surfactant widely in the bath and aids rapid dissolution.

While these fluorinated spray suppressants do not affect the structure or physical properties of the chromium deposits, they can increase any tendency for pitting to occur in thick plate and so are not generally recommended for addition to baths used for plating more than 25 μm of hard chromium.

ANODES

Anodes of chromium metal would neither be economical to manufacture nor technically advisable for chromium plating baths, for they would dissolve at approximately 100% efficiency and hence rapidly increase the chromium content of the baths. Insoluble anodes are therefore used. These are lead alloys containing 6–8% of either antimony or tin[21]. The former alloy was once used exclusively, but when fluoride-type catalysts were introduced it was found that it was corroded rather severely. The lead–tin alloy was found to be more resistant, and being also suitable for chromic acid baths containing sulphate only, has now largely superseded the antimonial lead type as the common anode material.

When operating correctly, these anodes should be coated with a chocolate brown film of lead peroxide (PbO_2)[71, 72]. In this condition, they will reoxidise much of the Cr(III) formed at the cathode back to Cr(VI). This will result in an equilibrium being set up between the cathodic reduction of Cr(VI) and its anodic oxidisation with the resultant concentration of Cr(III) depending on the relative cathode and anode surface areas. This feature can be used to advantage to re-oxidise excessive quantities of Cr(III) by using a piece of scrap metal having a small surface area as cathode while maintaining a large anode area. Kagan and Bonnimay[73] have studied the effect of anodic oxidation with the use of radioactive tracers.

If the lead alloy anodes are left idle for long periods in the chromium plating solution, an electrically insulating layer of lead chromate forms on them. This must be removed either by physical means or chemical dissolution before the desired peroxide film can be re-established[74].

Pure lead and iron have been used as anodes in the past. However, both tend to dissolve too rapidly, with the former becoming readily coated with lead chromate and the latter leading to an increase in the iron content of the solution. While the rate of dissolution of iron anodes depends on their purity, Armco iron being relatively slowly attacked, no ferrous anode is as efficient as lead for the re-oxidation of trivalent chromium. The use of lead and iron is therefore now almost entirely confined to that of auxiliary anodes. Platinised titanium has also been suggested as a more suitable material for this purpose, and indeed does remain uncorroded, but it can only be used in conjunction with a much larger area of lead alloy, since no oxidation of Cr(III) occurs at its surface.

REFERENCES

1. MORRISET, P., OSWALD, J. W., DRAPER, C. R and PINNER, R., *Chromium Plating* Robert Draper Ltd., Teddington (1954)
2. GREENWOOD, J. D., *Hard Chromium Plating*, 2nd edn, Robert Draper Ltd., Teddington (1971)
3. GREENWOOD, J. D., Chapters 11 to 13 of *Heavy Deposition*, Robert Draper Ltd., Teddington (1970)
4. VAGRAMYAN, A. T. and KUDRUAVTSEV, N. T., *Theory and Practice of Chromium Electrodeposition*, Akademiya Nauk SSSR, 197, translated from the Russian by Israel Program for Scientific Translations, Jerusalem (1965)
5. Proc. of a Conference on the Theory of Chromium Plating, Akademiya Nauk Litovskoi. SSR, Vilna (1958) translated from the Russian by Israel Program for Scientific Translations, Jerusalem (1965)
6. RYAN, N. E., *Metal Finishing*, **63** No. 1, 46 (1965); **63** No. 2, 73 (1965)
7. SULLY, A. M. and BRANDES, E. A., Chapter 6 of *Chromium* (2nd edn), Butterworths, London (1967)
8. MORRISET, P., *Chromage Dur et Decoratif*, Centre D'Information Du Chrome Dur, Paris (1961)
9. WEINER, R., *Die Galvanische Verchromung*, Eugen G. Lenze Verlag, Saulgau (1961)
10. SILVERMAN, L., *Metal Finishing*, **48** No. 2, 46 (1950)
11. KASPER, C., *J. Res. Nat. Bureau Stand.*, **9**, 353 (1932); **11**, 515 (1933); **14**, 693 (1935)
12. OGBURN, F. and BRENNER, A., *Trans. Electrochem. Soc.*, **96**, 347 (1949)
13. SNAVELY, C. A., *Trans. Electrochem. Soc.*, **92**, 537 (1948)
14. KNODLER, A., *Metalloberflache*, **17**, 161, 331 (1963)
15. RAUB, E. and MULLER, K., *Fundamentals of Metal Deposition*, Elsevier Publishing Co., Amsterdam, 138 and 139 (1967)
16. ROGERS, R. R., *Trans. Electrochem. Soc.*, **68**, 391 (1935)

17. WEINER, R., *Metal Finishing*, **64** No. 3, 46 (1966)
18. SARGENT, C. J., *Trans. Electrochem. Soc.*, **37**, 479 (1920)
19. HARING, H. E. and BARROWS, W. P., *Technol. Pap. U.S. Bureau Stand.*, **21**, 413 (1927)
20. FISCHER, J., *Wiss. Veroffent. Siemens-Werken*, **19**, 138 (1940)
21. PARTHASARADHY, N. V., *Metal Finishing*, **65** No. 9, 63; **65** No. 10, 70; **65** No. 11, 64 (1967)
22. CHELLAPOR, R. and PARTHASARADHY, N. V., *Metal Finishing*, **68** No. 2, 38 (1970)
23. FRY, H., *Trans. Inst. Metal Finishing*, **32**, 107 (1955)
24. *Electroplated Coatings of Chromium for Engineering Purposes*, BS 4641:1970
25. UNITED CHROMIUM INC., U.S. Pat. 2640022 (26.5.53)
26. SMART, A., *Electroplating and Metal Finishing*, **12**, 3 (1959)
27. SUCH, T. E., *Corrosion Prevention and Control*, **8** No. 8, 29 (1961)
28. SEYB, E. J., *Product Finishing*, **23**, 64 (1959)
29. LOVELL, W. E., SHOTWELL, E. H. and BOYD, J., *Proc. Am. Electroplaters' Soc.*, **47**, 215 (1960)
30. SAFRANEK, W. H. and HARDY, R. W., *Plating*, **47**, 1027 (1960)
31. SUCH, T. E. and PARTINGTON, M., *Trans. Inst. Metal Finishing*, **42**, 68 (1964)
32. DENNIS, J. K., *Trans. Inst. Metal Finishing*, **43**, 84 (1965)
33. JONES, M. H. and SAIDDINGTON, J., *Proc. Am. Electroplaters' Soc.*, **48**, 32 (1961)
34. CHESSIN, H. and SEYB, E. J., *Plating*, **55**, 821 (1968)
35. BEACOM, S. E., HARDESTY, D. W. and DOTY, W. R., *Trans. Inst. Metal Finishing*, **42**, 77 (1964)
36. PRESTON, JOHN & CO. (CHEMICALS) LTD., U.K. Pat. 1089629 (1.11.67)
37. KAMPSCHULTE, W. & CO., U.K. Pat. 1098066 (3.1.68)
38. SOCIÉTÉ D'ELECTRO-CHIMIE D'ELECTRO-METALLURGIE ET DES ACIÉRIES ELECTRIQUES D'UGINE, German Pat. 608757 (17.1.35)
39. TAYLOR, F., *Electroplating*, **5**, 109 (1952)
40. WEINER, R., *Galvanotechnik*, **47**, 438 (1956)
41. ROGGENDORF, W., *Galvanotechnik*, **56**, 158 (1965)
42. TWIST, R. D. L., *Prod. Fin.*, **25** No. 2, 20 and **25** No. 3, 37 (1972)
43. OLLARD, E. A., *J. Electrodepositors' Tech. Soc.*, **12**, 33 (1937)
44. GRAHAM, A. K., *Proc. Am. Electroplaters' Soc.*, **46**, 61 (1959)
45. SHENOI, B. A., GOWRI, S. and INDIRA, K. S., *Metal Finishing*, **64** No. 4, 46 (1966)
46. BRANCIAROLLI, J. P. and STUTZMAN, P. G., *Plating*, **56**, 37 (1969)
47. LONGLAND, J. E., *Metal Finishing Journal*, **14**, 224 (1968)
48. DIAMOND ALKALI CO., U.K. Pat. 1175461 (23.12.69)
49. CORILLIUM CORP., U.S. Pats. 3414492 (3.12.68) and 3418221 (24.12.68)
50. CORILLIUM CORP., U.S. Pat. 3442777 (6.5.69)
51. CAROSELLA, M. C. and METTLER, J. D., Chapter 6 in '*Ductile Chromium*'. American Society for Metals, Cleveland (1957)
52. PARRY, R. W., SWANN, S. and BAILAR, J. C., *Trans. Electrochem. Soc.*, **92**, 507 (1947)
53. YOSHIDA, T. and YOSHIDA, R. J., *J. Chem. Soc. of Japan* (Ind. Chem. Section), **58**, 89 (1955)
54. MACHU, W. and ELI-GHANDOUR, M., *Werkstoffe u. Korrosion*, **10**, 556, 617 (1959)
55. ZELL, M. R., *Metal Finishing*, **55** No. 1, 57 (1957)
56. CHISHOLM, C. U., *Trans. Inst. Metal Finishing*, **47**, 134 (1969)
57. BHARUCHA, N. R. and WARD, J. J., *Product Finishing*, **33** No. 4, 64 (1969)
58. BRITISH NON-FERROUS METALS RESEARCH ASSOCIATION, U.K. Pat. 1144913 (12.3.69)
59. WARD, J. J. B., CHRISTIE, I. R. A. and CARTER, V. E., *Trans. Inst. Met. Fin.*, **49**, 97 (1971)
60. BRENNER, A., BURKHEAD, P. and JENNINGS, C. W., *Proc. Am. Electroplaters' Soc.*, **34**, 32 (1947)
61. STARECK, J. E., SEYB, E. J. and TULUMELLO, A. C., *Proc. Am. Electroplaters' Soc.*, **41**, 209 (1954)
62. ZOSIMOVICH, D. P. and ANTONOV, S. P., *Ukr. Khim. Zhur*, **26**, 663 (1960)
63. GABE, D. and WEST, J. M., *Trans. Inst. Metal Finishing*, **40**, 6, 197 (1963)
64. CLEGHORN, W. H. and WEST, J. M., *Trans. Inst. Metal Finishing*, **44**, 105 (1966)
65. KONISHO, S., *Metal Finishing*, **61** No. 3, 54 (1963); **61** No. 10, 58 (1963)
66. HUME-ROTHERY, W. and WYLIE, M. R. J., *Proc. Roy. Soc.*, **A181**, 331 (1943)
67. HARDESTY, D. W., *J. Electrochem. Soc.*, **111**, 912 (1964)
68. GUENTHER, R. A. and VICTOR, M. L., *I & E. C. Products Res. & Dev.*, **1**, 165 (1962)
69. HAMMA, G. M., FREDERICK, W. G., MILLAGE, D. and BROWN, H., *Am. Ind. Hyg. Assoc. Quart.*, **15**, 3 (1954)
70. UDYLITE RESEARCH CORP., U.S. Pats 2750334 to 2750337 (12.6.56)
71. SALSER, T. M. H. and SHAMS EL DIN, A. M., *Electrochim. Acta*, **13**, 937 (1968)
72. HARDESTY, D. W., *Plating*, **56**, 705 (1969)
73. KAGAN, H. and BONNEMAY, M., *Chrome Dur.*, page 38, issue of 1958
74. GABE, D. R., *Met. Fin. J.*, **17**, 276 (1971)

Chapter 10

Decorative Nickel Plus Chromium Coating Combinations

Before the advent of chromium plating, no layer of any other metal was superimposed on nickel electrodeposits. The early developments are outlined in the historical survey given in Chapter 1, from which it will be seen that at first nickel was electrodeposited from solutions which gave matt deposits that required polishing to produce a bright appearance. Nickel quickly tarnishes on exposure to the atmosphere so that frequent polishing is necessary to retain a lustrous appearance. However, in the absence of a chromium top coat, there is obviously no possibility of the formation of local galvanic couples between nickel and chromium, which can lead to rapid penetration to the substrate. Single layer nickel coatings therefore provide reasonable protection of the basis metal particularly if they are of the dull or semi-bright types that require polishing, since polishing produces a smooth outer surface which has low porosity. There are now only a few applications of nickel coatings without superimposed chromium, and these are almost entirely for internal use in a dry environment, such as on telephone dials. However, today the nickel coatings would almost certainly be deposited in a bright state.

With the advent of chromium deposition, it was possible to prevent the rapid tarnishing of nickel by depositing an extremely thin layer of chromium onto it. At first, the resistance to tarnishing was thought to be the only effect of the chromium and it was not realised that a thin porous coating of chromium could affect the corrosion of the underlying nickel and the substrate. It is now also known that the underlying surface of the substrate affects the structure of the chromium overlay. Van Zuilichem et al.[1] have studied the effect of various undercoats on the quality of chromium deposits. They used electron microscopy and X-ray techniques to determine preferred orientation, internal stress and grain size. Dull unpolished nickel plus chromium is not significantly superior to bright nickel plus the same thickness of the same type of chromium as far as corrosion resistance is concerned. Although dull nickel is more electropositive than bright nickel, the structure and porosity of the chromium layer will be influenced by the surface conditions of the underlying coating and the rough or coarse-grained surface of dull nickel will have a greater tendency to retain foreign material than a smooth one, and so the rate of attack may increase. The only method of improving the corrosion resistance of a single layer nickel plus decorative chromium coating is to increase the thickness of the nickel layer, and this is usually uneconomical due to the prolonged plating time required and the consequent reduction in production rate. Nevertheless, if the thickness of nickel is doubled the life before penetration is quadrupled. The initial thickness of a dull nickel electrodeposit will be reduced by manual polishing so that this must be taken into consideration when complying with specifications. In tests carried out on flat panels having a coating of Watts nickel 30 μm thick, about 2·5 μm was removed by the polishing operation.

The introduction of bright nickel enabled polishing costs to be reduced, and this is now particularly important since skilled polishing labour is costly and not all components can be polished on automatic machines. The early bright nickels did not have levelling properties and so the substrate still required polishing to a reasonably high standard if a final bright and smooth surface was to be produced.

The reputation of nickel plus chromium coatings fell into disrepute soon after the introduction of bright nickels deposited from solutions containing both organic brighteners and levellers. Several factors contributed to this situation. It was not realised immediately that organic addition agents, particularly those of the second class, were decomposed by electrochemical reduction at the cathode or by chemical reaction in the bath, to form degradation products which could subsequently have a deleterious effect on the properties of the nickel deposit. Many of these 'break-down products' result in brittle deposits having such high tensile stresses that ultimately spontaneous cracking of nickel occurs. In some instances, if spontaneous cracking does not take place immediately it may be induced by slight deformation during assembly or early in the service life of a component.

Almost all bright nickel produced on a commercial scale contains sulphur and this inevitably makes the bright electrodeposit more electrochemically active[2] than dull nickel plated from a Watts bath. Consequently, attack in a corrosive environment on a bright nickel plus chromium coating occurs at

Figure 10.1. Cross-section through a corrosion pit in a bright nickel plus conventional chromium coating on a steel substrate after one year's static exposure on a roof in central Birmingham

a greater rate than in the case of dull nickel. Also, when the latter is polished. it can sometimes induce a type of micro-porous structure in the chromium plate, with subsequent benefits in corrosion performance. The reputation of decorative nickel plus chromium coatings also suffered in the early 1950s due to the nickel shortage which resulted from the Korean war, since it was not then possible to enforce thickness standards when insufficient nickel was available to meet requirements. A further period of world nickel shortage has now been experienced (late 1969 and early 1970) and once again the emphasis has been directed towards alternative coating systems or ones which involve the use of thinner nickel layers. The latter technique is now feasible without producing inferior coatings providing that the correct improved coating system is employed; these were not available twenty years ago. From about 1954[3] onwards, the shortcomings of the bright nickel plus decorative chromium system were realised, and attempts were made to eliminate poor quality plating by increasing the thickness of the nickel layer and ensuring that the physical properties were as near ideal as possible. The electrochemical properties of the nickel could not be improved, since sulphur incorporation is an inherent feature of most bright nickels. In spite of these improvements, this coating system was still not satisfactory for service in the most severe outdoor conditions such as those that are experienced by motor car components, due to pitting which was initiated at defects in the chromium and then rapidly penetrated the nickel coating to the substrate (Fig. 10.1). In 1965, the system was no longer included in the list of coatings for use in the most severe outdoor environments recommended in BS 1224, *Electroplated Coatings of Nickel and Chromium*[4].

MODIFICATION OF THE NICKEL PLUS CHROMIUM SYSTEM

The corrosion resistance of the nickel plus chromium coating can be improved by modification of either the nickel or chromium layer. Developments in nickel and chromium plating solution formulae and hence in the deposits obtained from them took place concurrently over a period of several years and the most satisfactory coatings in use today consist of those making use of developments in both metals.

A layer of copper is used underneath the nickel deposit in some instances, and is essential in the case of zinc alloy die-castings. This alloy cannot be plated directly in an acid nickel plating solution with a pH below ≈ 5 since the zinc would corrode and seriously contaminate the bath. Developments have also taken place in copper deposition and these will be discussed later in this chapter, but only in so far as they affect the performance of nickel plus chromium coatings. At the present time there are conflicting opinions on the use of copper as an undercoat for nickel on steel, although most authorities agree that it does not improve the corrosion resistance of standard bright nickel plus chromium coatings. Polishing of the copper deposit and subsequent re-jigging are sometimes necessary, and there is also evidence to suggest that the corrosion resistance of the composite coating can be inferior due to the presence of copper salts in the corrosion products. Copper undercoats tend to be popular at any period when there is a world shortage of nickel since most specifications permit the use of copper instead of part of the nickel but only by substitution of a thickness of copper greater than that of the nickel it replaces.

MULTI-LAYER NICKEL COATINGS

DOUBLE LAYER NICKEL

The major development as far as the nickel layer is concerned involves the use of a double layer or duplex system. This consists of an initial layer of semi-bright sulphur-free ($<0.005\%$ S). levelling nickel onto which is deposited a layer of conventional sulphur-containing bright nickel. The most satisfactory ratio has been found to be 70–80% semi-bright plus 20–30% bright nickel[5]. The chromium overlay can either be decorative bright chromium or any of the modified chromium coatings which will be discussed later. Pits form initially in the bright layer in the normal manner but when they penetrate to the semi-bright layer the latter is cathodically protected because the bright layer is more electronegative than the semi-bright layer[6]. Corrosion then spreads laterally instead of immediately continuing to penetrate the nickel layer. Characteristic 'flat-bottomed' pits

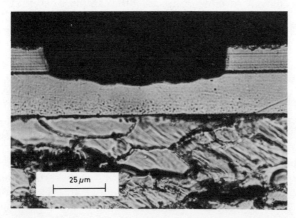

Figure 10.2. Cross-section through a corrosion pit in a duplex nickel plus conventional chromium coating on a steel substrate after three cycles of the CASS test

are formed as shown in Fig. 10.2. These pits do not appear as unsightly to the eye as those which penetrate to the basis metal causing blisters in the plate or allowing exudation of white or rust coloured corrosion products of the basis metal. However. this protective mechanism does not function well in a sulphur-containing atmosphere such as exists in large British cities. The more noble semi-bright nickel is activated by sulphur in the atmosphere and pits penetrate to the basis metal in the same manner as with a single layer of bright nickel. Once the pit reaches the substrate. a galvanic couple is formed between the substrate and the nickel coating. and a large pit is produced in the substrate below the coating (Fig. 10.3). After prolonged exposure to the corrosive environment. this results in the formation of blisters. particularly on zinc alloy die-castings and the eventual detachment

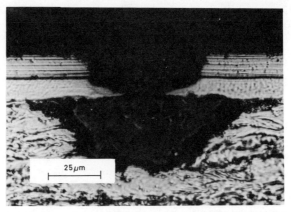

Figure 10.3. Cross-section through a corrosion pit in a duplex nickel plus conventional chromium coating on a steel substrate after one year's static exposure in central Birmingham, i.e. an industrial sulphur-containing atmosphere

of areas of the coating. When the dual nickel coating functions correctly, preferential attack of the bright nickel eventually leads to undermining of the chromium coating. In some instances, if the sample is carefully sectioned and mounted, microscopic examination reveals fragmented chromium, either in the pit or breaking away from the surface.

USE OF POTENTIAL MEASUREMENTS FOR EVALUATION OF DOUBLE LAYER NICKEL SYSTEMS

It has been indicated in the previous section that the magnitude of the difference in potential between the semi-bright and bright deposits must be sufficient to produce the correct type of flat bottomed pit. Since current flows in the local corrosion cell, the polarised potential of the anode is of greater significance than the open circuit or steady state potential. Experimental coatings can be assessed by carrying out service trials or accelerated corrosion tests, but these are time consuming and in the latter case the corrosion mechanism may not be exactly the same as in service. For the development of new coating systems it is therefore preferable to devise a more rapid sorting technique so that unsatisfactory systems can be discarded at an early stage.

Several authors[7,8] have quoted values for steady-state potentials determined in various electrolyte solutions and some have stated that certain minimum differences in potential must exist if a double-layer coating system is to be effective. These values can be misleading because they are influenced by a number of factors, such as the electrolyte employed, the surface condition of the nickel, the formation or absence of films on the nickel surface, the degree of agitation, the amount of oxygen dissolved in the electrolyte solution and the time of immersion. It is the experience of the authors[9] that an absolute steady state potential is never obtained, but consistent results can be achieved by reporting the appropriate value after a particular time of immersion. Two hours appeared to be a reasonable immersion time for the work in question (Fig. 10.4), although Du Rose[7] has reported that equilibrium is attained in approximately 5 min. For steady state potentials to be meaningful, all the test conditions must be specified precisely. The

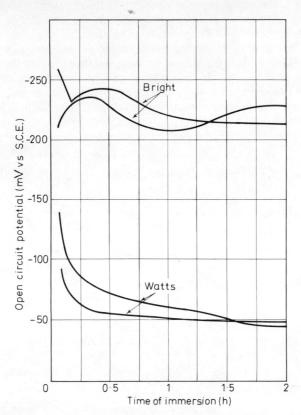

Figure 10.4. Effect of time of immersion on the open circuit potential of Watts nickel and a commercial bright nickel when immersed in an air-agitated nickel sulphate electrolyte solution at pH 2 (after Dennis and Such[9])

difference in potential between a particular bright and semi-bright nickel deposit is dependent upon the electrolyte used and so the relative order of nobility of a series of electrodeposits when immersed in different electrolyte solutions is not always the same. Some semi-bright deposits show intermittent passivity, i.e. fluctuating potential values are obtained from the same type of deposit when immersed in a particular electrolyte[9,10].

Agitation and aeration can be standardised for a particular series of experiments by specifying the air pressure and diameter of the inlet tube leading into the test vessel. The use of air for agitation results in saturation of the electrolyte solution with oxygen, but nitrogen can be employed to produce an environment free of oxygen. Although these procedures ensure uniformity of conditions in the solution, the environment may be completely different from that encountered in practice, even if the solution used is as natural as possible, such as collected rain or sea-water or artificially prepared near-equivalents to these natural waters.

Open circuit potentials do not necessarily provide a true indication of the behaviour of two metals forming a corrosion cell since polarisation phenomena must be considered. In the case of nickel plus chromium coatings, it is the nickel exposed at defects in the chromium which forms the anodic sites. A number of attempts have been made to devise experiments which enable corrosion currents to be determined or to permit polarisation curves to be plotted[7-11]. Anode potential measurements are susceptible to similar variations to open circuit measurements, e.g. agitation and surface conditions of the electrodeposit. Cleanliness of surface can be ensured by anodically polarising at a very low current density for several hours before commencing the experiment. A straight line relationship is obtained by

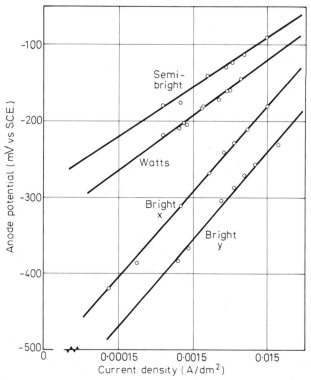

Figure 10.5. Relationship between current density and anode potential for electrodeposited nickel foils in nitrogen-agitated acetic acid–salt spray solution, pH 3·2 (after Dennis and Such[9])

plotting anode potential E against $\log_{10} i$, since from the Tafel relationship $E = a \pm b \log_{10} i$. If the Tafel relationships for semi-bright and bright deposits are plotted, it is possible from the slope b and the intercept a to estimate the relative dissolution rates of the two nickel deposits when in contact. The greater the ratio of the dissolution rates, the greater the tendency for the bright nickel in a duplex system to be preferentially corroded and therefore the greater the likelihood of the formation of flat-bottomed pits. Typical results for Watts, semi-bright and two commercial bright nickel deposits polarised in acetic acid–salt spray solution are shown in Fig. 10.5.

DECORATIVE NICKEL PLUS CHROMIUM COATING COMBINATIONS

Fairly simple equipment can be used to determine anode potentials. The cathode can be made either from a sheet of electrodeposited nickel identical to that of the anode or from a sheet of platinum. The anode current densities used should not greatly exceed 0·015 A/dm^2. The connections to the electrodes must be protected by lacquer or p.v.c. tape, so that bimetallic junctions do not come into contact with the electrolyte solution. The potential between a calomel electrode and the anode can be measured using a valve millivoltmeter. One hour must be allowed to elapse after each increment of current, so that stable conditions are established before the potential is measured.

THREE LAYER NICKEL SYSTEM

A three layer nickel system[12, 13] is a further development of the double layer system. The central layer has a very high sulphur content which is deposited from a non-air-agitated bath containing a specific organic compound. The object in this case is to arrange for preferential attack of the least noble nickel deposit to occur in a region in which it is not visible, so that it is not

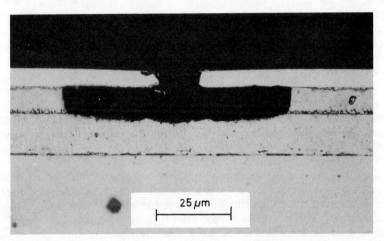

Figure 10.6. Corrosion pit in three layer nickel plus conventional chromium coating after exposure to the Corrodkote test. The proportion of the high sulphur nickel layer was much greater in this particular multi-layer coating than is usual, so that the lateral attack could be revealed more dramatically (courtesy International Nickel Co. Ltd.)

readily apparent to a casual observer that corrosion has occurred, for the large flat-bottomed pits that can form in double nickel are rather unsightly although not so detrimental as rust spots. The coating consists of an underlayer of semi-bright nickel, a thin central layer of nickel of high sulphur content (typically 0·7 μm containing 0·15%S) and a top layer of bright nickel. It has been suggested[14] that a bright nickel-cobalt alloy could be used for the central layer, since the presence of cobalt increases the electrochemical activity of the electrodeposit, but no commercial use has been made of this technique.

The mode of corrosive attack on a three layer nickel system, assuming an overlay of conventional decorative chromium, is shown in Fig. 10.6. In this way only comparatively small surface corrosion pits are visible, but after extensive exposure to a corrosive environment the undercutting becomes severe and gives rise to blisters or detachment of the top coat of bright nickel. This system probably functions more satisfactorily if the top nickel layer is somewhat more noble than a normal bright nickel. All the nickel systems described in this section extend the time required to penetrate to the substrate, not by totally preventing the dissolution of nickel but by allowing it to take place in a controlled manner. Three layer nickel systems can provide superior results to double layer systems but the complication due to the extra stage necessary in the plating sequence is not always considered to be justified by the increased benefit given. Nevertheless, these systems are in operation on a large scale in some organisations.

INDUSTRIAL PRACTICE

The double layer nickel plus decorative chromium system is one of the most popular coatings in use at the present time. The semi-bright nickel solution is comparatively cheap to maintain and control for it usually contains only one or two organic compounds, coumarin being the most common, together with a suitable wetting agent. It provides good levelling (65% for a coating thickness of 25 μm on an abraded substrate having a surface roughness of 0·375 μm CLA), so that expensive polishing is unnecessary. Bright nickels are usually more expensive to maintain and control, in addition to giving more brittle coatings, so that it is an advantage to be able to deposit the major part of the coating (70–80%) from the cheaper bath. Some semi-bright nickel solutions, in particular those based on coumarin, have to be purified regularly by treatment with activated carbon, but the break-down products formed in some bright solutions cannot always be so readily removed by this method. Usually the two baths are so formulated that drag-over of solution from the semi-bright to the bright solution does not cause deterioration of the latter. Many double layer coatings can be plated satisfactorily by transferring the work directly from the first solution into the second without rinsing, the adhesion between the nickel layers being sound, providing that the semi-bright nickel does not become momentarily anodic, as can happen if the plated component becomes bipolar as it leaves the semi-bright bath or enters the bright bath. Problems of poor adhesion between nickel layers rarely arise in commercial practice provided that good housekeeping of the plant is observed, so that neither organic nor metallic impurities, particularly zinc, accumulate. An automatic plating plant can be converted from one producing bright nickel plus decorative chromium to one producing double layer nickel plus decorative chromium without extensive rebuilding, often only two smaller tanks being required where one large one was formerly present. The length of the plating cycle will remain the same if the total nickel thickness of the double layer is the same as that of the single layer bright nickel, since the cathode efficiencies of the two nickel solutions and the plating current densities used are essentially the same.

MULTI-LAYER COATINGS OF DISSIMILAR METALS

Research has been carried out into the use of various sandwich coatings, but as yet none have been exploited on a commercial basis. In 1958, Knapp[15] investigated the effect of interposed layers of several metals on the weather resistance of electrodeposited nickel coatings. Even in this early survey most of the composite coatings tested were superior to a single layer of nickel. A coating in which the nickel deposit is plated in two layers separated by a thin layer of chromium has been evaluated in the USA[16]. With this coating system, the top nickel layer is penetrated but further penetration is halted at the intermediate chromium layer. Chromium is well known for the nature of its tenacious oxide film which ensures that chromium remains electropositive with respect to nickel. Corrosion is confined to the upper nickel layer, and thus, as in all multilayer nickel systems, attack of the substrate is greatly delayed (Fig. 10.7).

The most satisfactory results have been achieved by the use of dull, high

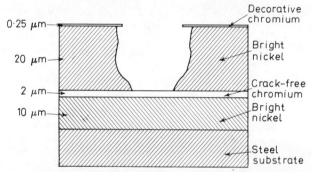

Figure 10.7. *Typical appearance of a corrosion pit in a bright nickel plus chromium plus bright nickel plus conventional chromium coating*

temperature-type crack-free chromium for the intermediate chromium layer. Articles of intricate shape should be plated at conditions which give maximum coverage, and chromium thicknesses up to 2 μm have been used to give maximum protection. Thorough rinsing is essential after the first chromium layer has been deposited and components should be dipped in a reducing solution such as alkaline sodium sulphite in order to reduce any residual chromium to the less troublesome trivalent state. Good adhesion of the top nickel layer is achieved by the use of a low pH nickel strike, the operating conditions for this being 2–5 A/dm^2 for up to 10 min. If the chromium is freshly deposited it is most satisfactory to load the work live into the nickel strike solution. The low pH strike solution is much more tolerant to chromium contamination than conventional nickel baths, particularly if this chromium is present in the trivalent state. The final nickel layer is usually deposited from a bright plating bath, and the thinner this layer the faster the lateral spread of the corrosion pit. The diagram shown in Fig. 10.7 indicates typical thicknesses of the various layers in a coating suitable for use in an outdoor environment and it illustrates the shape of a corrosion pit.

Turner and Miller[17] have examined the value of sandwich coatings consisting of bright nickel plus copper plus bright nickel plus micro-porous chromium. This system was claimed to have performed to a high standard on CASS and mobile testing. The results were less satisfactory on static roof testing, but it is not unusual for different behaviour to occur in mobile and static tests. The results were interesting in that copper did not act as a barrier layer, as might have been anticipated, but was sacrificially corroded. This corrosion mechanism occurred in all the corrosion tests employed, and so there was no obvious reason for the different degrees of protection afforded in various tests. It is apparent, as will be discussed in more detail in Chapter 11, that copper behaves in a more electronegative manner than all types of nickel when incorporated in sandwich coatings below micro-porous or micro-cracked chromium. Fig. 10.8 illustrates the mechanism of corrosion in which the lower bright nickel layer remains unattacked while

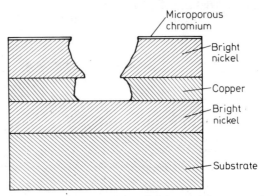

Figure 10.8. Typical appearance of a corrosion pit in a bright nickel plus copper plus bright nickel plus micro-porous chromium coating

considerable lateral corrosion of the copper layer takes place. This work of Turner and Miller[17] indicates that at least there are possibilities of making use of copper in decorative nickel plus chromium systems and yet still achieve a good corrosion resistance. This could reduce the overall cost of a coating by reducing the amount of nickel consumed and would be particularly useful at times of nickel shortages. However, further research is necessary before this type of sandwich coating could be considered for inclusion in national standards.

A patent[18] has recently been published which relates to the use of sandwich coatings characterised by their high ductility and protective action. A typical coating would consist of a lower layer of ductile sulphur-free semi-bright nickel, a second layer of either copper, silver or gold approximately 1·25 μm thick, a third layer of bright nickel and finally a top layer of conventional chromium. This system would be useful where duplex nickel plus decorative chromium fails too readily when subjected to high stress; e.g. at a bolt hole when the bolt is tightened the coating may crack and permit corrosion of the substrate to take place at an early stage.

All sandwich coatings involve greater complexity as far as the plating process is concerned, and are worthy of consideration only if they provide a definite advantage (either by way of cost saving or greater protection) over more simple coating systems. The main problems are associated with obtaining good adhesion between the successive layers of metal and preventing contamination of one plating solution with drag-out from an earlier solution in the plating line. The nickel plus chromium plus nickel plus chromium system, while providing excellent results on a pilot scale, would be particularly susceptible to these problems; the disastrous effect of chromium contamination in nickel baths has been described in Chapter 7. However, the plating of sandwich coatings should not be beyond the capabilities of a modern well-equipped plating shop, and so it is possible that in the future coating systems may become more complex.

SATIN NICKEL

A fully-bright highly-reflective finish is not required for all purposes and on some articles a 'softer' appearance has more aesthetic appeal. A lustrous surface is not permitted on certain internal components of American motorcars, since new legislation has been introduced to lessen the risk of drivers being dazzled by reflection from bright trim. A lustrous but satin finish provides a pleasing appearance on a wide range of articles, two of the most common uses being for camera components and door furniture.

A satin finish can be achieved by skilful polishing or scratch brushing of a dull nickel coating followed by deposition of a layer of conventional decorative chromium. The style of the finish produced in this manner is only limited by the skill of the polisher and the polishing media available. However, this can be a fairly expensive process due to the cost of the labour. An alternative process used to obtain a uniform satin finish is to incorporate inert particles of 1–5 μm in size, e.g. barium sulphate or silica, in a Watts nickel solution containing a brightener of the first class, such as *p*-toluene sulphonamide[19]. Decorative chromium is normally used as the overlay and this produces an attractive matt finish which does not finger mark too

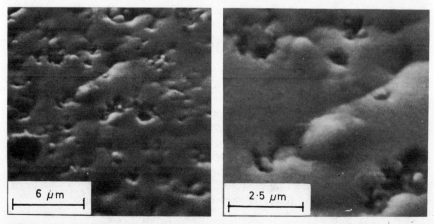

Figure 10.9. Scanning electron micrographs of the surface topography of a satin nickel deposit

severely, especially if the precaution has been taken to remove loosely adherent particles by a 'dedusting' process. The surface topography of a satin nickel deposit is shown in Fig. 10.9 and a cross-section of a satin nickel deposit showing the distribution of incorporated particles is illustrated in Fig. 10.10.

The incorporation of particles in electrodeposits has been investigated by

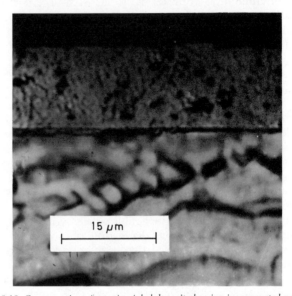

Figure 10.10. Cross-section of a satin nickel deposit showing incorporated particles

a number of authors[20, 21] and recently Tomaszewski et al.[22] have studied their behaviour in nickel and copper baths. Incorporation on vertical cathode surfaces from slightly acid nickel solutions can be achieved relatively simply, but incorporation from acid copper solutions does not occur so readily and is influenced by changes in the plating conditions. Experimental work has been carried out to evaluate the effect of certain additions to the plating bath which could encourage co-deposition.

Observations of the corrosion resistance of coatings consisting of satin nickel plus decorative chromium showed that it was far superior to bright nickel or polished dull nickel with a top coat of decorative chromium. This phenomenon was later exploited using much smaller particles to develop the special nickel layer which makes possible the deposition of micro-porous chromium. This will be discussed later since it provides a means of producing a bright finish which has very good resistance to corrosion on outdoor exposure. Therefore, while composite coatings consisting of particles embedded in an electrodeposited matrix have found, as yet, only limited applications for engineering purposes, they are already making a significant contribution towards producing improved decorative coatings.

Satin nickel deposits can also be produced by adding, to a Watts-type solution, a brightener of the first class and a special organic compound (typically a non-ionic surfactant) which is fully soluble when cold but which precipitates in fine droplets above a critical temperature, i.e. the 'cloud point'. These emulsified droplets influence the deposition mechanism at the cathode to produce an appearance described by Baker and Christie[23] as a matt velvet effect. For an additive to provide the desired pleasing appearance it must have the correct physical characteristics, but it must also have in its molecule groups which are known to be effective as brighteners in nickel solutions. If brightening groups are not present, an unattractive dull grey appearance results. The compounds used have wetting properties and so wetting agents are not required, and in any case these would be likely to interfere with the droplet formation. These satin nickel deposits have a hardness of about 500 HV, a ductility of 3·4% and contain up to 0·14% sulphur. Satin nickel deposits are also produced from baths containing organic or inorganic compounds either in true solution or present as colloidal suspensions. Since these additives influence the nickel structure by adsorption, close control of their concentration and the acidity and temperature of the bath is required for a consistent satin appearance, but the plant needed is only the same as for normal plating, without the modifications used for the droplet process.

Plant modifications are necessary to facilitate the formation of the fine droplets in solution. These coalesce and must be redissolved and reprecipitated by the alternate heating and cooling of the solution. The plant is therefore designed so that part of the solution continually passes over a weir, is then cooled below the cloud point so that the turbidity disappears, is subsequently filtered and finally reheated ready for recirculating to the plating tank. The droplets are very small when formed at the cloud point but grow in size until they reach dimensions at which they are cathodically active and therefore effective in producing the satin appearance. Above a certain size they are not effective and must be removed over the weir.

MODIFICATIONS OF THE CHROMIUM LAYER

Two techniques can be employed to prevent rapid penetration of the nickel plus chromium coating with the subsequent appearance on the surface of basis metal corrosion products. The first possibility is to ensure that the surface layer is pore free and covered by a passive film so that corrosion is completely prevented. Theoretically, this should be possible if a thick pore and crack-free layer of chromium could be deposited. This was the object of the development of 'crack-free' chromium. The alternative procedure is to deposit a chromium layer containing many discontinuities, such as cracks or pores, so that the galvanic attack of the nickel is spread out over a large area and therefore localised deep pitting prevented. This avoids the disastrous situation from a corrosion point of view of having a large cathode and a very small anode. This mechanism of reducing the corrosion current density is the basis of micro-cracked and micro-porous chromium. Saur[24] has investigated the influence of exposed nickel area on total nickel corrosion and on basis metal protection. A brief history and description of commercially available micro-discontinuous plating processes has recently been published[25].

CRACK-FREE CHROMIUM

As stated in Chapter 9, crack-free chromium[26] on nickel is a combination that gives better corrosion protection than does nickel plus decorative chromium, provided that the substrate is rigid, e.g. most zinc alloy die-castings. The thickness of crack-free chromium should be at least three times that of decorative chromium, i.e. 0·75 μm compared with 0·25 μm.

Unfortunately, if the plated component does get even slightly damaged, the position is then the worst possible from the corrosion aspect, i.e. a few macro-cracks inevitably result in the chromium, and produce a most unfavourable anode/cathode ratio. Crack-free chromium is similar in appearance to decorative chromium when deposited on the same nickel substrate, but to the experienced eye it usually appears to have a bluer colour. In service, corrosion usually takes place at isolated sites and each pit becomes large for the aforementioned reasons.

Crack-free chromium is unsuitable for use as the top coating of a nickel plus chromium system deposited on an aluminium substrate, even when the nickel coating is bonded satisfactorily to the aluminium using one of the techniques discussed in Chapter 12. Due to the large difference in galvanic potential between aluminium and nickel, large blisters occur in service around each pit. The adhesion of nickel to aluminium is inherently much lower than that of nickel to most metallic substrates, and after a certain degree of undermining of the coating, the tensile stress in the chromium is sufficiently high to detach the coating from large areas. Although the mean stress in crack-free chromium is approximately the same as that in decorative chromium, the total stress is higher as the layer is much thicker. As the layer is almost free of defects it is not stress relieved by cracking as in the case of micro-cracked chromium. The total stress even after the formation of a few pits is still high. This high tensile stress also mitigates against the use of crack-free chromium on plastics to which the undercoat has low adhesion, for it will cause the underlying electroless nickel or copper and the subsequent electroplate to blister.

MICRO-CRACKED CHROMIUM

Two layer system

The mechanism of protection by micro-cracked chromium has already been outlined in Chapter 9. This type of coating was first achieved by using a dual chromium system in which the undercoat was deposited from a solution giving crack-dree deposits and the upper coat from a solution producing micro-cracked deposits[27-30]. The total thickness of dual chromium is usually at least 0·75 μm and the crack pattern formed in the upper layer initiates cracking in the underlying previously crack-free layer. It is essential to obtain good adhesion between the two chromium layers, but this is not difficult to achieve in practice. The main operating precaution is to prevent drying of the first chromium plate and in any case not to have a time lapse of more than a few minutes between each plating process. It is best to apply a potential of 2 V to the chromium plated article before it enters the second chromium bath. Lindsay et al.[30] have shown that both crack pattern and chromium thickness play an important role in increasing the corrosion resistance of nickel plus chromium coatings.

Single stage micro-cracked chromium

Improvements in solution composition have made possible the deposition of a satisfactory micro-cracked chromium deposit from single solutions[31]. The minimum thickness necessary to obtain a fully micro-cracked deposit even over a complex shaped component is of the same order as that required using dual chromium. Nickel corrosion takes place at numerous sites and the surface becomes stained with nickel corrosion products, but basis metal corrosion only occurs after extensive exposure. Surface stains can be readily washed off using only water, if cleaning is done frequently, but if allowed to harden can be removed with mild abrasive or commercial chrome cleaners. The surface pitting is barely visible after cleaning until drastic undermining of the chromium has occurred.

Fig. 10.11(a) and (b) show the cross-section of a duplex nickel plus micro-cracked chromium coating after 10 cycles of CASS testing. Fig. 10.11(b)

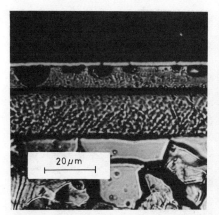 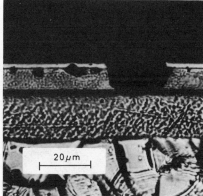

Figure 10.11. Two optical photomicrographs of corrosion pits in a duplex nickel plus micro-cracked chromium coating on a steel substrate after exposure to 10 cycles of the CASS test

illustrates the added benefit of duplex nickel underneath a chromium layer containing micro discontinuities; the largest pit shown in the photograph has not penetrated into the semi-bright layer. The selection of scanning electron micrographs illustrated in Fig. 10.12 indicates the various types of defects which can occur in nickel plus micro-cracked chromium coatings. Only certain cracks function as corrosion sites, as shown in Fig. 10.12(a). After extended exposure the nickel layer is corroded away at localised areas and chromium either collapses into the pits or is completely removed [Fig. 10.12(b)]. A greater proportion of cracks appear to serve as corrosion sites when a component is exposed to the CASS test, presumably because the surface is thoroughly wetted throughout the exposure period. whereas on outdoor exposure continual wetting and drying occurs. At the region shown in Fig. 10.12(c), many cracks have opened slightly to permit attack of the underlying nickel and so provide excellent protection of the basis metal. A few pits of the type shown in Fig. 10.12(d) have been observed to form on

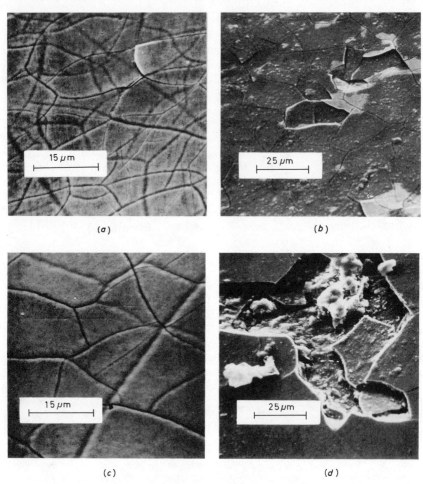

Figure 10.12. Scanning electron micrographs of the appearance of coatings having an overlay of micro-cracked chromium, after exposure to corrosive environments. (a) *2 years' mobile exposure of copper plus bright nickel plus micro-cracked chromium,* (b) *3 years' static roof exposure of bright nickel plus micro-cracked chromium and* (c) *and* (d) *10 cycles of CASS testing of duplex nickel plus micro-cracked chromium* (c) *and* (d) *are micrographs of different portions of the same specimen*

specimens during CASS testing, even when the majority of the surface behaved as in Fig. 10.12(c). A study of the morphology of corrosion pits formed in nickel plus micro-cracked or micro-porous chromium coatings has been published by Dennis and Fuggle[32].

The most satisfactory performance should be obtained when the distance between effective cracks is about twice the thickness of the nickel plus chromium coating, assuming the pits are uniform in size and hemispherical in shape (Fig. 10.13). The distance between cracks at the limiting conditions is equivalent to the pit diameter and the thickness of the coating is equivalent to the pit radius. If the number of effective cracks is too great, numerous small pits will be formed and the chromium completely undermined before the substrate is exposed. If the crack density is too low, the converse will be true and penetration to the substrate will occur too rapidly. On the basis of these assumptions, an effective crack density of 20/mm would be ideal for a nickel coating 25 μm thick. In service, not all cracks form sites for attack and a crack density of about 80/mm provides a suitable number of effective sites

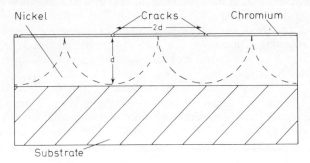

Figure 10.13. Schematic diagram illustrating the assumption that the optimum distance between cracks in micro-cracked chromium is twice the thickness of the nickel layer

in coatings of thickness between 25 μm and 40 μm. The observation and counting of cracks can usually be carried out without resource to special chemical etching or electrochemical techniques. Crack counting can be accomplished easily using a projection microscope at a magnification of about ×200. Higher magnifications or more sophisticated techniques such as replica or scanning electron microscopy are available for more detailed examination of the crack pattern. Discontinuities in chromium deposits, e.g. micro-cracks, macro-cracks or pores can be revealed by several techniques[33]. The most useful of these is that of Dubpernell which involves making the specimen cathodic in an acid copper sulphate solution. Typically this contains 150 g/l of $CuSO_4.5H_2O$ and 50 g/l of H_2SO_4. The chromium plated specimen is copper plated for about 5 min at a very low current density maintained by controlling the potential difference across the cell at between 0·1 and 0·2 V. A copper deposit is formed on most areas not covered by chromium as discrete spots at pores and in continuous lines on cracks, although the proportion of discontinuities revealed decreases with the time of standing after deposition of the chromium.

The coating of micro-cracked chromium required is thicker than for decorative chromium and so the plating time must be longer—approximately twice as long as for conventional chromium. This is due to the deposition efficiency of micro-cracked chromium being greater than for conventional chromium and so the proportional increase in time is less than the three times increase needed for thickness. Longer plating times necessitate the modification of most plating plants (if these were designed for use with a conventional chromium process) when it is desired to replace this by a modified form of chromium. In some instances, less change of equipment is required if the chromium is to be deposited in two layers; the dwell times in individual tanks would not need changing much and possibly an extra tank or the insertion of a chromium plating vat in place of a rinse tank is all that would be required.

Micro-cracked chromium produced by modification of the nickel underlay

An alternative method for obtaining micro-cracked chromium coatings has been described in the patent literature[34]. This technique was devised by workers at the Renault motorcar factory in France and is based on the idea of superimposing a thin layer of a highly stressed and brittle nickel electrodeposit over a conventional nickel undercoat. Either this nickel will crack spontaneously or the top coat of chromium subsequently applied from a conventional bath will crack so as to have a suitable micro-crack pattern. The stressed and brittle layer, which is normally about 1 µm thick, attains these properties by being deposited from a bath based on nickel chloride, the chloride ion intrinsically conferring a high tensile stress on nickel electrodeposits. The presence of a large concentration of a carboxylic acid, usually the acetate ion added in the form of its nickel, ammonium or sodium salt, enhances this effect. In addition, the nickel is plated at a higher current density (7 A/dm^2) from a bath maintained at a much lower temperature (25°C) than is normal for modern nickel plating[35] and this would assist in producing a brittle deposit. A typical basic solution would appear to contain 250 g/l of $NiCl_2.6H_2O$ and 100 g/l of $CH_3COONa.3H_2O$ at pH 4. To this solution can be added brighteners of the first class (typically saccharin) and of the second class (typically 2-butyne-1, 4-diol). Other processes based on nickel fluoborate[36] or on additions of amino acids have also been patented[37].

The corrosion performance of this micro-cracked nickel plus chromium system is said to be equivalent to that afforded by standard micro-cracked chromium coatings, which is not surprising, since the number of cracks produced by this more recent technique is also claimed to be around 40/mm. At the time of writing, few results of comparative corrosion tests have been published, but these[38] indicate that the use of such a micro-cracked nickel layer superimposed over a bright nickel electrodeposit with a top layer of decorative chromium is equal in corrosion resistance to the same total thickness of duplex nickel plus decorative chromium.

MICRO-POROUS CHROMIUM

This type of chromium achieves the same effect as micro-cracked chromium but by a somewhat different means. As in the technique just described, this process is based not so much on a modified chromium layer as on a modified layer of nickel. The special nickel solution contains a suspension of extremely small inert particles whose diameter is approximately 0·02 µm. Numerous types of water-insoluble materials are sited in the patent literature[39] as being suitable, but fine silica powder is probably the most common material used. Inert particles are incorporated in the same way that satin nickel is produced using larger particles. The special nickel can be deposited over any suitable nickel coating such as bright nickel or double layer nickel with no more precautions than are required for applying bright over semi-bright nickel. The special nickel is only about 2·5 μm thick and is bright in appearance, since the particle size is too small to affect the appearance. The inert particles embedded in the surface prevent subsequent deposition of chromium at these regions, hence a porous chromium layer is formed. Conventional decorative chromium is used and the thickness necessary is only of the same order as that used for decorative purposes. The particle size and distribution must be controlled and agglomeration must be avoided, otherwise the effective particle size will increase and produce visible effects.

Initially it was thought that the porosity of the chromium was significantly reduced due to bridging-over of the particles if thicker chromium layers were used. Oderkerken[40] plotted the relationship between pore density and chromium thickness. He claimed that 200 pores/mm^2 gave reasonable corrosion resistance, 800 pores/mm^2 good performance and 4000 pores/mm^2 very good performance. However, if the pore density is too high, excessive surface deterioration occurs, although the coating still affords excellent corrosion resistance to the basis metal[41]. The British Standard[4] demands a minimum of 100 pores/mm^2 in this type of coating. Recent investigations[42] have indicated that the tendency for surface dulling during corrosion is reduced if the chromium thickness is increased to 0·75 μm without having adverse effects on basis metal protection. Carter found that in the case of the thin micro-porous coatings (0·25 μm) circular pits formed at active corrosion sites and tended to coalesce. Thicker chromium deposits (0·75 μm) resulted in pits surrounded by what was described as a micro-pattern of crow's foot cracks.

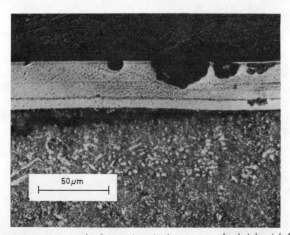

Figure 10.14. Optical photomicrograph of corrosion pits in a copper plus bright nickel plus micro-porous chromium coating on a zinc alloy die-casting after two years' static roof exposure

The optical photomicrograph of a cross-section shown in Fig. 10.14 illustrates the type of corrosion pits formed in a copper plus bright nickel plus micro-porous chromium coating on a zinc alloy die-casting; the pitting

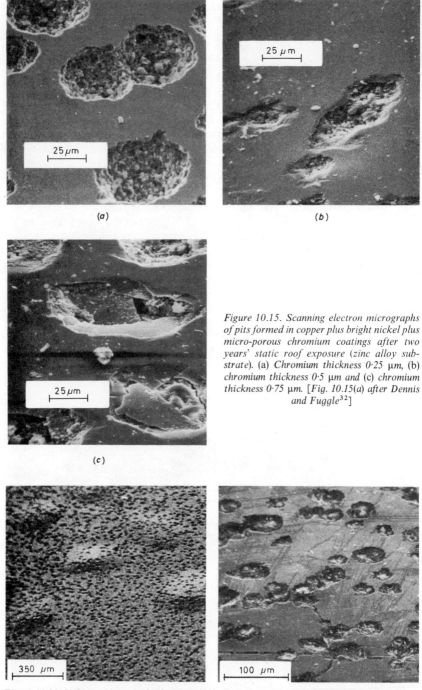

Figure 10.15. Scanning electron micrographs of pits formed in copper plus bright nickel plus micro-porous chromium coatings after two years' static roof exposure (zinc alloy substrate). (a) Chromium thickness 0·25 µm, (b) chromium thickness 0·5 µm and (c) chromium thickness 0·75 µm. [Fig. 10.15(a) after Dennis and Fuggle[32]]

Figure 10.16. Defects in copper plus bright nickel plus micro-porous chromium coatings after two years' static roof exposure (zinc alloy substrate). Chromium thickness 0·25 µm (after Dennis and Fuggle[32])

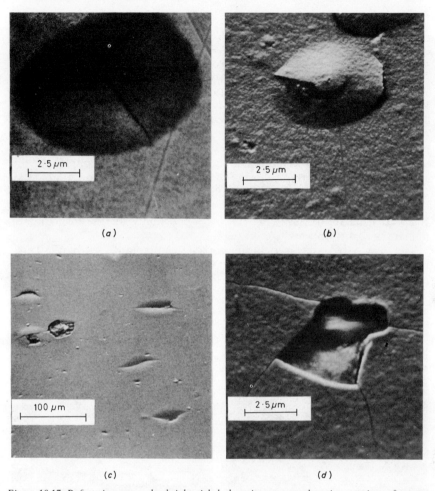

Figure 10.17. Defects in copper plus bright nickel plus micro-porous chromium coatings after one winter's mobile outdoor exposure (zinc alloy substrate). Chromium thickness (a) 0·25 μm, (b) 0·25 μm, (c) 0·5 μm and (d) 1 μm *(after Dennis and Fuggle[32])*

is similar to that occurring below a micro-cracked chromium overlay. As in the case of micro-cracked chromium, different types of corrosion pits develop in different environments and the chromium thickness also influences the morphology of pits, as already mentioned. A variety of defects are shown in Figs. 10.15, 10.16 and 10.17. When the chromium layer is thin (0·25 μm) hemispherical pits are formed and chromium is completely removed from the region of the pit [Fig. 10.15(a)] after two years' static outdoor exposure. However, at greater chromium thicknesses, small cracks develop in the chromium and it tends to collapse into the pits [Fig. 10.15(c)]. Blister formation is a common cause of failure of nickel plus chromium coatings on zinc alloy substrates even when the chromium layer is of the micro-porous type. Fig. 10.16(a) shows that fine pitting is superimposed on the blister formation. Eventually the small pits are joined together by cracks

[Fig. 10.16(b)] and finally large areas of the underlying metal are exposed. Fig. 10.17 illustrates the types of defects which occur after one winter's mobile outdoor exposure, this of course represents a much milder corrosion test than two years' static outdoor exposure. At this earlier stage of deterioration of the coating it is possible in many cases to detect the original defect which initiated the formation of the pit, as shown in Fig. 10.17(b) in which the chromium layer has been pushed up in concentric discs by nickel corrosion products formed underneath. By contrast, in Fig. 10.17(a) the chromium has been undermined and then cracked as it collapsed into the pit. The low magnification scanning electronmicrograph shown in Fig. 10.17(c) reveals the occurrence of several types of defects on the same sample. As the chromium was 0·5 μm thick, most of the defects were of the crow's-foot type. Fig. 10.17(d) illustrates that when thick chromium layers are employed (1 μm), small cracks develop and tiny pieces of chromium are subsequently removed from the surface; there is less opportunity for the formation of numerous corrosion sites.

The density of pores in a chromium layer can be indicated by the Dubpernell test referred to earlier. Fig. 10.18 shows the density of copper spots on a micro-porous deposit 0·75 μm thick. This test was carried out on a

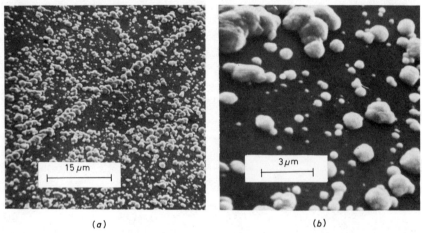

(a) (b)

Figure 10.18. Surface of a micro-porous chromium deposit, 0·75 μm thick, after copper plating using the Dubpernell technique to show the presence of pores. At the lower magnification [(a)] it can be seen that the copper has deposited along a scratch line

sample which had been plated some time earlier, so the modified Dubpernell test was employed as recommended in BS 1224[4]. It appears that this procedure indicates an artificially high pore density. By comparison with earlier figures in this chapter it is obvious that only certain pores function as effective corrosion sites, in the same manner that only a certain proportion of cracks in micro-cracked chromium function as corrosion sites. In Fig. 10.18(a) it can be observed that a row of copper nodules has formed indicating the presence of a fine scratch in the chromium layer.

The equipment for deposition of the special nickel must be designed with a view to the special features of the process. Agitation pipes must be arranged so that the air flow is directed downwards in order that the particles are prevented from settling on the bottom of the tank. Efficient agitation is necessary to prevent the heater coils from being buried in particles. as this could lead to local overheating and failure of the heaters. If the anodes become coated with a layer of particles. this could cause polarisation and unsatisfactory dissolution. The solution cannot be filtered by normal means for obvious reasons. However, this is not a major disadvantage since the plating time is limited to approximately 1 min, and in this time it is impossible to co-deposit particles of such a size as to cause roughness. For the same reason it has been found unnecessary to bag the anodes, whereas at first this presented some problems since the fine particles were readily trapped in anode bags of normal texture.

The advantage claimed for this process over micro-cracked chromium is that the number of particles per unit area over a complex shaped article will be constant if the agitation is adequate to provide a uniform suspension, whereas the micro-cracked chromium pattern varies with current density over a similar object; also the plating time for micro-cracked chromium must be extended so that the correct type of crack pattern is obtained in the low current density regions. A change to the micro-porous chromium system need not necessitate an increase in the total plating time. as the chromium is the conventional decorative type and the special nickel replaces part of the original nickel coating. Plant modification necessary would be the inclusion of a tank and ancillary equipment to operate the special nickel solution.

COPPER UNDERCOATS

The main recent development in copper plating has been the introduction of solutions capable of depositing levelled deposits from acid sulphate baths. Organic brighteners are used and the degree of levelling possible is at least comparable and often superior to that obtained with a similar thickness of bright nickel[43]. The use of this type of copper, which is deposited over a thin layer of copper plated from a cyanide bath to form the greater part of the copper layer. removes the necessity for de-jigging and polishing after dull copper plating and also greatly reduces the amount of polishing required by the basis metal. Some die-castings can thus be plated without any polishing. and others only need processing in bulk by a technique such as vibratory finishing. It is still essential to deposit a thin initial layer of copper from a cyanide solution onto both steel and zinc alloy die-castings for reasons already discussed. prior to their being acid copper plated. Alternatively. a thin initial nickel layer deposited from any standard bath is also quite satisfactory on ferrous substrates. In both cases. the thin initial layer is often termed a *strike plate*. The use of copper or nickel deposited by galvanic displacement in an adherent manner onto zinc alloy or steel substrates has been investigated as a means of forming a 'strike plate' even in deeply recessed areas[44].

A limited amount of experimental work has been carried out on zinc plating of zinc alloy die-castings prior to nickel and chromium plating. but this has not been pursued on a commercial scale. It was thought that zinc plating would improve the surface of die-castings by eliminating some of the pores and also by providing a purer and more uniform skin. A levelling acid zinc bath has been shown to possess good pore filling characteristics[45] and this should eliminate blistering due to entrapment of plating solution in surface pits. However, it has been found that acid copper baths have good micro (as distinct from macro) throwing power and so can serve the same purpose of infilling pores.

Copper pyrophosphate baths containing organic addition agents can also be used to deposit bright plate with levelling properties as an alternative to the copper sulphate based solution[43]. Although they have not the same levelling powers they are less corrosive both to the substrate and the plating plant. A modified type of pyrophosphate bath, without organic additives, when used for electrodeposition of the strike plate on zinc alloy die-castings with ultrasonics as the means of agitation, is said to give an excellent basis for subsequent additional copper or nickel electroplate, the plate having excellent macro coverage and yet being claimed to be more effective than copper cyanide deposits for filling fissures and pores[46].

Pointon[47] has studied the surface topography of copper deposits plated from cyanide and pyrophosphate copper baths onto zinc alloy die-castings, using various means of agitation. The effect of substrate surface finish (as cast, buffed, coarsely abraded and finely abraded) was also investigated. Obviously, a rough or nodular initial layer of copper should be avoided as it would cause problems in subsequent plating operations. Pointon observed that when plating onto abraded substrates using a pyrophosphate solution

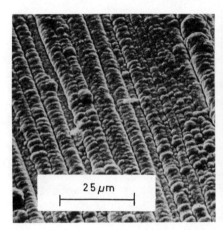

Figure 10.19. Scanning electron micrograph of the surface topography of a deposit from a pyrophosphate copper bath, plated onto an abraded zinc alloy substrate. Plating conditions: $1 \cdot 3$ A/dm^2, $30°C$, ultrasonic agitation $v = 38 \cdot 5\,kHz$

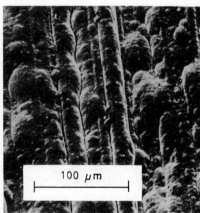

Figure 10.20. Scanning electron micrograph of the surface topography of a deposit from a cyanide bath plated onto an abraded zinc alloy substrate. Plating conditions: $3 \cdot 3\,A/dm^2$, $60°C$, ultrasonic agitation $v = 38 \cdot 5\,kHz$

the surface topography became finer as the current density was increased until the onset of burning. This mode of behaviour occurred regardless of the degree of agitation, but however, deposits were also coarsest in the absence of agitation and finest when using air agitation. Ultrasonic agitation resulted in nodular growths, but unlike the behaviour in the absence of agitation, nodules tended to grow in the valleys as well as at peaks.

It would appear that ultrasonic agitation is more effective in facilitating deposition in micro-depressions (Fig. 10.19) probably due to its greater efficiency in disrupting concentration gradients. On substrates in the as-cast and buffed condition, agitation had the opposite effect on the surface topography to that on abraded finishes; the structure was finer at low current density and became coarse and ridge-like as this was increased.

Deposits from cyanide copper solutions tended to be more 'rounded' (Fig. 10.20) than those from pyrophosphate solutions and these features coarsened as the current density increased, whatever the condition of the substrate surface or the degree of agitation. At low current densities, deposits from a cyanide solution became smoother in appearance as the degree of agitation was increased. This work did not indicate any significant advantages for the use of pyrophosphate instead of cyanide solutions, unless it is necessary to avoid cyanide in effluent.

The many varieties of nickel and chromium coatings that are now available allow the choice of different coating combinations to provide the same corrosion protection in any environment[48], providing their relative resistance to corrosive attack is taken into account. It is pointless to use an expensive coating which will far outlast the expected functional life of the manufactured articles of which it forms a part. Therefore, it is important to have some idea of the relative behaviour of the combinations available, particularly in the severe corrosive conditions encountered during exposure to industrial atmospheres, for marine applications or external vehicle usage. Knowledge of the comparative performance of various nickel plus chromium electrodeposits as protective coatings would enable either the service life of the article to be increased by substitution of an improved system for bright nickel plus decorative chromium, or, if preferred, a thinner total coating thickness to be applied, so as to give the same protection as previously, but at a lower cost. With this in mind, the Battelle Memorial Institute has produced corrosion performance ratios or indices based on the results of their extensive work on the behaviour of plated zinc alloy die-castings during long term exposure in different situations. Their results are obviously applicable to other substrates, and as other workers have obtained similar comparative values, these have influenced the recommendations of the International Organisation for Standardisation[49], which, in turn, has lead to changes in national standards, such as the recent British, American and German ones[4,50,51]. The corrosion resistance of nickel plus chromium coatings will be dealt with further in the next chapter.

REFERENCES

1. VAN ZUILICHEM, A. G., REIDT, M. J., VON ROSENSTIEL, A. P. and VERBRAAK, C. A., *Metalloberflache*, **19**, 1, 3 (1965)
2. SAMPLE, C. H., *Plating*, **47**, 297 (1960)
3. SUCH, T. E., *Trans. Inst. Metal Finishing*, **31**, 190 (1954)
4. *Electroplated Coatings of Nickel and Chromium*, BS 1224 (1965 and 1970)
5. WATSON, S. A., *Trans. Inst. Metal Finishing*, **39**, 91 (1962)
6. BECKWITH, M. M., *Plating*, **47**, 403 (1960)
7. DU ROSE, A. H., *Proc. Amer. Electroplaters' Soc.*, **47**, 83 (1960)

8. SAFRANEK, W. H., HARDESTY, R. W., and MILLER, H. R., *Proc. Amer. Electroplaters' Soc.*, **48**, 156 (1961)
9. DENNIS, J. K. and SUCH, T. E., *Trans. Inst. Metal Finishing*, **40**, 60 (1963)
10. MELBOURNE, S. H. and FLINT, G. N., *Trans. Inst. Metal Finishing*, **39**, 85 (1962)
11. FLINT, G. N. and MELBOURNE, S. H., *Trans. Inst. Metal Finishing*, **38**, 35 (1961)
12. UDYLITE RESEARCH CORP., U.S. Pat. 3 090 733 (21.5.63)
13. BROWN, H., *Electroplating and Metal Finishing*, **15**, 398 (1962)
14. EDWARDS, J., Research Report A.1483, *Corrosion Resistance of Nickel-Cobalt Deposits as Undercoats for Chromium*, British Non-ferrous Metals Research Association (1964)
15. KNAPP, B. B., *Trans. Inst. Metal Finishing*, **35**, 139 (1958)
16. BROWN, H. and WEINBERG, M., *Proc. Amer. Electroplaters' Soc.*, **46**, 128 (1959)
17. TURNER, P. F. and MILLER, A. G. B., *Trans. Inst. Metal Finishing*, **47**, 50 (1969)
18. M. AND T. CHEMICALS INC., British Pat. 1 188 350 (15.4.70)
19. UDYLITE RESEARCH CORP., U.S. Pat. 3 152 971 (13.10.64)
20. CLAUSS, R. J. and KLEIN, R. W., *Proc. Seventh Int. Met. Fin. Conf.*, Hanover, 124 (1968)
21. MARTIN, P. W., *Metal Finishing Journal*, **11**, 399 and 477 (1965)
22. TOMASZEWSKI, T. W., TOMASZEWSKI, L. C. and BROWN, H., *Plating*, **56**, 1234 (1969)
23. BAKER, R. A. and CHRISTIE, N., *Trans. Inst. Metal Finishing*, **47**, 80 (1969)
24. SAUR, R. L., *Plating*, **48**, 1310 (1961)
25. *Metal Finishing Plant and Processes*, **6** No. 1, 15 (1970)
26. SEYB, E. J., JOHNSON, A. A. and TUOMELLO, A. C., *Proc. Amer. Electroplaters' Soc.*, **44**, 29 (1957)
27. SEYB, E. J., *Proc. Amer. Electroplaters' Soc.*, **47**, 209 (1960)
28. SAFRANEK, W. H. and HARDESTY, R. W., *Plating*, **47**, 1027 (1960)
29. LOVELL, W. E., SHOTWELL, E. H. and BOYD, J., *Proc. Amer. Electroplaters' Soc.*, **47**, 215 (1960)
30. LINDSAY, J. H., LOVELL, W. E. and HARDESTY, D. W., *Proc. Amer. Electroplaters' Soc.*, **48**, 165 (1961)
31. M. AND T. CHEMICALS INC., British Pat. 1 070 685 (1.6.67), WILMOT BREEDEN LTD., British Pat. 1 087 613 (18.10.67) and W. CANNING & CO. LTD., British Pat. 1 091 526 (15.11.67)
32. DENNIS, J. K. and FUGGLE, J. J., *Trans. Inst. Metal Finishing*, **49**, 54 (1971)
33. TURNS, E. W. and BROWNING, M. E., *Proc. Amer. Electroplaters' Soc.*, **49**, 53 (1962)
34. RÉGIE NATIONALE DES USINES RENAULT, French Pat. 1 447 970 (27.6.66). British Pats. 1 122 795 (7.8.68) and 1 187 843 (15.4.70)
35. LONGLAND, J. E., *Electroplating and Metal Finishing*, **22** No. 12, 35 (1969)
36. M. AND T. CHEMICALS INC., U.S. Pat. 3 474 010 (21.10.69)
37. UDYLITE RES. CORP., U.S. Pat. 3 471 271 (7.10.69)
38. *Engineering Production*, **1**, 409 (1970)
39. N.V. RESEARCH, HOLLAND, British Pats. 1 020 285 (16.2.66), 1 039 741 (24.8.66), 1 041 753 (7.9.66) and 1 056 222 (25.1.67)
40. ODERKERKEN, J. M., *Electroplating and Metal Finishing*, **17** No. 1, 2 (1964); *Tijd. Oppervlakte Technieken*, **7**, 196 (1963)
41. TURNER, P. F., *Product Finishing*, **19** No. 12, 61 (1966)
42. CARTER, V. E., *Trans. Inst. Metal Finishing*, **48**, 19 (1970)
43. CHALKLEY, B., *Engineers Digest*, **30** No. 3, 55 (1969)
44. CLAUSS, R. J. and ADAMOWICZ, N. C., *Plating*, **57**, 239 (1970)
45. SAFRANEK, W. H. and FAUST, C. L., *Plating*, **45**, 1027 (1958)
46. SAFRANEK, W. H. and MILLER, H. R., *Plating*, **55**, 233 (1968)
47. POINTON, R. B., Project Report, Metallurgy Dept., University of Aston in Birmingham (1970)
48. SILMAN, H., *Industrial Finishing and Surface Coating*, **22** No. 267, 8 (1970)
49. ISO Recommendation R.1456 *Electroplated Coatings of Nickel Plus Chromium*; ISO Recommendation R.1457 *Electroplated Coatings of Copper Plus Nickel Plus Chromium on Steel (or Iron)* and ISO Recommendation R.1458 *Electroplated Coatings of Nickel* (all 1970)
50. ASTM B456–71 *Electrodeposited Coatings of Nickel Plus Chromium*
51. DIN 50 967: 1970, *Nickel Plus Chromium Coatings on Steel, Copper and Zinc Materials and Copper Plus Nickel Plus Chromium Coatings on Steel and Zinc Materials* (1970)

Chapter 11

Corrosion Resistance and Testing of Nickel Plus Chromium Coatings

In addition to its attractive appearance, good corrosion resistance is the most important requirement of a decorative coating. To ensure satisfactory performance, it is first necessary to specify a suitable coating combination and then to carry out adequate testing and inspection of the articles to ensure that they are plated to specification. Physical and mechanical testing has been discussed in Chapter 8, but two other testing procedures remain for consideration—thickness and corrosion testing. Electrochemical behaviour is obviously important in ensuring good service performance, but an adequate thickness of coating is essential to obtain protection by any particular combination of deposits. Inadequate thicknesses, particularly in low current density areas, are still a too common cause of service complaints. Methods of determining thickness fall into two categories—destructive and non-destructive. These have been reviewed elsewhere[1, 2] and only methods and problems associated with decorative nickel plus chromium coatings will be discussed in the present text.

THICKNESS TESTING

Non-destructive methods are obviously desirable, but in most instances the accuracy is not as great as that attained with destructive tests unless frequent calibration is carried against the latter; nevertheless, they are most useful for production quality control. Methods of determining average thickness over a relatively large area are not particularly useful, since they are usually destructive, do not provide information about plate distribution and also because corrosion failure usually takes place in the region of minimum thickness. However, they may be useful for small, irregularly shaped objects, particularly those which have been barrel plated, such as screws, nuts and bolts[3].

Direct measurement of the thickness of the coating at a suitable magnification by means of a microscope is the method recommended in BS 1224:

1970[4] for nickel layers. In this technique, the component is carefully sectioned so that the cut is made at right angles to the surface. mounted in Bakelite and then prepared in the same way as an ordinary metallographic specimen. This is the only way in which the thickness of each layer in a multi-layer nickel system can be determined. The boundaries between layers can be distinguished by etching with a suitable reagent and, for example. the demarcation between the lamellar structure of a bright deposit and the columnar

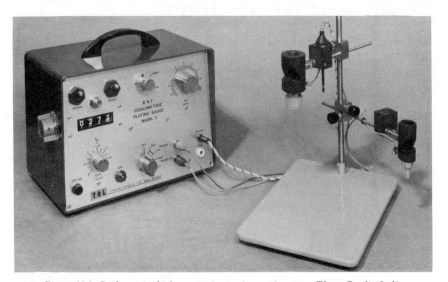

Figure 11.1. Coulometric thickness testing equipment (courtesy Thorn-Bendix Ltd.)

structure of a semi-bright nickel can be distinguished easily. Chromium layers are usually too thin to be measured accurately by this method.

The coulometric technique (Fig. 11.1) is probably the most useful of the destructive methods as it can be used both for a thin chromium layer and the total nickel layer. although it does not differentiate between the various types of nickel in a multilayer coating. The essential feature of the method is the measurement of the total quantity of charge required to anodically strip each metal from a clearly defined area. The test is stopped after each metal is penetrated and the appropriate solution introduced for stripping the next metal, so that, for example, the thicknesses of chromium. nickel and copper on a steel substrate can all be determined at the same spot. No skill is required to detect the end point of each test when using the commercially available instruments because these are designed so that the change in potential, which occurs on exposure of the underlying metal. results in the instrument being switched off. The thickness value is not influenced by the differing electrochemical activities of the various nickel deposits. but the electrolyte solutions used for each metal must be chosen so that dissolution takes place at 100% efficiency.

The B.N.F.M.R.A. Jet test (Fig. 11.2) is the only other destructive test still used to any extent, but this has been largely superseded by the coulometric test. its main use now being for components which are too small or complex

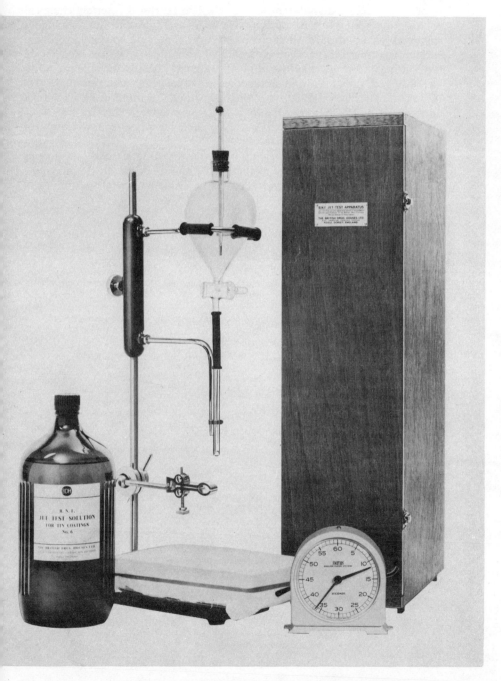

Figure 11.2. Jet test apparatus for determination of coating thickness (courtesy The British Drug Houses Ltd.)

to accommodate the coulometric test cell. The Jet test is a purely chemical solution method; a stream of corrosive liquid is allowed to impinge. under standardised conditions of temperature and flow rate. onto the coating surface. Coating thickness can be evaluated from the time taken to penetrate the coating. but it is sometimes difficult to detect visually when penetration has occurred. Its main disadvantage lies in the fact that the rate of penetration is influenced by the nature of the nickel layers. Standardisation against another method is therefore required; for accurate work this must be done for each type of nickel deposit, but even then the same nominal types of nickel plate can give different results.

Non-destructive thickness testing procedures can be divided into groups in a number of ways and one of the most useful is based on the magnetic properties of the substrate and coating. In Table 11.1 some examples of

Table 11.1 CLASSIFICATION OF COATING-SUBSTRATE COMBINATIONS WITH REGARD TO MAGNETIC PROPERTIES

A Magnetic coatings on non-magnetic substrates	B Non-magnetic coatings on magnetic substrates	C Non-magnetic coatings on non-magnetic substrates
Nickel on brass Nickel on copper Nickel on aluminium Nickel on zinc alloy die-castings Any magnetic coating on non-metallic materials such as plastics	Any non-magnetic metal on steel Any paint, plastics or vitreous enamel on steel Nickel on steel*	Copper on zinc alloy die-castings Anodised films on aluminium Paint, plastics and enamel coatings on non-magnetic materials Non-magnetic metals on plastics. etc.

* This system can be included in this group since nickel is only weakly magnetic.

Table 11.2 NON-DESTRUCTIVE METHODS OF THICKNESS MEASUREMENT APPLICABLE FOR THE COATING-SUBSTRATE GROUPINGS LISTED IN TABLE 11.1

A Magnetic coatings on non-magnetic substrates	B Non-magnetic coatings on magnetic substrates	C Non-magnetic coatings on non-magnetic substrates
Back-scatter of β particles Eddy current Thermoelectric Magnetic attractive force Magnetic inductive	Back-scatter of β particles Eddy current Thermoelectric Magnetic attractive force Magnetic inductive	Back-scatter of β particles Eddy current — — —

coating-substrate systems are given for the three categories. i.e. magnetic coating on non-magnetic substrate, non-magnetic coating on magnetic substrate and non-magnetic coating on non-magnetic substrate. Table 11.2

lists non-destructive tests which can be used to measure the thickness of the coating in each of these categories.

Variations in physical properties cause inconsistencies in results obtained by most non-destructive methods. The magnetic methods are the most useful as far as nickel coatings are concerned, but need careful calibration using coatings from the same baths and on the same substrate as standards for later measurements[5]. The 'pull-off' method can be used, with suitable calibration, for either magnetic or non-magnetic substrates, since nickel is only weakly magnetic. The magnetic-inductive technique uses a probe in which two contact heads are juxtaposed and between which passes the magnetic flux generated by a coil excited by a current source. The resultant effect on a measuring coil is affected by the permeability of the coating, and so when the instrument has been calibrated against any material (metal or plastic) coated with nickel electroplate of known thickness, it can be used for repetitive measurements of nickel coatings on the same basis material.

The thermoelectric method has a sound theoretical basis but has been found to have considerable disadvantages in industrial use for quality control, appearing to need frequent calibration. The β ray back-scatter method is a very accurate technique since it is not influenced by physical properties of the deposits and very thin coatings can be measured. Its main disadvantage is that the atomic numbers of the coatings and substrate should differ by at least four. Unfortunately, the metals which are the subject of this text are very close together in the Periodic Table (see Table 11.3) and the instrument

Table 11.3 TABLE OF ATOMIC NUMBERS OF ELEMENTS RELEVANT TO THICKNESS TESTING, BY THE β BACK-SCATTER METHOD, OF NICKEL PLUS CHROMIUM COATINGS ON COMMON SUBSTRATES

Metal	*Atomic number*
Aluminium	13
Chromium	24
Iron	26
Nickel	28
Copper	29
Zinc	30

is of no use for such combinations. It is also fairly expensive and is more suited for the measurement of the thickness of precious metal deposits, where the value of the coating justifies greater expenditure on test equipment.

Instruments have been designed based on inducing eddy currents in the surface layers of metals by the use of high frequency currents, whose magnitude depends on the conductivity of the 'skin'. If the coating and basis metal have conductivities which differ sufficiently, the eddy current passing through the composite surface layer depends largely upon the thickness of the coating. This principle can be used for the measurement of many different combinations of electrodeposited coatings and basis metals, including most of those of commercial importance. It is said to be unnecessary to use precisely the same type of deposit for calibration, e.g. different bright nickels have much the same conductivity. The instrument appears to be the most versatile of the non-destructive ones, although it is doubtful if it will measure the individual thicknesses of different metals in multi-layer deposits, and it is rather costly.

The real weakness in thickness testing is the lack of a reliable non-destructive test which employs a cheap instrument and yet is rapid, direct-reading, simple to carry out by unskilled personnel and will give accurate results on a sample of unknown origin. Destructive tests are more reliable, but, by the nature of the test, the percentage of components sampled must be far lower than can be checked in the case of non-destructive tests.

CORROSION TESTING

Apart from service trials of plated specimens in the actual environment in which they will be exposed, two types of test, viz. accelerated and outdoor exposure tests, are used to evaluate the corrosion resistance of coatings. However, even outdoor tests, although prolonged, are usually somewhat accelerated compared with average service conditions. Three accelerated tests are commonly used in the United Kingdom, and a fourth, the most accelerated of all (E.C. test), is used to some extent in America. They are:

(*a*) Acetic acid–salt spray test.
(*b*) CASS test (copper–acetic acid–salt spray).
(*c*) Corrodkote (slurry) test.
(*d*) E.C. test (electrochemical test).

Other tests have been used and are still employed in some countries, but they have been largely discarded for a variety of reasons. The neutral salt spray was the first of the spray tests, but it is too slow for decorative nickel plus chromium coatings unless they are extremely thin. The acetic acid–salt spray test was developed to overcome this difficulty and is much more rapid, although it is itself slow compared with some of the more recent methods. The once popular B.N.F.M.R.A. sulphur dioxide test[6] has now been discredited as a valid technique for comparing different nickel plus chromium systems[1,7]. It is really a means of detecting discontinuities in the chromium layer, and therefore an unfavourable result is obtained when testing micro-cracked or micro-porous chromium because the chromium is rapidly undermined and flakes off. Conversely, this test also exaggerates the beneficial effects of crack-free chromium. The sulphur dioxide test also fails to indicate the improved corrosion resistance of duplex nickel as compared with bright nickel. The more noble semi-bright layer is activated by the sulphur dioxide and preferential attack of the bright nickel does not occur. The authors consider that if the test was modified by decreasing the sulphur dioxide content, using 0·01% or less instead of 1%, it would probably provide useful information, particularly with regard to performance in industrial atmospheres, since these usually contain sulphur compounds. However, it would make the test much slower and it might lose any value it may have for acceptance testing of production parts, although it could still be useful for research purposes. The Kesternich test[8] which uses an atmosphere containing either 0·7 or 0·07% sulphur dioxide combined with cyclic condensation is still popular in Germany[9] but suffers from similar disadvantages to the B.N.F.M.R.A. test.

ACETIC ACID–SALT SPRAY TEST

In this test the spray solution is a 5% w/v NaCl solution with the pH adjusted to 3·2 by acetic acid. The test is carried out in a suitable non-corrodible cabinet at 35°C, and the spray rate must be controlled so that its collection rate over an area of 0·008 m^2 is between 1–2 ml/h. The collected spray solution must have a concentration of 5 ± 1% NaCl and its pH should be unchanged from the original. The spray solution should not be recycled as

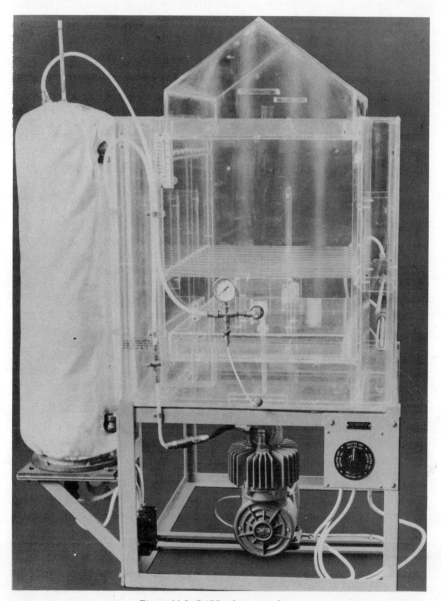

Figure 11.3. CASS salt spray cabinet

it contains products of corrosion which may accelerate subsequent attack. Test samples must be free from grease and dirt; swabbing with magnesium oxide slurry is the normal procedure for cleaning. Spray tests are directional and since the spray falls on the upper surfaces of articles the undersides will not be affected to the same extent. Significant surfaces are exposed at an angle of 15° to 30° to the vertical whenever possible but obviously this may not always be feasible with articles of complex shape. The period of exposure to the corrosive fog should be as continuous as possible, but it may be necessary to inspect the specimens periodically in order to assess the extent of corrosion occurring at times less than the full test period.

CASS TEST[7, 10, 11]

As the name implies, a copper salt (0·26 g/l of $CuCl_2$) is added to the spray solution, which otherwise is the same as that used in the acetic acid–salt spray test. The operating temperature is higher (49°C) but the collection rate of the sprayed solution and preparation and positioning of samples in the cabinet are the same as for the acetic acid–salt spray test. The cabinet (Fig. 11.3) is similar to that used for the previous test, but a larger humidifying unit is required so that the compressed air supplied to the jet is sufficiently humid to prevent evaporation of the sprayed liquid; this modification enables the correct collection rate and concentration to be maintained. This is a more accelerated test than the acetic acid–salt spray test and so is more useful as a routine inspection tool, particularly since the mode of corrosion is closely related to that which occurs in service.

The method of cleaning the surface before CASS testing can influence the degree and form of corrosion. The usual practice is to swab with magnesium oxide slurry and rinse thoroughly[10, 12], but there have been some suggestions that this treatment may be too severe.

CORRODKOTE TEST

This test, whose name is sometimes abbreviated to 'CORK', uses a different technique to the previous two in that the corrosive medium is applied in the form of a slurry, i.e. a paste. It can be brushed onto any surface and so the test is not directional. The paste is prepared as indicated in Table 11.4. These salts are claimed to simulate corrosion resulting from exposure in the Detroit area of the U.S.A., and the formulation was specifically devised for testing motorcar components. Details of this test are given in Appendix G of BS 1224:1970[4] and the appropriate A.S.T.M. specification[13].

Plated surfaces are not cleaned with magnesium oxide before applying the slurry, which should be brushed on using a circular motion to ensure that the surface is wetted[14]. The test is carried out by placing the slurry coated samples in a cabinet with controlled temperature and humidity, the specified conditions being 38°C and 90–95% r.h. When the basis metal corrosion products are coloured (e.g. iron hydroxide), the extent of corrosion can be assessed either from the appearance of the paste *in situ* or from the coating's appearance when the paste is removed. In cases where coloured

Table 11.4 PREPARATION OF STANDARD SOLUTIONS FOR THE CORRODKOTE TEST

Standard solutions and formulae of constituents used	Quantity of constituent
Cupric nitrate, $Cu(NO_3)_2 \cdot 3H_2O$	5 g/l
Ferric chloride, $FeCl_3 \cdot 6H_2O$	5 g/l
Ammonium chloride, NH_4Cl	100 g/l

Preparation of slurry
  Mix 7 ml cupric nitrate, 33 ml ferric chloride and 10 ml ammonium chloride solutions and add to 30g kaolin. Stir until slurry is obtained.

Data taken from BS 1224:1970[4]

corrosion products are not formed, the paste must be removed using running water and the metal surface examined for pits and/or blisters.

E.C. TEST[15]

The electrochemical test is very rapid and a test time of 2 min is reputed to correspond to one year's outdoor exposure. The success of this test depends on choosing an electrolyte solution in which the relative rates of electrolytic dissolution of copper, nickel and chromium are the same as the relative rates of corrosion in service. When a nickel plus chromium coating corrodes, the cathodic reaction

$$H_2O + \tfrac{1}{2}O_2 + 2e \rightarrow 2OH^-$$

takes place at the chromium surface and the anodic reaction

$$Ni \rightarrow Ni^{2+} + 2e$$

takes place at nickel exposed at defects in the chromium layer. By using a potentiostat to control the anode and cathode potentials, the rate of corrosion can be controlled. A suitable electrolyte solution and operating potential can be selected by plotting the relationship between current density and potential for the electrodeposits concerned in a particular coating system. The two electrolyte solutions given in Table 11.5 have been recommended by Saur and Basco[16] as having the correct characteristics (i.e. corrosion rates in the same order as in service) for the copper plus nickel plus chromium

Table 11.5 SOLUTIONS USED FOR THE E.C. CORROSION TEST‡

Composition (g/l)	Solution A*	Solution B†
Sodium nitrite	10	10
Sodium chloride	1·3	1
1,10-phenanthroline hydrochloride	—	1
Nitric acid	5	5

* For non-ferrous substrates
† For ferrous substrates
‡ Data taken from Saur[15]

system. The operating potential for these electrolytes is $+ 0{\cdot}3$ V (with respect to the S.C.E.). Sodium nitrate facilitates the corrosion of copper, and sodium chloride that of semi-bright nickel. Nitric acid is present to retain the corrosion products of copper and nickel in solution, while 1.10-phenanthroline hydrochloride indicates the presence of iron in the solution by the formation of a pink colouration. The first appearance of this colour indicates penetration of the coating and attack of the substrate. This colour change is obscured by the build-up of Ni^{2+} ion concentration in solution B, which should be discarded periodically as the pH rises, or salt precipitation may occur. The material used for the cathode must be inert with respect to the solution. The plated area to be tested must be determined and masked off by suitable means; the anode current density should not exceed $0{\cdot}32$ A/dm^2. Surface preparation is the same as for the salt spray tests, i.e. swabbing with magnesium oxide. It is recommended that electrolysis be carried out intermittently (electrolyse for 1 min—halt for 2 min). This test may grow in importance in the future, since it is very rapid and is reported to result in pits of the same morphology as those occurring in service. Probably its main drawback is that it requires a potentiostat as a current source and at the present time this precludes its use as a repetitive shop-floor technique, particularly in smaller organisations[17].

OUTDOOR TESTS

Outdoor tests can be divided into two categories, i.e. static and mobile. Mobile tests, in which panels are attached to various parts of road vehicles or boats, most closely simulate service conditions experienced by those modes of transport, but usually the test panels are positioned so that they are subjected to the most corrosive conditions encountered by a particular vehicle. Static roof or beach tests also have their uses and are easier to organise; panels are usually mounted in racks at about 30° to the vertical (Fig. 11.4). These static tests can give information which may be misleading when applied to the behaviour of coatings when in service on vehicles, e.g. the improvement afforded by crack-free chromium is exaggerated because the test panel is not subjected to vibration, deformation or bombardment by road chippings. Double layer nickel exposed on a roof site in an industrial region often does not show the improvement over bright nickel that has been observed in service. This is because, as in the case of the sulphur dioxide test, the semi-bright nickel is activated by the sulphur compounds present in the atmosphere. Outdoor tests are far more difficult to control than accelerated tests since it is inevitable that while the environment can be selected as industrial, urban, rural, marine, etc., the weather and pollution variations are out of the control of the investigator[18]. Therefore, comparisons between results obtained on batches of specimens exposed at different sites and at different dates offer considerable problems[19]. It is usually necessary to include in the exposure programme coating systems whose behaviour is well known, so that these can be used as controls and other results interpreted with respect to these controls. In this country, static tests are usually more severe than mobile tests, while in North America the reverse is true. This is caused by two factors. British industrial atmospheres are much more

Figure 11.4. Roof exposure site showing panels in position on racks

aggressive than American due to the higher concentration of sulphur compounds and the occurrence of more condensation and fog. Greater quantities of salt are used on the roads in Canada and the northern states of the U.S.A. than in the U.K. and so panels exposed on vehicles will be subjected to a very effective salt spray test for prolonged periods during the winter. Although the actual dosage rate for each period of snowfall is about the same in both countries (typically between 30 and 55 g/m^2 of road surface) the climate in North America, particularly in the snow-belt area, is such that salt spreading is required on many more days, resulting in a total annual usage which varies between 6 and 20 kg/m of each two-lane carriageway treated. Only in an unusually severe winter is the same amount used on British roads, except on the motorways where the amount exceeds, in most winters, the maximum American figure.

Degree of acceleration of corrosion tests

It is impossible to assign a precise figure to the degree of acceleration of both accelerated and outdoor tests, but the factors which will be quoted are based on information obtained from several sources as well as the authors' own experiences. As mentioned previously, the E.C. test is by far the most accelerated test, 2 min being equivalent to one year's service in the Detroit area. An approximate estimate for the rate of acceleration of the Corrodkote and CASS test compared to mobile exposure throughout the year for average British conditions would be about 400 to 500 times. One 16 h Corrodkote or CASS cycle is therefore usually assumed to be equivalent to about one year in mobile service on the roads. The acetic acid salt spray is less severe, 96 h being equivalent to one year. The ratio between static roof tests in Birmingham (October to March) and mobile exposure in an average year is about 10:1 (roof tests being the more accelerated). However, in a severe winter, such as that experienced in 1962–63, the road conditions are very similar to those usually experienced in the U.S.A., the factor then being reduced from about 10:1 to 3:1, as judged from an exposure programme carried out by the authors during that winter[7].

Choice of test

The choice of test method is influenced by the particular requirement[20, 21]; a particular test may be adequate as an acceptance test, but of little value for research and development purposes. Table 11.6 shows the order of merit of various coating systems when subjected to different corrosion tests. This illustrates how it could be misleading to use a single test without knowing the limitations of the test. For example, bright nickel plus crack-free chromium has a high rating in most tests, but a low rating in the mobile test. Many tests indicate that this system is superior to bright nickel plus microcracked chromium, although this is not the case in the majority of applications. Most tests can provide useful information, particularly for routine testing, when a certain amount of prior knowledge is available of the performance expected from coatings which are known from experience to be satisfactory in service.

Table 11.6 ORDER OF MERIT OF DIFFERENT NICKEL PLUS CHROMIUM SYSTEMS AFTER VARIOUS CORROSION TESTS*

Type and thickness of coating		Types of corrosion test and exposure time										
Nickel	Chromium	Static outdoor roof exposure	1 year (cleaned with tripoli paste)	Mobile outdoor exposure 10 months	Acetic acid salt spray		CASS		Sulphur dioxide		Corrodkote	
		1 year			72 h	216 h	24 h	72 h	24 h	72 h	One 20 h cycle (paste removed)	Three 20 h cycles (paste removed)
Bright 30 μm	Crack-free 0·75 μm	1	1	5	1	1	1	2	1	1	1	1
Watts 30 μm	Conventional bright 0·25 μm	2	3	1	1	1	2	2	4	2	4	4
Bright 30 μm	Micro-cracked 0·75 μm	2	1	3	4	4	4	4	5	5	1	1
Duplex 80/20 30 μm	Conventional bright 0·25 μm	4	4	1	1	3	2	1	2	2	1	1
Bright 30 μm	Conventional bright 0·25 μm	5	4	4	4	5	4	5	3	4	5	5

* Data taken from Dennis and Such[7].

METHODS OF EVALUATION OF CORROSION RESULTS

Numerous methods have been devised in an attempt to evaluate accurately the extent of corrosion [4, 12, 22–25] and these have been reviewed critically by Lowenheim[26]. Results obtained at different sites and times can be compared only if samples are evaluated systematically and reported quantitatively. Two systems are used frequently in the U.K. for evaluating nickel plus chromium coatings, these are the British Standard and the A.S.T.M. systems.

THE A.S.T.M. METHOD

This depends on assessing the total area that has been corroded[27]. The area of a particular type of corrosion pit or defect on the surface of the sample is assessed by comparison with standard charts or photographs. Depending on which of these types of illustrations resemble most closely the form of corrosion on the specimens, the appropriate series are placed alongside these and the extent of their corrosion matched as nearly as possible to one

Table 11.7 RELATIONSHIP BETWEEN A.S.T.M. PROTECTION RATING NUMBER AND PERCENTAGE TOTAL AREA CORRODED*

Rating number	% total area defective
10	0
9	0·0–0·1
8	0·1–0·25
7	0·25–0·5
6	0·5–1
5	1–2·5
4	2·5–5
3	5–10
2	10–25
1	25–50

* Data taken from ASTM B537-70

Table 11.8 LIST OF THE MOST IMPORTANT WEIGHTING FACTORS USED IN THE PROPOSED A.S.T.M. RATING METHOD*

Defect	Factor
Blistering	2
Pin-point corrosion	1
Basis metal corrosion stain	0·1
Moderate stain from coating metals	0·02–0·05
Light stain from coating metals	0·01
Surface pitting (light)	0·01
Surface pitting (heavy)	0·1
Cracking	1

* Not included in published ASTM standard

CORROSION RESISTANCE AND TESTING 249

of the standards and the specimen given the rating number applicable to that standard illustration. The table relating number (1–10) to percentage total corroded area (see Table 11.7) is based on the relationship

$$R = 3(2 - \log_{10} A)$$

where R is the rating number and A is the percentage total defective area. This system was adopted because the human eye is more sensitive to changes in ratio than to absolute intensity. The difference between 0·05% and 0·1% area defective has the same significance as that between 5% and 10%. Fourteen weighting factors were originally suggested by its devisors. The more important ones are shown in Table 11.8 and their use gave more emphasis to defects which cause rapid deterioration of surface finish.

A modified form of the A.S.T.M. system enables one value to be placed on protection and one on appearance. The protection rating is assessed solely on the manifestation of basis metal corrosion. For example, a bright nickel plus micro-cracked chromium coating on steel after outdoor exposure for several years could have a protection rating of 10 (no rust showing), but a much lower appearance rating (extensive nickel corrosion and staining). Alternatively, a steel sample having a bright nickel plus decorative chromium coating exposed in similar conditions would probably have a rating of zero for both protection and appearance (rust showing at all pits).

A further modification of this system has been devised to expand the scale between rating numbers 9 and 10 since this is the most important range[28]. The argument in favour of this is that precise rating of badly corroded

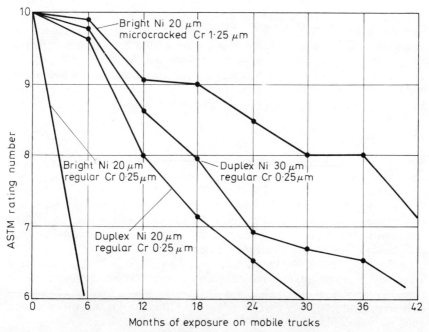

Figure 11.5. Corrosion resistance of various nickel plus chromium coating systems after mobile exposure (after Safranek and Faust [29])

specimens is of little value, but that it is important to have an accurate rating method to distinguish between the best coatings. The minimum rating for acceptance of a coating system after any particular test is often a matter for agreement between customer and plater but it is usually agreed that to specify a rating of 10 is unrealistic.

A useful method of reporting corrosion resistance is to record the time of exposure taken for deterioration to a specified rating number as assigned by the A.S.T.M. or any other system. The results shown in Fig. 11.5 are taken from a paper by Safranek and Faust[29] and show the performance of four coating systems on exposure on mobile truck trailers. This diagrammatic representation clearly indicates the superior performance of bright nickel micro-cracked chromium and double layer nickel plus conventional chromium. As mentioned in Chapter 10, corrosion endurance indices can be calculated to permit comparison of performance between coatings by allotting an index of 1 to a particular coating system. For example, if the standard coating system chosen is bright nickel (25 µm) plus decorative chromium (0·25 µm), indices for most other coatings will be greater than 1, since the time taken for deterioration to a particular rating number will be longer than for the standard coating.

BRITISH STANDARDS INSTITUTION METHOD

The method recommended in BS 3745: 1970 *Evaluation of results of accelerated corrosion tests on metallic coatings*[30] demands less skill and experience on the part of assessor than the A.S.T.M. method. However, it is a severe rating system, a small number of corrosion spots resulting in a low rating number. It differs from the A.S.T.M. system in that it only takes into consideration sites at which visible penetration of the coating has occurred. Surface staining or discolouration is not assessed. A transparent graticule divided into squares of 5 mm × 5 mm is placed on the surface of the sample and the number of squares containing one or more corrosion pits are counted. Frequency of corrosion spots is given by the expression:

$$F_I = \frac{100n}{N}$$

where F_I is the initial frequency, n is the number of squares containing corrosion spots and N is the total number of squares on the significant surface considered. Table 11.9 shows the relationship between frequency and initial rating number. If corrosion sites are greater than a critical size or too numerous in a particular region, the extent of corrosion is considered too disastrous to warrant assessment of the specimen and its rejection is then mandatory. These types of failure are classified as follows:

(*a*) Spots greater in size than 2·5 mm² or long cracks such that the area of crack in one 5mm square is greater than 2·5 mm². (These areas refer to the actual corrosion spot or crack, not to the area which may be covered by corrosion products. After the Corrodkote test this also refers to the corrosion spot, not the diffuse spot in the Kaolin paste.)

(*b*) 10 or more spots in any two adjacent squares.

The initial frequency rating is lowered if corrosion occurs in a localised

CORROSION RESISTANCE AND TESTING

Table 11.9. RELATIONSHIPS BETWEEN INITIAL FREQUENCY (%) AND INITIAL RATING NUMBER (BRITISH STANDARD METHOD OF EVALUATING THE EXTENT OF CORROSION)*

Initial frequency (%)	Initial rating number
0 (no corrosion spots)	10
Not more than 0·25	9
Not more than 0·5	8
More than 0·5 but not more than 1·0	7
More than 1·0 but not more than 1·5	6
More than 1·5 but not more than 2·0	5
More than 2·0 but not more than 3·0	4
More than 3·0 but not more than 6·0	3
More than 6·0 but not more than 12	2
More than 12 but not more than 25	1
More than 25	0

* Data taken from BS 3745:1970[30]

area of the sample. The corrosion spots are counted in the area 50 × 50 mm having the greatest number of spots, and the initial rating reduced by 1 for each 10 squares occupied by spots. Unlike the A.S.T.M. system, an acceptance level is specified and the relevant standard, BS 1224: 1970[4], sets this at 8. This rating method is probably less selective for research purposes than that published by the A.S.T.M., since it does not take into consideration the type of deterioration resulting from the use of some of the modern nickel plus chromium coatings, such as double layer nickel or micro-porous and micro-cracked chromium. In these cases, basis metal attack is delayed but nickel corrosion occurs with subsequent loss of brightness and so a rating system which distinguishes between attack on the substrate and attack on the coating has certain advantages.

TREATMENT OF SURFACE BEFORE RATING

The amount of washing and cleaning carried out before assessing the rating can influence its value, but this is more important in the case of the A.S.T.M. system. Removal of rust from around corrosion spots may be advisable when using the B.S.I. method to ensure that a particular pit is not so large as to cause rejection of a specimen. On the other hand, when rust is removed, pits may be difficult to see through a plastics graticule and a high rating number may be given.

When using the A.S.T.M. system, it is essential to standardise the cleaning procedure to achieve comparable assessment. After spray tests, streaks of corrosion product run down the test panel and mask the actual corrosion pits. This type of failure is easier to rate by comparison with the standard photographs than by comparison with the charts. For example, typical charts and photographs relating to rating numbers 3, 4, 7 and 8, are shown in Fig. 11.6. The pits would appear to be much smaller if the rust streaks were removed. Similarly, if nickel corrosion products and stain are removed from coatings having a top layer of micro-cracked or micro-porous chromium, the rating number would be higher. Rating of panels subjected to the Corrodkote test requires special consideration, for unless coloured corrosion products are formed, corrosion sites are not indicated in the Kaolin paste. For example, considerable nickel corrosion may occur, as in the case of duplex nickel coatings or micro-cracked or micro-porous chromium overlays, without penetration to the substrate taking place. In these instances, the only indication of corrosion, on examination of the kaolin paste, would be a pale green colouration due to the nickel corrosion products, and this would not be very noticeable in the off-white paste. White corrosion products from zinc or aluminium substrates would not be visible at all in the paste. If specimens are rated with the paste on, it is therefore possible in some instances to obtain a much higher rating number

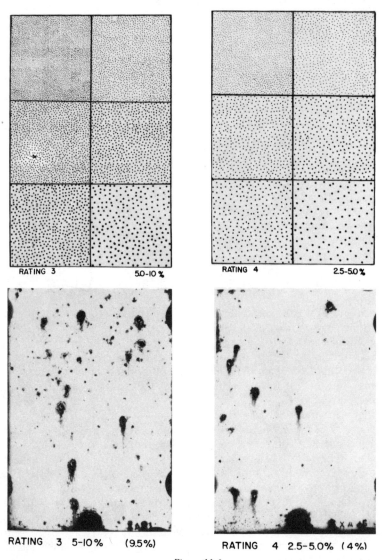

Figure 11.6.

CORROSION RESISTANCE AND TESTING

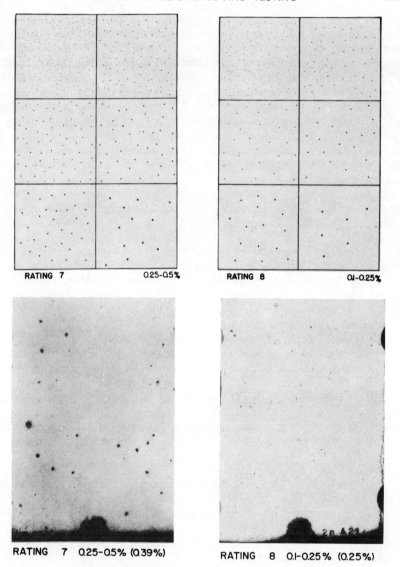

Figure 11.6 (Continued). A.S.T.M. rating charts and photographs relating to rating numbers 3, 4, 7 and 8 (after Graham)

than if they are rated with the paste removed, when surface pits, stains and blisters are revealed.

ASSESSMENT OF CORROSION BY EXAMINATION OF THE MORPHOLOGY OF CORROSION PITS

The surface of a corroded specimen can be examined using a simple optical microscope. This technique can provide information such as the density of pits per unit area, the size distribution and possibly the depth of penetration, but is unlikely to provide much evidence on the corrosion mechanism and shape of corrosion pits. However, examination of the cross section, etched when necessary, does illustrate the shape of the pit and this is an ideal method of revealing the preferential attack on bright nickel in a double layer nickel or the formation of numerous small pits in a nickel layer underneath a micro-cracked or micro-porous chromium layer [Figs. 10.2, 10.11(a) and (b) and 10.14]. It is important to remember that the exact location of a particular section through a pit can influence the shape and depth of the effects observed, for obviously it is difficult to ensure that the section is cut precisely through the centre of any pit.

Saur[31] has used the interference microscope to measure the average penetration into the semi-bright nickel and the average pit radius after E.C. testing. From this information, he calculated average corrosion velocity, which in the semi-bright layer is a measure of basis metal protection and in the bright layer a measure of visible pitting. If this procedure is adopted, exposure to the corrosive environment need not be continued until basis metal corrosion occurs, and the lengthy test times needed for evaluation of 'good' coatings can be avoided. Both factors assessed have significance as far as surface deterioration is concerned and the procedure can be employed regardless of the type of basis metal.

For research purposes, the electron probe microanalyser can be used to show which metals have been exposed in a corrosion pit formed in a multi-layer coating, although it cannot distinguish between bright and semi-bright nickel. If the characteristic X-ray emission is recorded for each metal concerned, the distribution of each metal can be observed. The most recent development in the field of electron microscopy—scanning electron microscopy—is particularly suitable for the examination of surface topography and so is ideal for the examination of corroded samples. For this particular purpose, it is most useful at the lower end of the magnification range, i.e. up to about ×5000. Since the depth of focus is far greater than that obtainable with optical microscopy, both the surface and the bottom of the pit are in focus at the same time. Fig. 10.12(d), showing the undermining of micro-cracked chromium, is a good example of the capability of the technique. It is possible to examine directly corrosion pits at a wide range of magnifications and to do so while preserving the three dimensional effect associated with visual observation of an object.

EVALUATION OF CORROSION RESISTANCE OF MULTI-LAYER NICKEL PLUS CHROMIUM COATINGS

Extensive corrosion testing programmes have been carried out in recent years in the U.K. and overseas (particularly in the U.S.A.) to assess the corrosion resistance of each modification to the nickel plus chromium system as it has been developed[7, 21-23, 32-36]. Steel or zinc alloy die-castings are the substrates used in most investigations, but brass[37] and aluminium[38, 39] substrates have not been neglected entirely. Outdoor and accelerated tests have each played an important role in the development of superior coatings. The development of the CASS and Corrodkote tests has made possible a fairly rapid assessment of coatings; only those selected as being most promising need be subjected to the time consuming operation of mobile and static outdoor exposure programmes. La Que[40] has reviewed the developments in America and gives a graph illustrating the relationship between deterioration occurring in the CASS test and after service in Detroit. This is a straight line relationship and so CASS test results have particular significance for that part of the world. although they are also relevant to many more environments.

All the exposure programmes have emphasised the inadequacy of bright nickel plus decorative chromium to meet the requirements of service in severe conditions. such as more or less continuous outdoor use. in regions where precipitation and air pollution are both heavy. This is particularly important for motorcar trim now that its finish is guaranteed in most cases for a specified period of time. sometimes two years.

Due to variations in test and service conditions it is impossible to enumerate categorically an order of merit for corrosion resistance of nickel plus chromium coatings. Detailed results have been published by many authors; these are too complex to review here but most of the modified. i.e. improved coatings. provide satisfactory protection when correctly applied. BS 1224 entitled *Electroplated Coatings of Nickel and Chromium* was revised in 1965 and 1970 with the object of persuading platers to use the improved systems, particularly for severe conditions. The 1965 edition of the standard did not permit the use of micro-porous chromium because it had not been evaluated fully at that time. However. its use is now permitted by the 1970 edition. since this coating system has been shown to provide excellent resistance to corrosion. The scope of the 1965 standard was extended to include nickel plus chromium coatings on aluminium. The latter has gained importance as a substrate with the introduction of a number of proprietary solutions which enable good adhesion of electrodeposited coatings to be achieved on aluminium and its alloys (see Chapter 12). The latest form of the standard requires that. when specifying the quality of plating required of the plater, the customer should state the basis metal and the service condition number, which denotes the severity of the conditions. but the classification number, which indicates the precise coating required. need only be specified if desired. There are four service condition numbers corresponding to degrees of severity of the conditions in which the plated components will be used. the most severe (No. 4) being for those to be almost continually exposed outdoors and the mildest (No. 1) for those used indoors in warm dry atmospheres. such as offices and living rooms. The classification numbers describe. in code form. the basis metal and the type and thickness of the nickel and chromium deposits. Only those combinations which have been found most satisfactory are allowed to be used for service condition number 4. no matter what their thickness. while for the other grades. single layer bright nickel and conventional chromium may be used.

It is suggested that the coating shall be deemed to be satisfactory. if. after the appropriate test for the period specified for the particular grade used. the rating. as determined using BS 3745: 1970[30] is at least 8. The level of deterioration due to stain and loss of lustre after testing is subject to agreement between the parties concerned. Tests are also recommended for assessing ductility, coating thickness and adhesion. This edition of the standard is an attempt to ensure the production of articles plated with high quality nickel plus chromium coatings, which will maintain their attractive appearance throughout their useful life. It sets a much higher standard than its predecessors for coating systems to be used in severely corrosive conditions.

The International Lead Zinc Research Organisation has carried out a survey on nickel plus chromium plated die-castings fitted on cars in two areas of America[29]. Components having different types of coating were selected to provide comparisons between the corrosion resistance of a number of nickel plus chromium systems. Some typical findings are shown

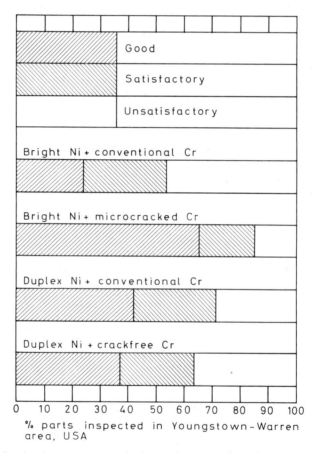

Figure 11.7. Results of an assessment of the degree of corrosion of zinc die-castings with various nickel plus chromium coatings. The components were in service for two years on motorcars in the Youngstown-Warren area in the U.S.A. (after Safranek and Faust[29])

in diagrammatic form in Fig. 11.7. The appearance of components was classified as good if the rating, as designated by the A.S.T.M. method, was 9·5 or 10. A rating of 8 or 9 was regarded as satisfactory and less than 8 unsatisfactory. These results clearly indicate the advantage to be gained by using one of the more recent coatings systems. Even the double layer nickel plus crack-free chromium performed fairly well, but it should be noted that the substrate was a rigid die-casting. Other workers have also reported the results of their extensive mobile testing of plated specimens. e.g. Bush[41] of the Ford Motor Co.

EFFECT OF COPPER UNDERCOATS

It has been proved fairly conclusively that inferior corrosion resistance results when part of the bright nickel layer in a bright nickel plus decorative chromium coating is replaced by an equal thickness of copper[42]. Copper undercoats can affect the surface appearance, once penetration through the nickel layer has occurred. La Que[40] has illustrated this point by showing the results of corrosion tests on two different coating systems applied to steel substrates. One steel panel was plated with a bright nickel layer 37·5 μm thick while the other had 25 μm of copper plus 12·5 μm of bright nickel. Before cleaning, the panel having the copper undercoat appeared superior to the other, although both were severely corroded. After cleaning, the appearance of the panel having only the bright nickel undercoat was better, while the crater-shaped rust spots on the panel having the copper plus nickel coating could not be removed by conventional means.

However, copper has been reported to be advantageous under micro-cracked chromium and micro-porous chromium[43–48], where it behaves as a cathode with respect to the nickel layer adjacent to it. Flint and Melbourne[49] have shown that the corrosion mechanism is quite different in the copper plus bright nickel plus decorative chromium system; the pits penetrate straight through the nickel and copper layers and the copper is preferentially attacked as illustrated in Fig. 11.8. The relevant standard electrode potentials indicate that copper should be more noble than nickel. but this is not the case under the conditions existing in a pit formed in a copper plus bright nickel plus decorative chromium coating. Anodic polarisation measurements on copper deposits in acetic acid–salt spray solution, sulphur dioxide solution and simulated industrial rain water[50] show that copper behaves in a more noble manner than sulphur-containing nickel electro-deposits. The difference in behaviour of copper in coatings having decorative and micro-cracked or micro-porous chromium overlays is thought to be due to the difference in magnitude of the corrosion currents operative in the corrosion pits. In the case of a chromium coating having micro-discontinuities, the area of nickel exposed is large and so the current density in each pit is very low. The corrosion current is almost entirely directed to the nickel layer, leaving the copper unattacked. If the corrosion current density is high (decorative chromium overlay) the copper becomes anodic with respect to the nickel. Flint and Melbourne[51] reported a similar mechanism of corrosion for the duplex nickel system. The latter functioned more satisfactorily when the corrosion current in the pit was relatively low. This is due to the shape of the polarisation curves. Clauss and Klein[45] have reported that copper is only advantageous, even when the chromium overlay is of the micro-porous type, when the copper is adjacent to, or in the close proximity of, a nickel layer containing sulphur. Copper afforded no improvement when deposited below a semi-bright nickel plus micro-porous chromium; the nickel was not preferentially attacked. The same authors carried out a limited investigation of the effect of copper under duplex nickel plus micro-cracked chromium. They used thin nickel layers in which a 1:1 thickness ratio of semi-bright to bright nickel existed. Usually in duplex systems a greater proportion of semi-bright is employed. The copper remained cathodic to the combined nickel layer and the coating system performed satisfactorily. However, the authors suggested that further work is required to determine the ideal

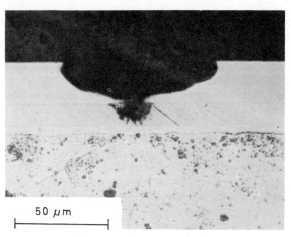

Figure 11.8. Corrosion pit in copper plus nickel plus conventional chromium coating showing preferential attack of the copper layer (courtesy International Nickel Co. Ltd.)

thickness ratio of sulphur-free and sulphur-containing nickels in this coating system. It is of interest to compare the behaviour of copper when employed as an undercoat in a nickel plus micro-porous system to that when it is used in the sandwich coating described in Chapter 10.

Due to the recent scarcity of nickel, emphasis has been directed towards the use of thinner nickel coatings without reducing the corrosion resistance. Davies[44] has reported that replacement of 16% of the bright nickel layer by copper, as is allowed by BS 1224: 1970, gave satisfactory performance when the top coat was micro-cracked chromium. His results when replacing the same proportion of nickel by copper in the presence of a top layer of decorative chromium confirmed that the corrosion resistance was inferior to that when using only a layer of bright nickel. Bache and Turner[46] have carried out a similar investigation to Davies, except that they employed micro-porous chromium instead of micro-cracked. They claimed that if a copper undercoat is used, a reduction in total nickel thickness as great as 40% can be made without lowering the corrosion resistance below that specified in BS 1224.

EFFECT OF CATHODIC DICHROMATE TREATMENT

The Battelle Memorial Institute has developed a technique for the production of protective, but almost invisible, chromate films on the surface of nickel and chromium plated articles. A film having a thickness of approximately 50 Å is formed on the chromium electrodeposit and at any nickel exposed in the discontinuities of that deposit. Such films consist of hydrated chromium oxide and, having an impedance value of 80–90 Ω/cm^2 after drying, greatly reduce the galvanic corrosion that otherwise results when conventional nickel plus chromium plate is subjected to corrosive environments and so decreases the rate of attack. This can be seen from the work of Safranek et al.[52-53] at Battelle and also from the work of Davies[54] in the U.K., who investigated the effect of this treatment on thin nickel electrodeposits, usually recognised as being unsuitable for exterior service. Both the American and British results demonstrate the benefits that can be achieved by the use of this post-chromium plating treatment, although the film does slowly lose its initial efficacy with time. The greatest improvements given by this treatment are obtained when bright nickel plus conventional chromium is the electroplated coating; much smaller benefits are observed if, for example, micro-cracked chromium is present, since this type of coating already affords such substantial improvements in corrosion performance that any additional benefit is not so noticeable. Chromate films of this type are claimed to enhance the adherence of paint to all chromium plated surfaces on which paint is otherwise often poorly adherent.

The chromium oxide films are cathodically precipitated from solutions containing 50 g/l of sodium dichromate ($Na_2Cr_2O_7.2H_2O$) and 1 g/l of chromic sulphate $[Cr_2(SO_4)_3]$ maintained at pH 2·5. This treatment is performed subsequent to rinsing after chromium plating but before the articles are dried. A current density of 0·3 A/dm^2 is employed for approximately 1 min using a solution temperature of 95°C, followed by standard water rinsing and drying off.

CORROSION RESISTANCE OF DEFORMED NICKEL PLUS CHROMIUM COATINGS

Electrodeposited chromium is brittle and inevitably cracks on slight deformation; nickel however can be deposited in either a ductile or brittle state, the former being usually dull and the latter bright. Clearly the ductile form would be preferred if the coatings were likely to undergo deformation; cracks formed in the chromium layer are less likely to propagate into a ductile nickel undercoat than they are into a brittle nickel layer. Attempts have been made to evaluate the corrosion resistance of deformed nickel plus chromium coatings, but the results vary in detail, probably because deformation has been carried out in different ways. Flint[55] used a technique in which a round, tapered bar was pressed to a constant depth of 5 mm into the plated surface, near the centre of a test panel and parallel to one of its edges. The present authors have used two techniques, i.e. indentation and extension. The indentation test[7] involved the use of the Hounsfield cupping equipment as used in an Olsen or Erichsen test. A standard depression 1·6 mm deep was made in the test panels, but in one series the indentation was made in the plated surface and in the other it was made in the opposite (unplated) side. The Hounsfield Tensometer was employed to obtain deformation by extension of steel strip-type test pieces plated with various nickel plus chromium coatings[56]. Outdoor and accelerated tests were used to provide the subsequent corrosive environment.

In the work carried out by Flint, polished semi-bright nickel plus micro-cracked chromium proved the most satisfactory system, but either double layer nickel or copper plus bright nickel both overlaid with conventional decorative chromium, also gave reasonably good performance. Crack-free chromium was most unsatisfactory, since not only did this thick and brittle layer crack easily, but it also increased the tendency for crack penetration into the underlying layers. The semi-bright nickel coatings were the only

Table 11.10 ELONGATION BEFORE PROPAGATION OF CRACKS THROUGH COATING TO SUBSTRATE*

Type and thickness of coating		% Elongation at which rust was first observed after exposure for 8h in acetic acid–salt spray cabinet
Nickel	Chromium	
Polished Watts, 30 μm	None	30
,,	Crack-free, 0.75 μm	4.5
,,	Conventional bright, 0.25 μm	5.0, rust only at isolated pores
,,	Micro-cracked, 0.75 μm	5.0, rust only at isolated pores
Bright, 30 μm	None	1.0
,,	Crack-free, 0.75 μm	0.2
,,	Conventional bright, 0.25 μm	0.2
,,	Micro-cracked, 0.75 μm	0.5, some original cracks opened up
Semi-bright, 30 μm	None	15
,,	Crack-free, 0.75 μm	1.0, rust in edge cracks only
,,	Conventional bright, 0.25 μm	2.5
,,	Micro-cracked, 0.75 μm	>5.0
Polished semi-bright, 30 μm	Micro-cracked, 0.75 μm	>2.5 Nickel corrosion only at 2.5%
Duplex, 30 μm	None	2.0
,,	Crack-free, 0.75 μm	0.2
,,	Conventional bright, 0.25 μm	0.4
,,	Micro-cracked, 0.75 μm	0.9, rust in edge cracks only

* Data taken from Dennis [56]

ones which did not crack at the time of deformation and these were found to have been reduced in thickness in the region of the indentation. On semi-bright nickel undercoats, the crack density which occurred in conventional chromium after deformation was about 200 cracks/mm while in micro-cracked chromium it was only about 100 cracks/mm, an increase of about 40/mm over the initial crack density. The original cracks in micro-cracked chromium widen in preference to new crack formation. After exposure to a corrosive environment, loss of lustre occurred in the deformation region of panels plated with semi-bright nickel plus decorative chromium. Numerous small pits were formed, the corrosion mechanism presumably being somewhat similar to that pertaining to any nickel plus micro-cracked chromium system. Bright nickel coatings cracked readily, the number of cracks penetrating to the substrate being greatest when thick, initially crack-free chromium was used as the top coat. In the case of double layer nickels plus

Table 11.11 TYPE AND FREQUENCY OF CRACK FORMATION ON EXTENSION*

Type of coating		% Elongation	Type of cracks and number per millimetre after extension
Nickel	Chromium		
Polished Watts	Crack free	1	80P, cracks in chromium only
,,	Crack free	5	80P
,,	Conventional bright	0·5	80P, cracks in chromium only
,,	Conventional bright	5	140P
,,	Micro-cracked	1	80M
,,	Micro-cracked	5	80M, severe opening up and propagation of cracks
Bright	Crack free	0·2	5P
,,	Crack free	1	5P
,,	Conventional bright	0·2	2P
,,	Conventional bright	1	3P
,,	Micro-cracked	0·5	120M
,,	Micro-cracked	1	120M
Semi-bright	Crack free	1	80P, cracks in chromium only except at edge
,,	Crack free	5	100P
,,	Conventional bright	1	160P
,,	Conventional bright	2·5	160P
,,	Micro-cracked	1	80D
,,	Micro-cracked	5	120D
Duplex	Crack free	0·2	20P
,,	Crack free	1	20P
,,	Conventional bright	0·4	14P
,,	Conventional bright	1	14P
,,	Micro-cracked	0·5	100M
,,	Micro-cracked	1	100M

P Parallel cracks at right angles to the direction of the load.
D Discontinuous cracks approximately at right angles to the direction of the load.
M Opening up of some original micro-cracks
* Data taken from Dennis[56]

conventional chromium, most of the cracks halted at the interface between the ductile semi-bright and the brittle bright nickel. In the most severely deformed region, cracks had propagated by the notch effect through the whole of the nickel coating, in spite of the high ductility of the semi-bright nickel; basis metal was therefore exposed. When the substrate was more electrochemically active than the nickel coatings, examination of cross sections of pits revealed that corrosion took the form of hemispherical pits in the substrate at the bottom of cracks in the coating.

The work by Dennis and Such[7] using the indentation attachment on the Hounsfield Tensometer also indicated that ductile nickel undercoats and micro-cracked chromium were the types of deposit which gave the most satisfactory performance and that crack-free chromium was the most disastrous. However, the improvement shown by the former coatings was not as marked as in Flint's experiments. Watts nickel plus decorative chromium gave the best results, but micro-cracked chromium did not perform as well as expected. A few of the original zone boundaries in the crack pattern opened up, but the intervening block of micro-cracked chromium remained undeformed; the strain was not taken up in small increments at each crack. Safranek and Faust[29] have shown that composite coatings of copper plus double layer nickel plus micro-cracked chromium

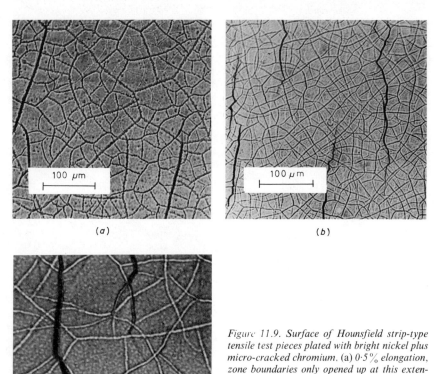

Figure 11.9. Surface of Hounsfield strip-type tensile test pieces plated with bright nickel plus micro-cracked chromium, (a) 0·5% elongation, zone boundaries only opened up at this extension, (b) 1% elongation and (c) 1% elongation (after Dennis[56])

exhibit better ductility than do bright nickel plus micro-cracked chromium coatings. Cracking, as a result of deformation using a ball indenter (15 mm d), occurred in the latter case at load levels that were resisted completely by the copper plus double nickel plus micro-cracked chromium coating.

Indentation of a plated panel from the unplated side was a particularly severe test, local extension of the coating taking place on the convex surface. When the indenter was applied to the plated side, compressive forces were operative at the bottom of the indentation. Fortunately, this denting is the type of deformation most likely to result from accidents occurring to plated articles and in particular to components of motor vehicles; however, flexing of components leads to both tensile and compressive stresses. The notch effect of chromium and the embrittling effect of hydrogen were illustrated clearly by this technique. When the nickel coatings were indented in the

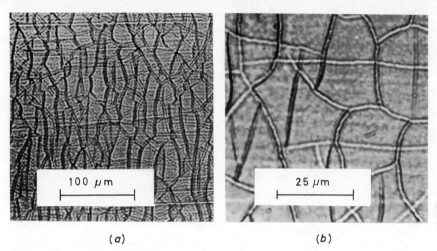

Figure 11.10. Surface of Hounsfield strip-type tensile test pieces plated with polished semi-bright nickel plus micro-cracked chromium, after 2·5% elongation (after Dennis[56])

absence of chromium, cracks were not visible in any type of coating, whether on the convex or concave surface. However, if chromium was deposited and subsequently stripped off anodically in sodium hydroxide before indentation, cracking still occurred in several instances unless the nickel was baked to remove the absorbed hydrogen.

Deformation by extension[56] showed the same general trend of performance as did the indentation tests. The ductility of composite nickel plus chromium coatings was usually increased when micro-cracked chromium was used as the overlay. When a ductile nickel, such as Watts nickel, was used as the undercoat, the ductility of the composite coating was approximately the same whichever type of chromium was used. Double layer nickel undercoats were not shown to be too successful by this technique. Table 11.10 shows the elongation necessary before cracks propagated to the substrate. For the purpose of this investigation, the limit of ductility was assumed to be the point at which cracks did propagate to the substrate, this being indicated by the presence of red rust from the substrate after eight hour's exposure in an acetic acid–salt spray cabinet.

Cracks at right angles to the direction of application of the load formed in all cases where crack-free and decorative chromium top coats were deposited. The frequency of cracking was greater on ductile nickel undercoats than on brittle bright nickel, and in the latter case there was a greater tendency for penetration to the substrate. Frequency of cracking after specified elongation is shown in Table 11.11. On deformation of bright nickel plus micro-cracked chromium, only certain favourably oriented cracks opened up (Fig. 11.9). The zone boundaries were usually the weakest points as shown in the same figure. Induced cracks predominantly followed original cracks, but could propagate into initially uncracked regions; however, all cracks were discontinuous, unlike those occurring in crack-free and decorative chromium deposits. Zoning does not occur when micro-cracked chromium is deposited on polished ductile nickel (Watts or semi-bright) and the crack pattern formed on extension was different from that occurring in micro-cracked chromium on bright nickel. The cracks were finer and closer together as shown in Fig. 11.10; this again resulted in less chance of propagation to the substrate.

All the experimental work reported emphasises the need to choose a suitable coating, if a component is liable to sustain deformation in service and be exposed to a corrosive environment. Micro-cracked chromium, preferably over a ductile or double layer nickel, is likely to provide the most satisfactory service. Results reported[45] for the behaviour of micro-porous chromium indicate that this coating may behave similarly to micro-cracked chromium since the porous layer also contains many points of weakness.

REFERENCES

1. DENNIS, J. K., section in *Proc. Symposium on Decorative Nickel-Chromium Plating*, Institute of Metal Finishing, London (1966)
2. NICHOLSON, A. H., Chapter 3, and DENNIS, J. K., Chapter 4 of *Quality Control in Metal Finishing*, G. ISSERLIS (Editor), Columbine Press, Manchester (1967)
3. *Electroplated Nickel or Nickel Plus Chromium Coatings on Steel, Copper or Copper Alloy (Including Brass) Threaded Components*, BS 3382: Parts 3 and 4 (1965)
4. *Electroplated Coatings of Nickel and Chromium*, BS 1224: 1970
5. *Measurement of Coating Thicknesses by the Magnetic Method: Electrodeposited Nickel Coatings on Magnetic and Non-magnetic Substrates*, A.S.T.M. Method B530-70
6. EDWARDS, J., *Trans. Inst. Metal Finishing*, **35**, 55 (1958); *Proc. Amer. Electroplaters' Soc.*, **46**, 154 (1959)
7. DENNIS, J. K. and SUCH, T. E., *Trans. Inst. Metal Finishing*, **40**, 60 (1963)
8. KESTERNICH, W., *Stahl u. Eisen*, **71**, 587 (1951)
9. DIN 50 018-63
10. NIXON, C. F., THOMAS, J. D. and HARDESTY, D. W., *Proc. Amer. Electroplaters' Soc.*, **46**, 159 (1959)
11. *Copper-accelerated Acetic Acid–Salt Spray (Fog) Testing (CASS Test)*, A.S.T.M. Method B368-68
12. SUKES, G. L., *Metal Finishing*, **57** No. 12, 59 (1959)
13. *Corrosion Testing of Decorative Chromium Plating by the Corrodkote Procedure*, A.S.T.M. Method B.380-65
14. BIGGE, D. M., *Proc. Amer. Electroplaters' Soc.*, **46**, 149 (1959)
15. SAUR, R. L., *Product Finishing*, **29** No. 11, 142 (1965)
16. SAUR, R. L. and BASCO, R. P., *Plating*, **53**, 35 (1966)
17. KWAPISZ, R. A., *Product Finishing*, **34** No. 12, 74 (1970)
18. STANNERS, J. F., *Br. Corros. J.*, **5**, 117 (1970)
19. GOETHNER, G. A., *Metal Finishing*, **68** No. 5, 45 (1970)
20. HEILING, H. M., *Metall.*, **14**, 549 (1960)
21. KOJUCHAROV, W. K., *Werkstoffe Korros.*, **21**, 573 (1970)
22. SAFRANEK, W. H., MILLER, H. R. and FAUST, C. L., *Proc. Amer. Electroplaters' Soc.*, **40**, 133 (1959)
23. EDWARDS, J. and CARTER, V. E., *Trans. Inst. Metal Finishing*, **40**, 48 (1963)
24. MELEKIAN, C., TIERNEY, J. and LINDSAY, J. H., *Metal Progress*, **84**, 167 (1963)
25. BLUM, W., *Nickel/Chromium Plating Corrosion Data*, International Nickel Co., London (1962)
26. LOWENHEIM, F. A., *Plating*, **52**, 121 (1965)
27. *Standard Recommended Practice for Rating of Electroplated Panels Subjected to Atmospheric Exposure*, A.S.T.M. Method B537-70

28. NASEA, J., TIFFANY, B. E. and BUSH, G. F., *Plating*, **49**, 989 (1962)
29. SAFRANEK, W. H. and FAUST, C. L., *Trans. Inst. Metal Finishing*, **40**, 217 (1964)
30. *Evaluation of Results of Accelerated Corrosion Tests on Metallic Coatings*, BS 3745:1970
31. SAUR, R. L., *Plating*, **52**, 663 (1965) and **58**, 1075 (1971)
32. SAFRANEK, W. H., MILLER, H. R., HARDESTY, R. W. and FAUST, C. L., *Proc. Amer. Electroplaters' Soc.*, **47**, 96 (1960)
33. MILLAGE, D., ROMANOWSKI, E. and KLEIN, R., *Proc. Amer. Electroplaters' Soc.*, **49**, 43 (1962)
34. SAUR, R. L. and BASCO, R. P., *Plating*, **53**, 981 (1966)
35. SEYB, E. J., *Plating*, **54**, 1135 (1967)
36. DU ROSE, A. H., *Trans. Inst. Metal Finishing*, **42**, 57 (1964)
37. KUBACH, G., *Galvanotechnik*, **61**, 788 (1970)
38. WITTROCK, H. J., Chapter 9 in *The Finishing of Aluminium*, G. H. KISSIN (Editor), Reinhold Publishing Corp., New York (1963)
39. SUCH, T. E. and WYSZYNSKI, A. E., *Plating*, **52**, 1027 (1965)
40. LA QUE, F. L., *Trans. Inst. Metal Finishing*, **41**, 127 (1964)
41. BUSH, G. F., *Automotive Engineering Congress*, Preprint 650A Society of Automotive Engineers (1963)
42. SAMPLE, C. H., Section in *Properties, Tests and Performance of Electrodeposited Metallic Coatings*, A.S.T.M. Special Technical Publication No. 197 (1956)
43. TURNER, P. F., *Product Finishing*, **19**, 61 (1966)
44. DAVIES, G. R., *Electroplating and Metal Finishing*, **21**, 3 (1968)
45. CLAUSS, R. J. and KLEIN, R. W., *Proc. 7th International Metal Finishing Conference*, Hanover, 124 (1968)
46. BACHE, H. J. and TURNER, P. F., *Electroplating and Metal Finishing*, **20**, 312 (1967)
47. TURNER, P. F. and MILLER, A. G. B., *Trans. Inst. Metal Finishing*, **47**, 50 (1968)
48. CARTER, V. E., *Trans. Inst. Metal Finishing*, **48**, 19 (1970)
49. FLINT, G. N. and MELBOURNE, S. H., *Trans. Inst. Metal Finishing*, **38**, 35 (1961)
50. SAFRANEK, W. H., HARDY, R. W. and MILLER, H. R., *Proc. Amer. Electroplaters' Soc.*, **48**, 157 (1961)
51. MELBOURNE, S. H. and FLINT, G. N., *Trans. Inst. Metal Finishing*, **39**, 85 (1962)
52. SAFRANEK, W. H., MILLER, H. R. and FAUST, C. L., *Plating*, **50**, 507 (1963)
53. SAFRANEK, W. H. and MILLER, H. R., *Plating*, **52**, 873 (1965)
54. DAVIES, G. R., *Electroplating and Metal Finishing*, **21**, 393 (1968)
55. FLINT, G. N., *Trans. Inst. Metal Finishing*, **40**, 98 (1963)
56. DENNIS, J. K., *Electroplating and Metal Finishing*, **18**, 376 (1965)

BIBLIOGRAPHY

PLOG, H. and CROSBY, C. E., *Guide to Coating Thickness Measurement*, Robert Draper Ltd., Teddington (1971)

Chapter 12

Recent Developments

HIGH SPEED NICKEL PLATING

The rate of electrodeposition, being governed as it is by Faraday's law, depends directly upon the current density applied on the cathode, but as already explained in Chapter 2, this current density has a limiting value above which acceptable plate is not obtained. The anion present in the nickel plating bath can affect this limiting current density, chloride and sulphamate ions being markedly beneficial in raising it. The use of nickel chloride solutions has been investigated by Wesley et al.[1] amongst others, claims being made that with a solution velocity of 23 m/min, sound deposits could be obtained at 450 A/dm^2. Sulphamate baths have been described in detail in many papers which Hammond[2] has recently reviewed.

The rate at which the electrolyte solution passes over the surface of the cathode has a considerable effect on the maximum current density at which satisfactory electrodeposits are obtained. This has been recognised for many years and was the reason for the introduction first of cathode movement, usually at the rate of about 0·1 m/s, and then air agitation, as means of providing solution movement over the cathode and thus reducing the thickness of the diffusion layer. The most recent paper on this effect is that by Gabe[3]. The application of ultrasonic energy as a means of agitating electroplating baths has been tried in the laboratory and its effects described[4,5], but this technique does not seem to have been adopted for commercial nickel plating, possibly because it appears that its benefits are little different from those obtained when using violent air agitation.

Other means of speeding up electrolyte solution movement have been tried, such as pumping, paddle rotation and impingement of jets, and some have been used on a production scale. In particular, claims have been made that current densities up to 10 A/dm^2 can be achieved by the use of paddles or impellers which are said to give solution speeds of about 0·5 m/s, even in plating baths of large volume. However, as yet no method has been so revolutionary as to make electroplating fit readily into the modern concept of high speed, continuous and automated production, as carried out in many metal forming shops. However, General Motors did apply the principle of rapid transfer between consecutive operations in what they termed their *Contour High Speed Plating Machine*[6]. In this machine, the total time required for depositing 30 μm of multi-layer nickel onto a car bumper was less than 2 min. The bumpers were mounted on fixtures and passed one at a time through an automatic plant containing 25 successive closed cells, one for each different cleaning, plating and rinsing treatment. In each electrolytic cell, the bumper was placed only 10 mm away from a conforming electrode. Through this gap the electrolyte solution was pumped very rapidly to give solution movement which was said to be 8 m/s when the equipment was first installed, although later it was lowered to 5 m/s. Nickel deposition was at first conducted at about 150 A/dm^2, but at the slower rate of solution movement this was reduced to about 100 A/dm^2. Not only did this technique result in a very high deposition rate but the use of a conforming, insoluble anode meant that the metal distribution was much better than when plating the same bumpers in the conventional manner. A ratio of 2:1 between maximum and minimum thicknesses of nickel plate was achieved, compared with the normal 8:1, thus saving 250 g of nickel on each bumper. The nickel plating solutions were conventional Watts type, but since lead anodes were used the bath did not contain chlorides; even the proprietary brighteners were standard ones. The pH of the nickel solutions fell as their nickel ion concentrations were depleted and so this was restored by additions of nickel carbonate[7], which dissolved readily in the acidic solution (pH 3 or less). Although the chemical and electrochemical operation of the plant was not without difficulties, it was mainly the inability to solve the many engineering problems associated with this machine and its ancillary equipment that led to this laudable pioneering effort being brought to an end after two years. However, this topic is still interesting electrochemists, as indicated by a recent conference[8] in Moscow.

PLATING ON ALUMINIUM

The need to plate aluminium and its alloys arises because their surface characteristics may not be suitable for a particular application, although some properties such as their excellent ratio of strength to weight or electrical conductivity per unit cost may be advantageous. This text is concerned with the application of nickel plus chromium electrodeposits, and aluminium coated with these metals finds many uses. Decorative coatings are applied to cast, stamped and extruded articles for use as furniture fittings, motorcar components, domestic hollow-ware and a variety of other consumer products. Coatings recommended for service in environments of varying degrees of severity are listed in BS 1224[9] and discussed by Bailey[10].

Nickel or chromium electrodeposits are also applied to aluminium substrates for engineering applications to provide wear resistant surfaces. A component can be machined relatively easily and cheaply from aluminium and the properties necessary for the service conditions obtained by depositing a suitable coating. The use of plated aluminium for large components results in a considerable weight saving and this may be essential for certain purposes. Typical applications of coatings for engineering purposes include their deposition on printing rollers, moulds, cylinder liners and spindles for textile machines. One recent and important use of plated aluminium is on the N.S.U. Wankel rotary engine[11], where the nickel deposit includes oxide or carbide particles so as to give a wear resistant composite coating. Direct hard chromium plating of the bores of standard internal combustion engines has been practised for many years. Nickel electrodeposits can also be used to build up worn or over-machined aluminium components. However. other metals are plated onto aluminium substrates for certain applications. The oxide film present on aluminium prevents solder from wetting its surface[12]. and so if soldered connections are to be made to it. the aluminium must be plated with a suitable metal such as tin. Gold is also deposited on aluminium for some electrical components. such as contacts. Another example of the use of aluminium plated for engineering applications is the deposition onto it of lead/tin alloy deposits to serve as bearing surfaces[13].

Aluminium is a relatively cheap metal; in terms of cost per unit volume it is the cheapest non-ferrous metal and the second cheapest of all commercially important metals. Consequently. there is a considerable economic incentive to use aluminium for many applications and a necessity to plate onto it in many instances. Aluminium is an extremely reactive element and always has a naturally occurring oxide film present on its surface. This confers a reasonable degree of corrosion resistance. but is not always adequate for either decorative or protective purposes. The surface can be modified either by the formation of a thick oxide film. as produced by anodising. or by the deposition of a plated coating. Anodised coatings are excellent for many purposes. but have certain shortcomings. The oxide layer is brittle and forms an insulating layer. Reflectivity is reduced unless high purity metal is used and anodised aluminium has limited corrosion resistance in some natural environments.

METHODS OF PLATING ON ALUMINIUM

Attempts have been made to plate aluminium since commercial production of the metal commenced. However. the bond strength achieved by the early procedures was very low and would not be acceptable today. A variety of techniques have been employed and these will be discussed briefly. but the processes of present commercial significance are based on the use of an immersion deposit prior to the application of the electrodeposited coating. The factor which makes the plating of aluminium difficult is the presence of the air-formed oxide film. which forms so quickly that good adhesion is not achieved even if the metal is treated in the conventional manner to remove the oxide and then transferred very rapidly to the plating bath.

Mechanical roughening

This is one of the oldest techniques used to prepare the aluminium surface for plating. Sand and grit blasting have been used but produce a surface which is much too rough for most purposes. The roughened surface permits some 'keying' but usually gives a low bond strength and hence mechanical treatments are rarely used today.

Chemical etching

Chemical etching can be used as an alternative technique to mechanical methods as a means of producing a roughened surface. Some of the acid etchants contain metal ions which form an immersion coating[14]. The object of chemical roughening is to produce etch pits so that the coating can be mechanically keyed to the substrate. The etching conditions are quite critical, but it is claimed that with experience many alloys can be plated successfully. Bond strength may be improved by post-plating heat treatment. Chemical roughening suffers from the same disadvantage as mechanical in that the surfaces produced are so rough that they cannot be smoothed by conventional thicknesses of decorative nickel, even by those possessing levelling properties.

Anodic oxidation

Anodising or deliberate thickening of the oxide layer has been referred to earlier as a means of protecting the surface. A modification of this procedure can be utilised as a basis for electrodeposition. It is well known that the anodic oxide film consists of two layers, the barrier and porous layer. The dimensions of the pores are dependent on the operating conditions and the electrolyte solution employed. Phosphoric acid solutions[15, 16] produce oxides having large pores and these are the type used to prepare the surface for electroplating. Deposition must take place in the pores so that the deposit is keyed into them; problems may arise with alkaline plating solutions if the oxide dissolves before the surface is completely covered with electrodeposit. This can be overcome by depositing an initial layer of copper from a pyrophosphate bath. Various commercial processes have been introduced, but all the phosphoric acid anodising processes are sensitive to alloy composition. No aluminium alloy die-castings can be plated by this method and only some sand casting alloys. Articles of complex shape are also difficult to plate. The use of chemical oxidation has also been investigated in an attempt to eliminate the relatively high cost of the electrical equipment necessary for anodising, but it has not yet proved to be of commercial importance.

Direct plating on aluminium

Most of the methods for plating directly onto aluminium have involved the use of specialised zinc plating baths[17-20], although information has been published on the deposition of chromium[21] and nickel[22] onto aluminium. The process which probably gained the greatest commercial success, particularly in the U.K., was the Vogt process which was first reported by Fyfe[18]. Wallbank[19] has discussed the solutions used and described its application for the plating of holloware. The process sequence of the Vogt method and a brief indication of the solutions used are shown in Table 12.1. It is claimed that most aluminium alloys can be plated by the Vogt process, and its advantage over immersion processes which depend upon a displacement reaction. is that a uniform zinc coating can be applied electrolytically in spite of variation in the surface composition of the substrate. If the surface of the basis metal is not homogeneous. an immersion coating of different thicknesses will be obtained on the matrix and on intermetallic constituents. The zinc layer deposited in the Vogt process is extremely thin (0.5 A/dm^2

Table 12.1 SEQUENCE FOR VOGT ELECTROPLATING PROCESS FOR NICKEL PLATING ALUMINIUM*

1. Cathodic alkali clean at room temperature using approx. 7V for 4 min.
2. Water rinse.
3. Dip in a mixture containing equal volumes of concentrated nitric and sulphuric acids at room temperature for 4s.
4. Water rinse.
5. Cathodic alkali clean at room temperature using approx. 7V for 20s.
6. Water rinse.
7. Zinc plate at 0.5 A/dm^2 for 20s at room temperature in a solution containing:
   zinc chloride. 0.5 g/l   sodium cyanide. 0.5 g/l
   sodium hydroxide. 10.5 g/l
8. Brass plate at 1 A/dm^2 for 8s at 30°C in a solution containing:
   copper acetate. 13 g/l   zinc chloride. 13 g/l
   sodium cyanide. 31 g/l   sodium carbonate. 9 g/l
   sodium bisulphite. 13 g/l
9. Water rinse.
10. Nickel plate at 1.5 A/dm^2 in Watts type solution containing magnesium sulphate.
11. Water rinse.
12. Heat treat at 230°C for 30 min.
13. Polish nickel. if a bright finish is required.

* Data taken from Wallbank[19].

for 20 s) and this is followed by a thin brass electrodeposit (1 A/dm^2 for 8 s). Finally. a dull nickel layer is deposited from a solution containing nickel sulphate. magnesium sulphate. sodium chloride and boric acid. Good adhesion is achieved only if the plated articles are then stoved for about 30 min at 230°C to cause interdiffusion to occur. If good adhesion has not been obtained. blistering will result at this stage. After stoving. the dull nickel must be nickel finished and reactivated before chromium plating. This sequence of operations does not lend itself to automation. Unjigging is necessary after nickel plating and two quite expensive operations remain to be carried out. i.e. heat treatment and polishing of the nickel. Advantage cannot be taken of bright nickel plating since a special nickel plating bath is necessary. Problems can also arise with the thin zinc and brass coatings deposited from unusual solutions. A variation of the Vogt process was introduced by Ore[20]. This made use of somewhat different cleaning and plating solutions and required two stoving treatments. one after nickel and the other after chromium plating. together with appropriate polishing of both deposits.

Immersion deposits

The use of immersion films on aluminium prior to electrodeposition of metallic coatings is the most widely used technique, as discussed in Vanden Berg's review paper[23]. A number of metal salts have been employed to give these galvanically deposited coatings, but present proprietary formulations are based on either zincate or stannate solutions. The use of the former process and the essential pretreatments are described in detail in the A.S.T.M. *Recommended Practice for Plating on Aluminium and its Alloys*[24].

The use of alkaline solutions for the deposition of immersion coatings of zinc and aluminium prior to electrodeposition was patented by Hewitson[25] as early as 1927, and in 1939 Korpium[26] patented a further development of the technique. The simplest solutions contain only sodium zincate and excess sodium hydroxide. Saubestre and Morico[27] have claimed that a concentrated solution containing 500 g/l of sodium hydroxide and 100 g/l of zinc oxide is preferable to a more dilute solution containing 125 g/l of sodium hydroxide and 25 g/l of zinc oxide, due to the deposits obtained from the more concentrated solution being thinner and more compact. Wyszynski[28] has compiled a graph showing the variation of the rate of film growth against time of immersion for zincate solutions of different formulations. These results, taken from the work of various authors, show some discrepancies, but the general trend was for faster growth from the simple solutions. Bailey[29] has also reported that dilute zincate solutions give coarse, thick deposits resulting in poor adhesion, whereas more concentrated solutions produce finer grained and more compact deposits giving much better adhesion.

Due to the inadequacies of immersion deposits from simple zincate solutions, many attempts have been made to incorporate other cations, anions and complexing agents in the solution in order to modify the growth mechanism[30-34]. To provide a high bond strength it is desirable that nucleation should take place rapidly at many centres but that subsequent crystal growth should be inhibited. Wyszynski[28] has listed nine requirements which must be satisfied in order that a commercial solution may be acceptable; the most important of these are as follows:

1. The solution must produce good, adherent zinc deposits on any aluminium alloy.
2. The deposit produced must be thin, yet coverage of the surface must be complete. Dendritic growth must be discouraged and rapid nucleation encouraged.
3. The solution should be dilute to minimise drag-out and carry-over problems, and also to assist its penetration into blind holes in complex components.
4. The solution must be economical to prepare and operate. Therefore, not only must the initial material costs be low but the solution should be stable both when in use or when standing idle.

Zinc alloy immersion deposits

Such and Wyszynski[30] have investigated the effect of operating conditions on the rate of film formation from an alloy-type solution. Fig. 12.1 shows the effect of temperature on the rate of film formation, and it is apparent that the type of alloy being plated has a significant effect. It can be seen that the behaviour of commercial aluminium is different from that of the alloys in that for the former the temperature coefficient decreases with temperature, whereas for the latter it increases with temperature. The rate of film formation

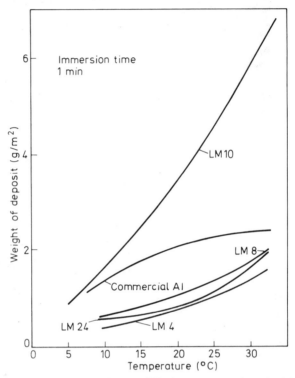

Figure 12.1. *Effect of temperature on the rate of film formation on various aluminium alloys using an alloy-type zincate solution. Percentage of main alloying elements in alloys coated:*

Alloy	Copper	Magnesium	Silicon
LM 4	2–4	0·15	4–6
LM 8	0·1	0·3–0·8	5–6
LM 10	0·1	9·5–11	0·25
LM 24	3–4	0·1	7·5–9·5

(after Such and Wyszynski[30])

depends on the alloy composition which in turn determines its electrochemical potential with respect to zinc. Optical microscopy has indicated that the immersion coating is not of uniform thickness on aluminium alloys; in Fig. 12.2 the light areas in the region of the intermetallic constituents suggest a thin coating. This can be confirmed by electron probe microanalysis as shown in Fig. 12.3. The zinc coating is extremely thin or completely absent in the region of the silicon particles. Fig. 12.4 shows scanning electron micrographs of immersion deposits on commercial aluminium and on an aluminium alloy.

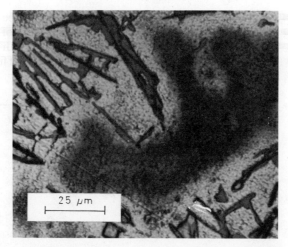

Figure 12.2. Optical photomicrograph of the surface of alloy LM 4 after immersion in an alloy-type zincate solution. The light areas in the region of the intermetallic constituents suggest the presence of a thin immersion coating (after Such and Wyszynski[30])

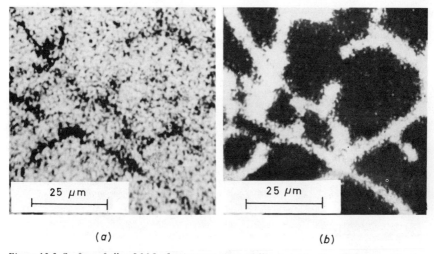

(a)   (b)

Figure 12.3. Surface of alloy LM 8 after immersion in an alloy-type zincate solution; examination carried out using an electron probe microanalyser. Light areas indicate high concentrations of the elements in question. (a) Zinc X-ray image showing that very little zinc is present in the region of the silicon and (b) silicon X-ray image (after Such and Wyszynski[30])

The mechanism of zinc immersion processes is reviewed by Wernick and Pinner[14], as are the variations in treatment that have been found beneficial. These variations are often employed in commercial practice, particularly when plating onto aluminium alloys; the most common of these is known as the *double zincate dip*. This technique was first devised by Korpium[26] and involves the dissolution of the first zinc coating in nitric acid and the formation of a second zinc coating in either the same or a different zincate solution. This procedure results in the deposition of zinc coatings having a fine-grained compact nature. which are the most satisfactory type. Table 12.2 gives a typical plating process using the zinc immersion coating method. However. the acid etching or pickling procedures vary according to the main constituent of the alloy (i.e. magnesium. silicon or copper) as described

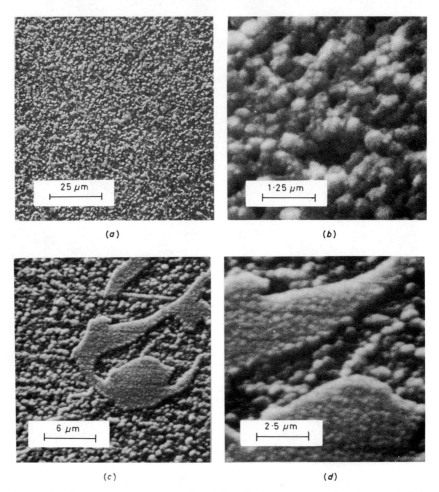

Figure 12.4. Scanning electron micrographs of zinc alloy immersion deposits on commercial aluminium and alloy LM 8. (a) *and* (b) *deposit on commercial aluminium;* (c) *and* (d) *deposit on alloy LM 8 (after Arrowsmith. Dennis and Fuggle)*

in the literature[14]. When simple zincate solutions are employed. special electroplating baths may be necessary. It should be emphasised that standard nickel baths are not satisfactory, the high pH, dull nickel type being preferred where direct deposition of nickel is required. More frequently. nickel is plated in the dull or bright form over a copper undercoat deposited from Rochelle or pyrophosphate baths. However, where the type of zinc alloy immersion deposit containing nickel and copper is used. as described by Such and Wyszynski[30]. deposition directly onto this from normal commercial

Table 12.2 SEQUENCE FOR SIMPLE ZINCATE IMMERSION PROCEDURE FOR NICKEL PLATING ONTO ALUMINIUM

1. Trichlorethylene degrease (if necessary).
2. Alkali soak clean at 65°C for 2 min.
3. Water rinse.
4. Dip in 50% (v/v) nitric acid* at room temperature for 30s.
5. Water rinse.
6. Dip in zincate solution at room temperature for 2 min.
7. Water rinse.
8. Copper plate using Rochelle cyanide or pyrophosphate solution.
9. Water rinse.
10. Nickel plate using bright or dull plating solution.

* Other acid mixtures may be used for alloys of aluminium.
Note. The use of a double zincate dip is often advantageous. This necessitates the repetition of stages 4 to 7, employing shorter immersion times.

Table 12.3 SEQUENCE FOR ALLOY ZINCATE IMMERSION PROCEDURE FOR NICKEL PLATING ONTO ALUMINIUM†

1. Trichlorethylene degrease (if necessary).
2. Cathodic alkali clean at room temperature for 2 min.
3. Water rinse.
4. Dip in 50% (v/v) nitric acid* at room temperature for 1 min.
5. Water rinse.
6. Dip in alloy zincate solution at room temperature for 2 min.
7. Water rinse.
8. Nickel plate using bright or dull plating solution.

* Other acid mixtures may be used for alloys of aluminium.
† Data taken from Such and Wyszynski[30].
Note. The use of a double alloy zincate dip is often advantageous. This necessitates the repetition of stages 4 to 7 employing shorter immersion times.

bright nickel plating baths gives an adherent coating of good appearance. A typical processing sequence is listed in Table 12.3.

Stannate process

Immersion coatings of tin have been deposited on aluminium from various types of solutions[35, 36] in addition to stannate baths, although the latter have proved the most successful. Bryan[37] has discussed the use of stannate solutions and described the plating procedure. Aluminium pistons have been plated using these solutions to provide a coating having lubricating properties.

Stannate solutions can also be employed as a means of enabling electrodeposited coatings to be soundly bonded to aluminium substrates[38]. A typical sequence of operation for a current commercial process is shown in Table 12.4. The acid dip is a particularly important stage since it should result in the formation of a uniform oxide layer on all kinds of aluminium alloys. The composition of the acid dip must be formulated to treat the particular alloy which is to be plated. In many instances nitric acid is satisfactory, but it may be necessary to add hydrofluoric acid to it in order to process some alloys such as those containing manganese or silicon. After acid dipping the work is swilled and transferred to the stannate solution. Jongkind[39] has stated that a continuous layer of tin is not deposited using the commercial process in question but that an electrodeposited bronze layer is plated directly onto the aluminium, not on top of a layer of tin. He indicates that the main reactions can be represented by the following equations:

$$Al_2O_3 + 2KOH \rightarrow 2\ KAlO_2 + H_2O$$

$$2Al + 2KOH + 2H_2O \rightarrow 2\ KAlO_2 + 3H_2$$

The first equation represents the dissolution of the oxide layer formed during acid dipping and the second the dissolution of aluminium with the liberation of hydrogen. It is believed that the reaction between aluminium and stannate to form an immersion deposit of tin only takes place to a very limited extent. After immersion in the stannate solution, work is transferred without swilling to a special bronze plating solution, electrical contact being made before entry into the bath.

It has been claimed that the stannate process is superior to the simple zincate one as a means of permitting the deposition of copper plus nickel

Table 12.4 SEQUENCE FOR STANNATE IMMERSION PROCEDURE FOR NICKEL PLATING ONTO ALUMINIUM†

1. Trichlorethylene degrease (if necessary).
2. Alkali soak clean at 60°C for 1 min.
3. Water rinse.
4. Dip in 30% (v/v) nitric acid solution* at room temperature for 30 s.
5. Water rinse.
6. Dip in stannate tin solution at 30°C for 30 s.
7. Plate in special bronze alloy deposition solution at 30°C for 3 min at 5 A/dm².
8. Water rinse.
9. Nickel plate using bright or dull plating solution.

* Other acid mixtures may be used for alloys of aluminium.
† Data taken from Jongkind[39].

plus chromium coatings on aluminium. When the simple zincate process is used, lateral attack of the zinc layer occurs resulting in blisters and exfoliation on corrosion. This is reputed not to occur when the stannate process is employed. However, modern solutions which result in the deposition of alloy zinc deposits can be used quite satisfactorily as a means of plating commercial aluminium and many of its alloys with copper plus nickel plus chromium or nickel plus chromium coatings which provide good corrosion protection in outdoor environments. For the most satisfactory results, duplex nickel or a chromium layer containing discontinuities should be used.

A new process for direct plating onto aluminium has been reported recently by Beyer[40], but no indication of any solution formulations have been given, although the strike bath is probably based on a bronze solution. The most essential features of the process are the use of a special electrolytic strike solution followed by transfer without swilling to a copper bath, the work being transferred 'live' into the latter. It is claimed that this process can be used where immersion processes fail, and an example has been quoted to illustrate its use for joining copper and aluminium tubes to make a connector device.

Chromium plating on aluminium

Chromium is plated onto nickel or other suitable electrodeposited undercoats, but as already mentioned, direct chromium plating onto aluminium is frequently undertaken on certain components. Several of the pretreatments used for nickel plating are satisfactory prior to chromium deposition. A wet blasting technique is claimed to be successful for preparing the surface of aluminium alloys for hard chromium plating. A slurry of quartz crystals and water is employed and, after abrading, the articles are transferred to the plating vat without washing the slurry from their surface. Oxidation of the surface before plating is prevented by the wet film. The most popular pretreatment before hard chromium plating is said to be chemical etching combined with galvanic deposition of heavy metals, e.g. nickel or manganese. This is carried out by dipping the aluminium in acidified solutions of metal salts. However, the use of zincate immersion processes for preparation of aluminium for the deposition of hard chromium has also been described; alloy zinc deposits have been found particularly applicable for this purpose.

PREPARATION OF UNCOMMON METALS TO BE NICKEL PLATED

Many metals other than the common and relatively cheap ones are nickel plated in comparatively small quantities for specific applications where their desirable properties such as lightness or high melting-points are advantageous and justify their extra cost. However, some of these less common metals have little resistance to corrosion or oxidation while others may have other inferior surface properties, such as poor wear resistance or a tendency to gall. To obtain an adherent coating of nickel or other metals on top of these substrates may present considerable difficulties, particularly if they come into the category of refractory metals, typical examples of these being titanium, tungsten, molybdenum and tantalum. Many widely different methods for dealing with these refractory metals have been devised, and the position at the end of the 1960s is well described in two publications[41,42]. More recent work on each metal is summarised in succeeding paragraphs. However, magnesium is probably the most important and the most frequently plated metal, apart from the common ones. It presents even more problems than does aluminium, since it oxidises far more readily, and much work has been carried out in attempts to overcome this difficulty.

Magnesium

The A.S.T.M. recommended practice for preparation of magnesium and its alloys for electroplating[43] describes two processes. One employs a zinc immersion deposit obtained from a pyrophosphate bath, while the other makes use of an electroless nickel/phosphorous alloy (see later section) deposited from specially formulated solutions. Both techniques were devised by the Dow Chemical Co.[44, 45], as were the essential prior pickling operations, these pickles being most frequently based on chromic acid and often containing fluorides.

The zinc immersion process makes use of a solution containing zinc sulphate and tetrasodium pyrophosphate with sodium carbonate to adjust the pH to between 10 and 11, but it also always contains a source of fluoride ion as a vital ingredient and it is significant that so do the recommended electroless nickel solutions[46, 47]. Review papers have been published fairly recently in the U.S.A.[48] and France[49].

Titanium

Because of its high strength to weight ratio, this metal has recently achieved widespread application, in spite of its tendency to seize when in contact with itself or other metals. Titanium is a difficult metal on which to produce adherent electroplated coatings due to the presence of a tenacious oxide film, and many widely differing techniques have been described for pretreatment of this metal and its alloys. Three methods are advised by the A.S.T.M. in their recommended practice[50]. These are respectively: a chemical etch in a solution containing sodium dichromate and hydrogen fluoride, an anodic etch in a solution of hydrofluoric acid in ethylene glycol containing only a small quantity of water and mechanical etching by liquid abrasive blasting. A bibliography attached to this Recommended Practice is a useful introduction to the extensive literature on this topic, but other papers[51-55] should also be consulted. It is noticeable that a common feature of most of the solutions used for chemical or anodic etching is that they contain chlorides or fluorides no matter how dissimilar the remainder of the process sequences are.

Tungsten

Although the A.S.T.M. have considered it justified to issue a recommended practice for plating of this metal[56], few other references are available[57]. The recommended practice states that two different techniques can be used, either electrochemical etching in hydrofluoric acid solution using a.c. or the electrodeposition of a chromium 'strike' plate after pickling in a mixture of hydrofluoric acid and nitric acids; both can be followed by a 'strike' from an acid nickel solution. Seegmiller et al.[57] claim that either striking in acid cobalt or nickel baths is satisfactory, provided it is preceded by an anodic etch in the same bath, or a strongly alkaline solution.

Molybdenum

This metal can suffer catastrophic oxidation at elevated temperatures and its protection from this by plating and methods for doing so have been described[53, 57]. One technique[58] for obtaining adherent nickel plate involves the application of an initial coating of chromium from a standard bath onto a previously etched surface. The etching can be done in several ways, but anodic treatment in a mixture of concentrated sulphuric and phosphoric acids is preferred. After conventional acid activation of the chromium, a very thick coating of nickel is deposited.

Tantalum

Because of its stable and protective oxide film, tantalum is very corrosion resistant and normally there is no need to apply other metals onto it, although its treatment has been mentioned in general papers dealing with difficult-to-plate metals[41, 42]. However, the nickel plating of tantalum may be required in order to confer solderability on its surface and one paper[59] has recently dealt with this topic in detail. Using anodic/cathodic etching in Seegmiller's[57] almost non-aqueous methanol solution of hydrochloric and hydrofluoric acids as the pretreatment, a very thin layer of nickel is deposited from a Watts bath and then diffused into the tantalum by heating in vacuo at temperatures between 500° and 650°C in order to form a diffused layer of nickel-tantalum, onto which is then deposited a thicker layer of nickel. Finally, a heat treatment at 450°C is carried out, again in vacuum.

Uranium

Recently, the protection of this metal by nickel has aroused some interest and methods for doing so have been described, together with the results of corrosion tests. One technique depends on the chemical etching of the uranium in a nitric acid solution containing nickel chloride[60] while another involves an anodic etch in a mixture of concentrated phosphoric and hydrochloric acids in order to obtain good adhesion[61].

ELECTROLESS NICKEL

PHOSPHORUS ALLOYS

The process termed *electroless* nickel plating was rediscovered by Brenner and Riddell in 1944, and their subsequent investigations were published in 1946 and 1947[62]. They observed that when electrodepositing nickel from a bath containing sodium hypophosphite, the cathode current efficiency was much greater than 100% and it was then found that in addition to the nickel being electrodeposited, additional metal was being plated out by means of a chemical reduction reaction which supplied the necessary electrons. It is of interest to note that the principle of this reduction was first noted by Wurtz[63] in 1845. However, although Roux[64] patented its use in 1916 as a nickel coating method, it was not exploited until a century later.

One important feature of this hypophosphite reduction of nickel ions is that only certain metals provide the necessary catalytic effect to initiate deposition. Fortunately, all those in Group VIII of the Periodic Table, including nickel, have this property. Therefore, once an initial coating of nickel has been established, the electroless process is self-perpetuating and thus has been better described as *auto-catalytic*. This differentiates this deposition reaction, which theoretically imposes no limit on the thickness of metal that can be deposited from those processes which operate only by a displacement reaction leading to thin 'immersion' coatings, which are due to the fact that the metal to be coated has a more negative potential than the potential of the metal ions of the coating metal. Chemical reductants other than sodium hypophosphite, in particular borohydrides, alkyl amine borane compounds and hydrazine, have been found to produce nickel coatings. Although the boron derivatives are utilised for production purposes, it is interesting to note that by far the greatest volume of modern electroless nickel baths are derived directly from those originally devised by Brenner and Riddell. Detailed descriptions of their state of development and applications at the end of the 1950s are given in two books[65, 66] and a comprehensive review paper[67]. One of the most important points that widened the usage of this process was the finding that non-catalytic metals, such as copper and copper alloys, could be electroless nickel plated satisfactorily, provided they were first coated with a thin layer by either making then momentarily cathodic in the electroless nickel bath or else bringing them into contact with a catalytic metal such as steel.

The mechanism involved in this hypophosphite reduction has given rise to much speculation, for although the reduction can be represented by the following equation:

$$Ni^{2+} + (H_2PO_2)^- + H_2O \rightarrow Ni + 2H^+ + H(HPO_3)^- \quad \ldots(12.1)$$

the details of how this reaction occurs have not yet been finalised[68, 69]. Gutzeit[67] has suggested that equations 12.2 and 12.3 indicate a possible route, with the nickel ions being catalytically reduced in equation 12.3 by means of the active atomic hydrogen, which is formed according to equation 12.2 and adsorbed onto the catalytic surface, with simultaneous formation of orthophosphite and hydrogen ions.

$$(H_2PO_2)^- + H_2O \rightarrow H(HPO_3)^- + 2H \text{ (on catalytic surface)} \ldots(12.2)$$

$$Ni^{2+} + 2H \text{ (on catalytic surface)} \rightarrow Ni + 2H^+ \quad \ldots(12.3)$$

Whatever the mechanism, the deposits are not pure nickel but a nickel/phosphorus alloy, whose limits of composition are 3–15% phosphorus and typically 4–10%. Hence, electroless nickel plate normally refers to a nickel/phosphorus alloy which, not unexpectedly, has very different properties from unalloyed nickel.

In a similar manner to the modifications wrought by additives to nickel plating processes, the basic solution containing nickel and hypophosphite ions has been improved by additions of small quantities of metals or organic compounds which result in benefits such as increasing the stability of the bath, the speed of deposition and the brightness of the deposit[70]. Typical compositions of such solutions are given in Table 12.5. These modifications give best results when made to solutions already formulated so as to have optimum properties. Therefore, some consideration of the basic solutions available is necessary, and so it is essential to distinguish between the two major types of electroless nickel baths containing hypophosphite ions as

Table 12.5 COMPOSITION AND OPERATING CONDITIONS OF TYPICAL BATHS FOR DEPOSITION OF ELECTROLESS NICKEL/PHOSPHORUS ALLOYS

Composition (g/l)	Type of bath		
	Acid (hot)	Alkaline (hot)	Alkaline (cool)
Nickel sulphate*	33	–	–
Nickel chloride*	–	30	21
Sodium hypophosphite	20	10	24
Malic acid	18	–	–
Sodium succinate	16	–	–
Lead (as Pb^{2+})	0·003	–	–
Ammonium citrate	–	65	–
Sodium citrate	–	–	45
Ammonium chloride†	–	50	30
pH	5–6	8–10	8–9
Temperature (°C)	85–95	80–90	30–40

* The nickel ion content can be provided either by nickel chloride or nickel sulphate, since their anions have only minor effects on the properties of the baths and the deposits therefrom.
† The above comment applies to the use of ammonium chloride or ammonium sulphate, either being suitable for supplying the content of ammonium ions.

reductant. These are the ammoniacal ones with pH lying between 8 and 10 and the acid type with pH ranging from 4 to 6. Although it was an ammoniacal bath that Brenner and Riddell first described in 1946, this type of solution has achieved very little commercial usage for applying electroless nickel onto metals. As with all electroless nickel baths, the rate of deposition from an ammoniacal bath is very temperature dependent. If a commercially acceptable rate is to be achieved ($\approx$ 15 μm/h), bath temperatures exceeding 80°C are essential, with 90°C and over being preferred. Ammonia is rapidly lost from such hot liquids giving rise to objectionable fumes and causing the bath to become unbalanced. For purposes where only a thin film of electroless nickel is required, cool ammoniacal baths, operated at temperatures less than 40°C, have been found to have certain desirable characteristics, particularly stability, and so are frequently used for applying the initial conductive coating onto prepared plastics prior to their being electroplated. Although relatively few results of investigations into alkaline electroless nickel baths have been published, one report[71] states that the internal stress of coatings from a bath containing pyrophosphate ions is very high in tension (350 N/mm^2), with claims being made that saccharin acts as a stress reliever.

Hot, acid electroless nickel baths are used almost exclusively for the deposition of relatively thick coatings onto metals. The essential components of these baths are nickel sulphate or chloride and sodium hypophosphite, but they invariably also contain a buffer, usually the sodium salt of a carboxylic acid, such as acetate, citrate or glycollate. Otherwise, the pH of the bath drops during use to values at which the rate of deposition is very low.

282   RECENT DEVELOPMENTS

These carboxylic acids also play another important role in the solution. As an electroless nickel bath is worked, orthophosphite is inevitably formed by oxidation of the hypophosphite and accumulates in solution until the solubility product of nickel phosphite is exceeded, and this has a low value at

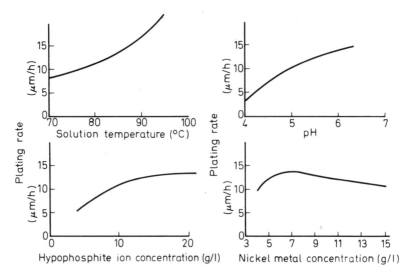

Figure 12.5. *Effect of variables on the rate of deposition from an electroless nickel solution (after Baldwin and Such[78])*

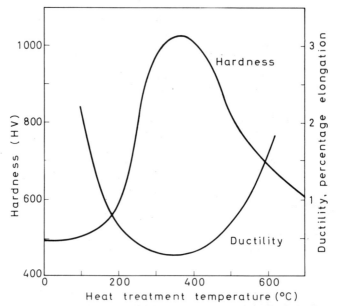

Figure 12.6. *Variation of hardness and ductility of an electroless nickel deposit with heat treatment temperature (after Baldwin and Such[78])*

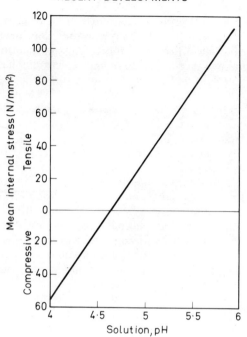

Figure 12.7. Variation of mean internal stress óf an electroless nickel deposit with solution pH (after Baldwin and Such[78])

the conditions of operation. However, many of the buffers used as complexants for nickel ions and other organic components are added specifically for this purpose and these greatly increase the tolerance of the bath to phosphite content and so prolong its life. The effects of carboxylic acids on various properties of the electroless nickel solutions and deposits have been investigated by many workers[69, 72, 73], often with particular reference to their effect on plating rates. In addition to these three main constituents of acid electroless nickel baths, many other compounds have been patented for use as accelerators of deposition, succinates and fluorides being important examples. Other compounds act as stabilisers of these baths. It has been found that minute additions of chemicals containing the thio group (—SH) whether organic or inorganic, heavy metal ions and certain polar surfactants, can be adsorbed onto the surface of tiny nickel particles formed in the bulk of the solution and so prevent them acting as nuclei for the uncontrolled deposition of more nickel which results in spontaneous decomposition of the bath. The concentration of these stabilisers is very critical, for even a slight increase above their stabilisation level slows down the rate of plating and a larger excess inhibits it completely.

The presence of these and other patented chemicals has made it possible to operate modern electroless nickel baths in plant and at conditions not too dissimilar from those used for conventional nickel electrodeposition. When electroless nickel baths were first introduced commercially, the chemical engineering equipment considered necessary for their satisfactory operation was quite extensive and expensive. The development of these more tolerant electroless nickel processes has enabled their operation to be treated as a standard metal finishing technique rather than an esoteric process, although the solutions still require more frequent and stringent chemical analysis and maintenance than do nickel electroplating baths. A good description of the plant and accompanying equipment needed for modern electroless nickel deposition on a commercial scale has been given by Greenwood[74], while the chemical control and maintenance of the plating bath itself has been described by Murski[75], with both aspects being dealt with by Brown and Jarrett[76].

If a reasonably consistent rate of plating is desired (which is usually between 10 and 20 μm/h), it is necessary to control operational variables and solution composition as closely as economically feasible. Fig. 12.5 shows the effect of four variables on the speed of deposition from a typical electroless nickel solution. It can be seen that only the nickel ion content has an insignificant effect on plating rate, with hypophosphite content, acidity and temperature all being important in this respect. The necessity for close control of the operating conditions and for making frequent replenishment additions to the solution is apparent from these relationships. These same variables also affect the phosphorus content of the alloy and so presumably its structure[77]. Surprisingly, they do not affect the hardness of the deposit, its most important property after its uniformity of thickness. The hardness of electroless nickel is 450–500 HV as deposited, but it can be increased to 900–1000 HV by suitable heat treatment. The time and temperature recommended for this treatment are 1 h at 400°C, although somewhat lower temperatures can be used for longer times to produce coatings of slightly lower hardness. The ductility of the as-plated electroless nickel is low—between 1 and 2% elongation[78] but, when heated, has the usual inverse relationship to hardness, as can be seen from Fig. 12.6, which relates hardness and ductility to temperature of heat treatment. Electroless nickel deposits usually have a low tensile stress (30–60 N/mm^2) but pH has an interesting effect on the stresses produced, changing these from low compressive to high tensile values, as shown in Fig. 12.7. This effect is believed to be due to the variation of phosphorus content with pH of solution[79]. Although electroless nickel is employed primarily as a corrosion protective coating on functional articles of complex shape, its high hardness compared with electrodeposited nickel resulting in good wear and abrasion resistance, has led to its being employed for engineering purposes, as discussed in many publications[80,81], including the reviews already referred to. It has been claimed that electroless nickel coatings are more corrosion resistant than are nickel electrodeposits of equal thickness, but recent comparative corrosion tests[82] have cast doubt on this and also indicated that post-deposition heat treatment has a considerable effect on their corrosion performance, as well as their hardness.

The microstructure of the electroless nickel/phosphorus alloy is lamellar as deposited and remains unchanged on heat treatment at temperatures up to 400°C. However, if it is heated at temperatures greatly exceeding 400°C, the laminations disappear and are replaced by a fine-grained structure. It has been postulated that electroless nickel deposits containing less than 7% phosphorus consist of a Ni_3P phase in a nickel matrix, with the Ni_3P acting itself as the matrix when higher phosphorus contents are present. Certainly, various workers[77, 78, 81] have demonstrated independently that there is a dramatic change in strength, hardness and ductility at a phosphorus concentration of 6–7% with alloys of that composition being relatively the most ductile and least hard. However, all electroless deposits are harder than electrodeposited dull nickel and this, together with their ability to coat uniformly even the most complex of components such as those with sharp corners, re-entrant angles and long small-bore holes, ensures that electroless nickel processes will maintain their established position in the nickel plating field, in spite of their cost being higher than for conventional electrodeposition.

BORON ALLOYS

The deposition of electroless nickel using borohydrides or alkyl amine boranes as reducing agents has achieved some commercial usage[83]. The former are mainly employed in hot alkaline solutions for plating onto many alkali-resistant metals[84] and the latter in cooler, weakly acid baths for

Table 12.6 COMPOSITION AND OPERATING CONDITIONS OF TYPICAL BATHS FOR DEPOSITION OF ELECTROLESS NICKEL/BORON ALLOYS

Composition (g/l)	Type of bath		
	Acid (hot)	Alkaline (hot)	Alkaline (cool)
Nickel chloride*	30	30	—
Nickel sulphate*	—	—	20
Ethylenediamine	60	—	—
Sodium borohydride	0·6	—	—
Diethylamine borane	—	3	—
Methanol	—	40	—
Dimethylamine borane	—	—	3
Thallium nitrate	0·07	—	—
Sodium acetate	—	20	—
Sodium succinate	—	20	—
Sodium citrate	—	10	—
Ammonium citrate	—	—	12
Ammonium chloride†	—	—	15
2-mercaptobenzothiazole	—	—	0·0002
Sodium hydroxide	40	—	—
pH	14	5–6	6·5–7·5
Temperature (°C)	90–95	50–60	25–35

* The nickel ion content can be provided either by nickel chloride or nickel sulphate, since their anions have only minor effects on the properties of the baths and deposits therefrom.
† The above comments apply to ammonium chloride and ammonium sulphate, either being suitable for supplying the content of ammonium ions.

plating onto light metals and plastics[85]. A comparison of the two types of plating baths are given in Table 12.6. Both types deposit a nickel/boron alloy, nickel boride (Ni_2B or Ni_3B) being co-deposited with nickel. The reaction mechanisms occurring with borohydrides have been quoted as those in equations 12.4 and 12.5, although these are only schematic representations since the details of the actual reactions are still unknown:

$$NaBH_4 + 4NiCl_2 + 8NaOH \rightarrow 4Ni + NaBO_2 + 8NaCl + 6H_2O \quad \ldots (12.4)$$

$$2NaBH_4 + 4NiCl_2 + 6NaOH \rightarrow 2Ni_2B + 8NaCl + H_2 + 6H_2O \quad \ldots (12.5)$$

These reactions result in the deposition of an alloy containing 5–7% boron at a rate of 10–30 µm/h. This deposit has a hardness of 500–750 HV as-plated, and 1000–1250 HV after heating at 400°C for 1h. While no quantitative results for ductility have as yet been published, it appears from qualitative tests that the as-plated nickel/boron alloy may be more ductile than the nickel/phosphorus electroless nickel, but becomes embrittled during heat treatment. The reduction of nickel ions by borohydrides has the advantage of being catalysed by a greater number of common metals (including copper and its alloys) than when hypophosphite is the reductant. However, a disadvantage of the borohydride process is that it needs the same amount of plant as originally required for electroless nickel baths containing hypophosphite, instead of the simplified equipment now used for the latter.

The alkyl amine boranes (otherwise named *N-alkyl amino borazanes*, $R_2NH.BH_3$), usually dimethyl or diethyl amine borane, are less rapidly hydrolysed in aqueous solutions than the borohydrides and so can be employed in weakly alkaline or acid solutions. They are therefore used for the plating of aluminium and other alkali-sensitive metals, but have achieved their major application for depositing the electroless nickel layer onto plastics which are subsequently to be electroplated. Alloys containing be-

Table 12.7 COMPOSITION AND OPERATING CONDITIONS OF A TYPICAL HYDRAZINE ELECTROLESS NICKEL BATH*

Nickel acetate	60 g/l
Glycollic acid	60 g/l
EDTA (tetrasodium salt)	25 g/l
Hydrazine	100 ml/l
Sodium hydroxide	30 g/l
pH	10·5–11·0
Temperature (°C)	85–90

* Data taken from Dini and Coronado[87].

tween 1 and 5% boron are deposited as a result of the reaction which has been represented as:

$$R_2NH.BH_3 + 3Ni^{2+} + 5OH^- \rightarrow 3Ni + (R_2NH_2)^+ + H_3BO_3 + 2H_2O \quad \ldots (12.6)$$

$$4R_2NH.BH_3 + 6Ni^{2+} + 8OH^- \rightarrow 2Ni_3B + 4(R_2NH_2)^+ + 2H_3BO_3 + 3H_2 + 2H_2O \quad \ldots (12.7)$$

The rate of deposition depends on the temperature and pH of the bath employed, but is usually between 0·1 and 0·15 µm/min, which is lower than in the much hotter borohydride baths, but quite adequate for the thin initial coatings required on plastics. The composition of an electroless nickel bath that can be used for this purpose at temperatures just above ambient is given in Table 12.6.

PURE ELECTROLESS NICKEL

Although the electroless deposits of nickel containing either phosphorus or boron have found many applications, a 'pure' electroless nickel would be useful for certain specialised purposes, such as for semiconductor applications, and it is claimed that this has been achieved by the use of hydrazine as a reactant[86, 87]. By use of baths similar to that given in Table 12.7, deposits

Table 12.8 TYPICAL SOLUTIONS USED PRIOR TO ELECTROLESS DEPOSITION OF COPPER OR NICKEL ON ABS PLASTICS

Etching	
Concentrated sulphuric acid (s.g. 1·84)	50% (v/v)
Water	50% (v/v)
Chromic acid to saturation	15 g/l*
OR:	
Concentrated sulphuric acid (s.g. 1·84)	55% (v/v)
Concentrated phosphoric acid (s.g. 1·75)	20% (v/v)
Water	25% (v/v)
Chromic acid	8 g/l
Sensitising	
Stannous chloride ($SnCl_2 \cdot 2H_2O$)	10 g/l
Concentrated hydrochloric acid (s.g. 1·16)	40 ml/l
Activating	
Palladium chloride ($PdCl_2 \cdot 2H_2O$)	0·5 g/l
Concentrated hydrochloric acid (s.g. 1·16)	10 ml/l
OR:	
Silver nitrate	2 g/l
Ammonium hydroxide	10 ml/l

* Approximate solubility at operating temperature of 60°C.

containing over 99% nickel have been achieved, with the main impurity being nitrogen (0·25–0·35%). The rate of deposition is ≈ 6 μm/h. The deposit as-plated is relatively hard (≈ 400 Knoop) and brittle, but after heating at 450°C for 1h becomes softer (≈ 150 Knoop) and much more ductile. Unfortunately, it is stated that these 'pure' nickel deposits are more porous than the electroless nickel/phosphorus alloy type and, probably due to the expense of hydrazine, no work on overcoming this effect has been published.

PLATING ON PLASTICS

The idea of plating onto plastics may seem rather strange to those who are not familiar with the subject, since in many instances, metals are protected by a coating of some plastics to prevent corrosion. Nevertheless, at the present time, plated plastics are providing competition to plated brass, zinc alloy die-castings and anodised aluminium. Various reasons for this are discussed in two recent comprehensive books on this topic of coating plastics with metals[88, 89]. One advantage of plastics is that mouldings of complex shape can be produced cheaply, accurately and with a high quality surface finish. Plastics have a much lower density than metals and have a good strength to weight ratio. Another important advantage is that the substrate does not corrode. The processing sequence required to prepare plastics for electroplating is more elaborate and more expensive than that needed for the preparation of metal substrates, but the costly polishing operation is unnecessary. Plating of plastics is still a relatively new process, but the weight of plastics being plated is increasing rapidly each year[90]. It is therefore difficult at the present time to make realistic comparisons between the cost of manufacturing an article by plating an ABS moulding or by plating a zinc alloy die-casting, particularly as the polishing costs for the latter depend so much on the complexity of their shapes and the standard of finish required. However, the two figures are probably quite similar and as the consumption of plating grade ABS increases and the moulding and plating processes are improved, the plating of ABS is likely to become more competitive.

The coating of plastics and other non-conducting materials by electroplated metals was an early development in the history of electrodeposition, although initially these non-conductors were used as mandrels for electroforming purposes, rather than as substrates permanently covered with electrodeposited metal. However, electroplated non-conductors were subsequently produced as articles in their own right[91], usually in comparatively small numbers, e.g. the plating of baby shoes for souvenirs, or natural leaves for jewellery. Until recently, the only plastics items plated in comparatively large quantities were buttons. This situation was due to the fact that the increased difficulty and cost of plating of the limited range of plastics then available more than offset the advantages that these materials possessed compared with most metals. It is self evident that any surface that is to be plated successfully in an aqueous solution must be both hydrophilic and have a conducting surface. Until lately, the first property was often obtained by mild mechanical abrasion and sometimes by chemical etching, while the second was conferred by a variety of methods, the most popular being the use of thin films of graphite applied from colloidal suspensions or silver produced by chemical reduction. Electrodeposition of a much thicker coating, almost invariably copper, then took place. While the simple acid copper sulphate bath was used for this purpose, some specialised and careful jigging techniques were required to avoid 'burning-off' the initial thin fragile conductive coating and to obtain fairly uniform deposition. A thick deposit (≈ 150 μm) was required to encapsulate the plastics substrate, if this were of appreciable size, since there was little or no adhesion between that and the copper plate. Hence, it was difficult to reproduce fine details, particularly if the copper plate had to be polished to transform it from its as-plated dull state to a lustrous condition. Because of these technical problems, the process of plating on plastics was generally limited to specialised applications of small total quantity, with the aforementioned plating of buttons, frequently by barrelling, forming a notable exception.

PLATING OF ABS PLASTICS

In the early 1960s it was discovered that a polymer formed by the combination of acrylonitrile, butadiene and styrene, usually referred to as ABS, could be chemically etched so as to enable an electroplated coating to be bonded to it. These chemical etchants consisted of strongly oxidising acid mixtures that attacked the surface of the ABS in a controlled manner, so that it became not only hydrophilic but enabled some adhesion to occur between that surface and an electroplated coating. This ability of ABS plastics to be etched in this specific manner enabled them to be coated with metal of thicknesses no greater than those required for metallic substrates, rather than necessitating the thick coatings required to encapsulate other plastics. The concurrent development of bright acid copper baths, which gave deposits also possessing levelling powers (see Chapter 11), and later the availability of stabilised electroless copper solutions and electroless nickel solutions active at near ambient temperatures, gave extra impetus to the possibilities opened up by this discovery, so that the amount of ABS plated quadrupled between 1967 and 1970. Since 1964, the effect of diverse etchants has been investigated on many other thermoplastic polymers, but only polypropylene and polysulphone have, as yet, been found to be practicable commercial propositions. Of these, the latter is too expensive to have more than limited applications, while the cheapness of polypropylene has so far not outweighed the problems encountered when processing the varying commercial grades of this polymer. Until polypropylene can be plated as consistently satisfactorily as can ABS, it will not make much impact on the growing plated plastics market, whose leader continues to be ABS, the current status of which is discussed in three recent review papers[92-94]. Hence, the following descriptions of procedures used to prepare plastics for plating will be specifically applicable to ABS, although the general principles apply to other plastics, as will be mentioned briefly later in this section (see Table 12.8 on p. 287).

ABS consists of polybutadiene (graft rubber) particles in a styrene-acrylonitrile copolymer matrix. The procedure essential to promote satisfactory adhesion is to etch the surface of the ABS so that the graft rubber particles are oxidised and attacked preferentially. This is achieved by immersion in solutions containing both chromic and sulphuric acids, formulated within limits of composition determined by experiments[95]. The presence of o-phosphoric acid is also beneficial in some cases when a milder etching solution is required for certain grades of ABS[96]. Fluorinated surfactants of the same type that are added to chromium plating baths are also claimed to be beneficial in ABS etching solutions[97]. However, there is no doubt that the most important ingredient is the Cr^{6+} provided by the chromic acid and this must be replenished as it is reduced to the trivalent state. In addition to rendering the ABS hydrophilic, this acid mixture enables the surface to provide adhesion to a metal coating. This adhesion is of a low value (90 g/mm peel strength) when compared with that of a metal-to-metal bonding, but adequate to prevent blisters occurring due to temperature changes in spite of the considerable differences between the coefficients of thermal expansion of plastics and metals. The use of a ductile rather than a brittle electroplate assists in this.

The reason why these particular acid mixtures have this property of producing adherence is still the subject of debate[96, 98–103]. When this effect was first observed, chemical bonding was considered to be mainly responsible, but subsequent investigations have indicated that mechanical 'keying' is probably primarily responsible. However, this physical attachment is possibly assisted by chemical linkages and the papers of protagonists of both theories should be consulted for their detailed arguments.

While, preferably, the surface of ABS should be perfectly clean before immersion in the etchant, this is not essential because these oxidising solutions are capable of coping with a reasonable amount of contamination, which is all that should be present if ABS components have been correctly handled after moulding. If, for some reason, the plastics have become excessively soiled, then, prior to etching, pretreatment in an alkaline cleaning solution or in another similar type of acid etchant may be necessary. A more likely cause of non-uniformity of the ABS surface lies in the method of forming these plastics components, which are, in most instances, produced by injection moulding. The moulding conditions normally satisfactory for ABS parts are not necessarily ideal for mouldings that are to be electroplated. Whether this is due to the formation of localised stresses or the deformation of the polybutadiene particles is not yet certain, but moulding conditions must be controlled closely, if optimum plating results are to be achieved, even in the case of grades of ABS known to give good adherence to electroplated coatings[96, 104–106].

When a suitable surface state of the ABS has been achieved by etching, the next operation is to render it conductive. While the older methods of graphite or silver deposition could be used, they are not readily adaptable to mass production techniques and so an 'electroless' deposit is invariably used for this purpose. This can be obtained either (a) from an electroless nickel solution of the types described earlier in this chapter, but adapted to operate at temperatures below the softening point of the plastics, or (b) from an electroless copper solution. The thin nickel or copper films deposited form an adequate basis for subsequent electrodeposits, although certain precautions such as striking in a copper pyrophosphate or nickel sulphamate bath have been found from commercial experience to be advisable.

As stated in the section on electroless nickel, deposition will occur only on certain catalytic metals and so it cannot proceed on plastics, unless these are 'activated' by adsorption of one of these catalytic metals on their surface. Palladium and silver are the two metals most commonly used with a similar two-stage process sequence being employed to obtain a film of either of these. Immersion in stannous chloride solution leads to an adsorbed film of this salt, which is sufficient to reduce to the metallic state the palladium ions contained in an acid chloride solution or the silver ions contained in an ammoniacal nitrate solution, either solution being suitable for the second stage. These metals are adsorbed sufficiently strongly on the plastics surface to resist removal during normal water rinsing and act as catalytic nuclei for the deposition of electroless nickel or copper. An alternative sequence feasible when using palladium is to immerse the plastics components first into the acidified palladium chloride solution and then into a solution of some chemical reductant, such as an alkylamine borane[85], to produce metallic palladium. A so-called one-stage process involves an immersion in a solution in which the stannous and palladium chlorides are combined. Since their interaction results in the formation of colloidal palladium metal in the presence of excess stannous chloride, this colloidal palladium is adsorbed directly onto the ABS surface rather than being produced *in situ*[107, 108]. A subsequent dip in hydrochloric acid is necessary to remove the palladium metal from the p.v.c. insulating coating of the jig holding the ABS parts, otherwise these also become coated with electrodeposits which accumulate with recycling. This aspect of plating on ABS plastics, the avoidance of plating on the plastics jig insulation, is particularly important where large scale production, especially when automated, is being carried out. Most electroless copper baths are more sensitive than electroless nickel ones and so readily deposit on most jig coatings, necessitating the use of uninsulated stainless steel jigs for the etching and activating stages and subsequent transfer of the ABS to the usual p.v.c. coated brass or copper jigs immediately prior to the electroless plating. This additional handling operation is not encountered with electroless nickel baths, and, as these are often more stable, they are now frequently preferred to electroless copper solutions, although stabilised versions of these are now available. One recent development which is claimed to shorten the pretreatment cycle by combining the etching and one of the activating stages, involves the dissolution of a palladium salt in the etching bath[109–111]. Even if this and other improvements are commercially adopted, it must be clear that intrinsically the preparation of a plastic for electrodeposition will always be more complex than it is for a metal, as is recognised by those experienced in the practical aspect of this field[112–115].

This dictum applies even to the electrodeposition itself, for as mentioned previously, an undercoat of a ductile and levelling electroplate is necessary before application of the usual fully bright finishes. If the plated plastics are to be exposed to considerable temperature variations, this ductile coating must have a certain minimum thickness. It should preferably be of copper since this electroplate is obtainable in a very ductile and low stressed state, even when in a fully bright and levelled condition. Bright copper plating processes based either on the acid sulphate or the ammoniacal pyrophosphate baths are suitable. Semi-bright, levelled nickel plate can be employed but does not confer so much resistance to the effect of fluctuations in temperature as does copper. Subsequent deposits of bright nickel and chromium or brass are a common finish. The bright nickel can be comparatively thin if the plated plastics are to be utilised only in dry atmospheres. However, for moist conditions, particularly in contaminated environments, better corrosion protection must be provided for the underlying copper. A duplex nickel plate, possibly plus a micro-discontinuous chromium topcoat may be necessary, in order to avoid rapid penetration to the initial copper electroplate, although controversy still continues as to the necessity for applying such coatings to plastics substrates, in view of their intrinsic corrosion resistance.

Mechanism of pretreatment processes

The acid etching of the ABS is probably the most crucial stage of the whole sequence. The composition of the solution is very important and this has been related to the adhesion properties conferred and to the surface topography of the etched plastics. From published work, it is possible to formulate solutions having the ability to cope with a range of surface conditions of

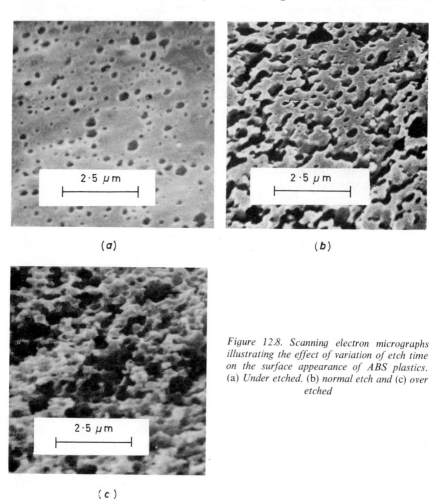

Figure 12.8. Scanning electron micrographs illustrating the effect of variation of etch time on the surface appearance of ABS plastics. (a) Under etched, (b) normal etch and (c) over etched

mouldings made from different proprietary grades of ABS. However, it is also vital to control the temperature of the solution and the time of immersion in order to avoid under or over-etching. Exemplary formulae, operation conditions for acid etchants and subsequent pretreatment solutions are given in Table 12.8 (page 287). However, there are many variants of these basic solutions, most of which are patented; the subject matter of these patents has been collated and critically reviewed by Lowenheim[116].

The electron micrographs shown in Fig. 12.8 illustrate the appearance of a particular grade of ABS after various etch times which represent under, normal and over-etching. The surface is considered to be under-etched when the matrix is unattacked and over-etched when considerable attack of the matrix has taken place resulting in deep penetration into the surface[103]. The ideal situation is that in which the acrylonitrile-butadiene matrix is attacked slowly, while the polybutadiene particles are dissolved quite rapidly. The dissolution of the latter leaves holes into which the metal can be 'keyed'. If a well-bonded coating is peeled off an ABS substrate, failure occurs in the surface layers of the plastic below the electrodeposited coating. Fig. 12.9(a) and (b) show plastics adhering to the underside of an electrodeposited coating after peel testing and Fig. 12.9(c) and (d) show the plastics surface

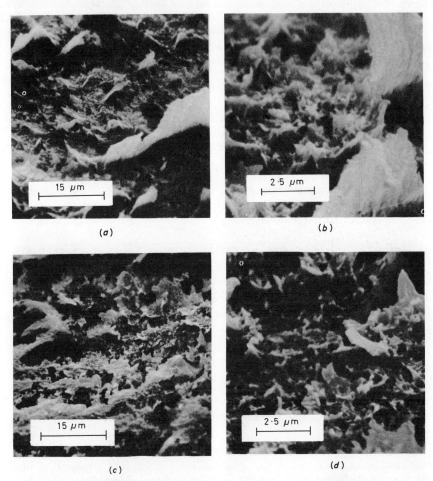

Figure 12.9. Scanning electron micrographs illustrating the appearance of the underside of the metal foil and the surface of the plastics after peeling off the plated coating. (a) and (b) underside of foil; (c) and (d) surface of ABS

from which it was detached. Several explanations have been put forward to account for the occurrence of the weak layer. It may be due to holes produced by etching which are situated too far below the surface to be filled by electroless metal. Alternatively, it has been suggested that the plastics in this region are embrittled in some manner by the etch solution. Adhesion, as determined by the peel test, is usually higher if the sample is stored at ambient temperature for a few days or heated for 1h at temperatures approaching the softening point of the plastics before testing. It has been suggested that this is due to water diffusing away from this critical region. Atkinson et al.[117] have shown that the thickness of the plastics adhering to the stripped metal foil is proportional to the thickness of the electrodeposited coating. Comparative tests should therefore be carried out with a standard coating, usually 50 µm of copper.

After etching, immersion of the well-rinsed plastics in stannous chloride solution at ambient temperature is often the next stage. This solution is a dilute one containing hydrochloric acid, and the adsorbed salts are claimed to consist of particles of 10–30 Å in diameter, which are aggregated in clumps. The nature of the adsorbed deposit is not certain, some form of stannous oxide or hydroxide being postulated[118].

After thorough rinsing, immersion in a dilute, acidified palladium chloride solution is often the next step. Various pH ranges and solution temperatures are successfully used but cannot be chosen at random since they are interdependent. The interaction of the adsorbed stannous compounds and the palladium chloride results in the deposition of a thin film (≈ 20–40 Å) of metallic palladium over the plastics surface, including the interior of the microscopic etched cavities.

These catalytic sites then function during subsequent immersion in electroless nickel or copper solutions to induce deposition of these metals in discrete thin spots (≈ 25–50 Å), which then grow until they coalesce with

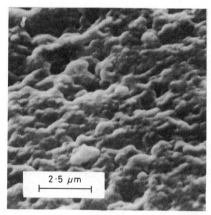 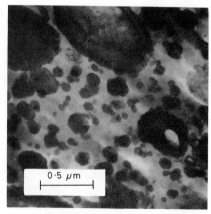

Figure 12.10. Scanning electron micrograph of the surface of etched ABS when coated with a typical thickness of electroless nickel

Figure 12.11. Transmission electron micrograph illustrating the structure of ABS plastics; the round spots represent the polybutadiene (graft rubber) particles

adjacent ones and cover the whole surface, a process that should take 20–60s. Sufficient thickness of film (≈ 1 µm) is produced in 5–10 min plating in electroless nickel and 20–30 min in electroless copper. Fig. 12.10 shows the appearance of an electroless nickel deposit after immersion for 7 min, this being the recommended processing time for the solution concerned. Results of further investigations into the nucleation and growth of electroless metals on plastics, together with a mathematical analysis, have been published by Rantell[119].

Solutions used for electroless nickel deposition have already been described. Electroless copper baths are of the Fehling type. which employ formaldehyde as the reductant and usually contain complexing agents other than Rochelle salts. For example. EDTA acts as an accelerator and 2-mercaptobenzothiazole as a stabiliser. The addition of these or similar compounds has resulted in the development of copper baths which can be operated over much longer periods. sometimes at elevated temperatures. than can the unmodified and unstable Fehling solutions.

Testing of plated plastics

The testing of these metal/polymer combinations present some differences from those encountered with electroplated metals. The corrosion of plated plastics in accelerated tests has been found to have different relationships to service corrosion. particularly out of doors, than those of plated metals. Accordingly. less emphasis is given to accelerated corrosion testing of plated plastics. partly because there is still disagreement as to what criteria to accept for failure and also because of the controversy regarding thickness and types of electrodeposits required. However. much more attention is paid to the testing of the adherence of metal coatings to plastics substrates than to metals on metals, since the former are more likely to fail during service due to poor adhesion than are the latter.

Adhesion

Two types of tests are used to assess the adhesion of plated coatings to plastics substrates; these are peel tests and thermal cycling tests. The essential features of these tests have been outlined in Chapter 8 and the specific details are quoted in various national standards[120-123]. A system which performs well in one test does not necessarily do so in the other and so at the present time there is still a certain amount of conflict as to which test correlates most closely to the behaviour of plated plastics in service. The peel test is incorporated in the A.S.E.P.[120] and A.S.T.M.[121] standards in the U.S.A.. while only thermal cycling tests are included in the British[122] and German[123] standards. Different criteria result in good performance in a particular test; e.g. a high adhesion value is achieved in the peel test when the ABS contains a certain size and distribution of graft rubber particles. while in the thermal cycling test the best results are frequently obtained when the difference in co-efficients of expansion of the metal and plastic is relatively small. Although reference has been made earlier to the normal or optimum etch time. the degree of etch has surprisingly little effect on the adhesion values obtained using the peel test for a particular plated coating on a specific grade of ABS, provided that a certain minimum etch time has been exceeded. this minimum time being quite short. The grade of ABS has a much greater influence on adhesion, so that to achieve the optimum performance from plated ABS it is necessary to evaluate the processing conditions for a particular grade of ABS in a particular commercial system. Atkinson et al.[117] summarised the factors that affect the depth of etch; these include the rubber content, the rubber particle size, the composition of the copolymer, orientation near the surface and the stress level. The shape, size and distribution of rubber particles can be revealed by the transmission electron microscopy technique described by Kato[124]. Fig. 12.11 shows the

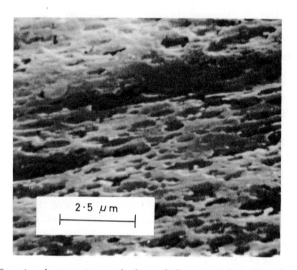

Figure 12.12. Scanning electron micrograph of an etched commercial moulding showing directional orientation

distribution of particles in a typical plating grade of ABS. Fig. 12.12 shows the directionality that is frequently revealed in commercial mouldings on etching. Most of the experimental work reported has been carried out on samples moulded under ideal conditions so as to be free from stress.

The thermal cycling test has the advantage that it can be carried out on shaped components, such as commercial mouldings; special test pieces are not required. Various types of failures develop on thermal cycling; these include surface sinking, blister formation and fine cracking[125, 126].

Corrosion testing

Carter[126] has carried out an extensive investigation into the behaviour of plated plastics on static and mobile outdoor exposure, and when exposed to the CASS and thermal cycling tests. His results showed that systems which behave well in a thermal cycling test are also likely to do so on outdoor exposure. Failure due to lack of adhesion rarely occurs as a result of normal atmospheric temperature variations on samples passing the British Standard thermal cycling test (BS 4601:1970, Appendix F, Test B).

It is clear that, no matter what the substrate, copper plus nickel plus chromium coatings will corrode initially in the same manner and at the same rate, but once any coating has been penetrated to the substrate, the latter's corrodibility and electrical conductivity will affect the degree of subsequent corrosion. The pits in metallic coatings on plastics substrates do not spread laterally as much as in the case of metallic substrates, consequently the pits do not become as wide as on plated metals, and in addition cannot exude basis metal corrosion products. For these reasons, plated plastics exposed out of doors retain an acceptable appearance for longer times than do metal articles plated with coatings of the same type and thickness. However, experimental evidence has been produced[127] that indicates that lateral anodic dissolution of an initial electroless nickel coating proceeds at a much faster rate than does that of an electroless copper layer, no matter what electroplated coatings have been subsequently applied. This lateral corrosion can lead to premature failure of the plated plastics due to loss of adhesion of the metallic coating. The CASS test can be used for assessing the likely corrosion performance of plated plastics when in outdoor service, but with some reservation, since although the total number of pits is often the same in both cases, they are much broader after CASS testing and so appear more unsightly.

Most decorative nickel plus chromium systems provide satisfactory corrosion resistant coatings when applied to ABS plastics. However, crack-free chromium should be avoided since the total stress in this type of chromium is so high that it results in the coating being detached from the ABS substrate during the chromium deposition. Micro-cracked or micro-porous chromium provide improved corrosion behaviour in a similar manner to that given on metal substrates.

Ductility and tensile strength of plated plastics

As would be expected, the tensile strength of any plated plastics is greater than that of the unplated article, being usually increased by a factor of between 10 and 30%, the actual gain in strength depending on the ratios of the combined thickness and tensile strength of the metal coating to those of the plastics substrate[128]. The flexural modulus or stiffness of the plated plastics is much increased, being approximately three times greater in the case of ABS and seven times greater for polypropylene. There is some doubt as to whether Izod impact strengths of plastics are improved or reduced by plating. This may be related to the embrittlement that is reported[129] to occur in some instances when plating onto ABS mouldings. Work has been carried out to discover which stage in the processing sequence is responsible for this phenomenon (particularly the etching, copper plating, copper plus nickel plating and copper plus nickel plus chromium plating operations). Tensile tests and three-point bend tests have been used to determine ductility, although the latter did not provide severe enough deformation to test etched ABS to failure. It was found that the effect of electroless deposition was not very great and etching had little effect on ductility, but after the electrodeposition of a ductile copper coating a serious reduction in ductility had occurred. The decrease in ductility on plating is due mainly to the presence of a relatively brittle metal coating. Since the latter supports a significant proportion of the load, once it fails the total load is applied to the ABS substrate causing it to fail instantaneously.

PLATING PLASTICS OTHER THAN ABS

Many plastics materials other than ABS have been successfully electroplated on a small scale as described at length by Goldie[89] and more briefly by Smith and Lewis[130] and Rantell[103]. However, only the plating of polypropylene and polysulphone has achieved significant commercial usage. The mechanism that enables adherent electrodeposits to be obtained on these and other thermoplastic polymers is still under discussion. with both mechanical[131,132] and chemical[133] bonding being postulated as responsible for the adhesion; their behaviour is therefore analogous to that of ABS.

Plated polypropylene. in particular. has achieved some popularity. especially in the U.S.A.[134,135] and Japan[136]. This is partly due to the lower cost of polypropylene as compared with ABS. but also because of the greater adhesion of the electroplated coating to the former polymer (double or treble that normally obtained on ABS. i.e. 400–700 g/mm compared with 100–300 g/mm peel strength). This latter property of polypropylene. combined with its intrinsically greater resistance to heat distortion than that of ABS. enables plated mouldings of the former polymer to withstand much higher temperatures without failure. 150°C being claimed for plated polypropylene as compared with 100°C for some plated ABS.

It has been reported that two different approaches have been undertaken in order to produce satisfactory electroplated polypropylene mouldings. One technique is based on the formulation of solutions capable of processing most grades of polypropylene moulding resins. while the other involves the use of special resin mixes. often containing fillers such as titanium dioxide. The process solutions needed to deal with a wide range of polypropylene types rely on the same stannous chloride sensitisation. palladium chloride activation and electroless copper or nickel sequence. as first used for ABS. However. the formulation for the etch. although basically a mixture of chromic and sulphuric acids. may vary greatly from that most suitable for ABS[137] and it often appears necessary to precede it by an organic solvent dip. e.g. dioctylphthalate or a turpentine emulsion. although an etch mixture containing certain organic hydrocarbon liquids has also been patented[138]. The solvent dip is said not to be necessary when mouldings are produced from grades of polypropylene especially developed for electroplating. However. the formulation of the acid dip is still critical and a mixture containing chromic. sulphuric and phosphoric acids is often preferred[139].

Comparisons of the resistance of ABS and polypropylene mouldings plated with varying metal combinations and exposed to corrosive conditions and temperature fluctuations. either artificially created or as experienced naturally in outdoor service. have indicated that an undercoat of copper electroplate is not necessary on polypropylene since a duplex nickel coating is adequate to provide resistance to thermal changes[140]. Plated polypropylene has been found to have superior corrosion resistance to ABS. allegedly due to the lower moisture absorption of the former polymer.

Polysulphone is a more expensive resin than either ABS or polypropylene and the processing conditions are critical. It has a lower coefficient of thermal expansion and a higher softening point than either plastics. and consequently it has found some applications where plated plastics are utilised for high temperature applications. Its preparation for electroplating is similar to that for ABS and polypropylene, inasmuch as it is etched in a mixture of chromic and sulphuric acids, sensitised in stannous chloride and activated in palladium chloride before metallising in electroless copper. However, the acid etching is preceded by two additional and vital treatments[141]; the polysulphone mouldings must first be annealed at 165°C and then immersed in an organic solvent, e.g. dimethylformamide.

REFERENCES

1. WESLEY, W. A., SELLARS, W. W. and ROEHL, E. J., *Proc. American Electroplaters' Soc.*, **36**, 79 (1949)
2. HAMMOND, R. A. F., *Metal Finishing Journal*, **16**, 169, 205, 234, 276 (1970)
3. GABE, D. R., *Metal Finishing Journal*, **16**, 340, 370 (1970)
4. WEINER, R., *Metall.*, **15**, 97 (1961)
5. SHENOI, B. A., INDIRA, K. S. and SUBRAMANIAN, R., *Metal Finishing*, **68** No. 7, 40; No. 8, 57; No. 9, 56 (1970)
6. *Product Finishing*, No. 11, 66 (1964)
7. LE NICKEL ASSOCIÉTÉ ANONYME, U.K. Pat. 1 213 267 (25.11.70); *Electroplating and Metal Finishing*, **24** No. 1, 33 (1971)
8. *Accelerating Electrolytic Processes for the Deposition of Metallic Coatings*, R. Zh. Mash. Moscow (1970)
9. *Electroplated Coatings of Nickel and Chromium*, BS 1224:1970
10. BAILEY, D. J., *Electroplating and Metal Finishing*, **23** No. 5, 42 (1970)
11. METZGER, W., et al., *Galvanotechnik*, **61**, 998 (1970)
12. WYSZYNSKI, A., *The Engineer*, **223**, 135 (1967)
13. MCFADDEN, M. F., *Products Finishing*, **34** No. 2, 76 (1969)
14. WERNICK, S. and PINNER, R., *The Surface Treatment and Finishing of Aluminium and Its Alloys*, 3rd edn, Robert Draper Ltd., Teddington (1964)
15. BENGSTON, H., *Trans. Electrochem. Soc.*, **88**, 267 (1945)
16. SPOONER, R. C. and SERAPHIM, D., *Trans. Inst. Metal Finishing*, **31**, 29 (1954)
17. WORK, H. K., *Trans. Electrochem. Soc.*, **60**, 117 (1931)
18. FYFE, R., *Metal Ind.*, **77**, 300 (1950)
19. WALLBANK, A. W., *J. Electrodep. Tech. Soc.*, **28**, 209 (1952)
20. ORE, R., *J. Electrodep. Tech. Soc.*, **29**, 97 (1953)
21. WORK, H. K. and SLUNDER, C. J., *Trans. Electrochem. Soc.*, **59**, 429 (1931)
22. MINISTRY OF NATIONAL DEFENSE, Ottawa, Canadian Pat. 590 840 (19.1.60)
23. VANDEN BERG, R. V., *Trans. Inst. Metal Finishing*, **45**, 161 (1967)
24. *Recommended Practice for Preparation of and Electroplating on Aluminium and its Alloys by the Zincate Process*, A.S.T.M. B253–68
25. EASTMAN KODAK CO., U.S. Pat. 1 627 900 (10.5.27)
26. SCHERING–KAHLBAUN AG., U.S. Pat. 2 142 564 (3.1.69)
27. SAUBESTRE, E. B. and MORICO, J. L., *Plating*, **53**, 899 (1966)
28. WYSZYNSKI, A., *Trans. Inst. Metal Finishing*, **45**, 147 (1967)
29. BAILEY, G. L. J., *J. Electrodep. Tech. Soc.*, **27**, 233 (1951)
30. SUCH, T. E. and WYSZYNSKI, A., *Plating*, **52**, 1027 (1965)
31. ZELLEY, W. G., *J. Electrochem. Soc.*, **99**, 513 (1952)
32. DIVERSEY CORP., U.S. Pat. 3 235 404 (15.2.66)
33. ALUMINIUM CO. OF AMERICA, U.S. Pat. 2 650 886 (1.9.53)
34. ALUMINIUM CO. OF AMERICA, U.S. Pat. 2 676 916 (27.4.54)
35. KRAMER, O., *Metal Industry*, **66**, 121 (1945)
36. HEIMAN, S., *J. Electrochem. Soc.*, **95**, 205 (1949)
37. BRYAN, J. M., *Metal Industry*, **83**, 461, 502 (1953)
38. SEYB, E. J., JONGKIND, J. C. and GOWMAN, L. P., *Proc. Amer. Electroplaters' Soc.*, **51**, 133 (1964)
39. JONGKIND, J. C., *Trans. Inst. Metal Finishing*, **45**, 155 (1967)
40. BEYER, S. J., *Plating*, **56**, 257 (1969)
41. BEACH, J. G. and GURKLIS, J. A., *Procedures for Electroplating Coatings on Refractory Metals*, Defense Metals Research Information Center Memorandum 35 (O.T.S. Publication 161185), Battelle Memorial Institute, Columbus (1969)

42. SAUBESTRE. E. B., *J. Electrochem. Soc.*, **106**, 305 (1959)
43. *Recommended Practice for Preparation of Magnesium and Magnesium Alloys for Electroplating*, A.S.T.M. B480–68
44. DELONG, H. K., *Proc. American Electroplaters'*, *Soc.*, **36**, 217 (1949)
45. DOW CHEMICAL CO. INC., U.S. Pat. 3 152 009 (6.10.64)
46. INNES, W. P., 'Plating on Magnesium' section in *Metal Finishing Guidebook and Directory*, Metals and Plastics Publications Inc., Westwood (1971)
47. GENERAL MOTORS CORP., U.S. Pat. 2 916 401 (8.12.59)
48. SPENCER, L. F., *Metal Finishing*, **68** No. 12, 32 (1970); **69** No. 2, 43 (1971)
49. BACQUIAS, G., *Galvano*, **39**, 621 (1970)
50. *Recommended Practice for Preparation of Titanium and Titanium Alloys for Electroplating*, A.S.T.M. B481–68
51. DOMNIKOV, L., *Metal Finishing*, **60** No. 3, 59 (1962)
52. MARSHALL, W. A., *Trans. Inst. Metal Finishing*, **44**, 111 (1966)
53. FRIEDMAN, I., *Plating*, **54**, 1035 (1967)
54. BROWN, E. E., NASA Technical Brief 67–10532, Marshall Space Flight Center, Huntsville (1967)
55. HANSEL, G., *Galvanotechnik*, **60**, 293 (1970)
56. *Recommended Practice for Preparation of Tungsten and Tungsten Alloys for Electroplating*, A.S.T.M. B482–68
57. SEEGMILLER, R., GORE, J. K. and CALKIN, B., *Proc. Amer. Electroplaters' Soc.*, **49**, 67 (1962)
58. COUCH, D. E., SHAPIRO, H., TAYLOR, J. K. and BRENNER, A., *J. Electrochem. Soc.*, **105**, 450 (1958)
59. YANIV, A. E. and LASZEWSKA, Z., *Trans. Inst. Metal Finishing*, **48**, 5 (1970); **49**, 42 (1971)
60. DINI. J. W. and CORONADO. P. R., *Trans. Inst. Metal Finishing*, **47**, 1 (1969)
61. OWEN. L. W. and ALDERTON, J. R., *British Corr. J.*, **5**, 217 (1970)
62. BRENNER, A. and RIDDELL, G., *Proc. Amer. Electroplaters' Soc.*, **33**, 23 (1946); **34**, 156 (1947)
63. WURZ. A., *Comptes Rendus*, **18**, 702 (1844); **21**, 149 (1845)
64. SOCIÉTÉ ALUMINIUM FRANÇAIS, U.S. Pat. No. 1 207 218 (5.12.16)
65. *Symposium on Electroless Nickel Plating (Catalytic Deposition of Nickel/Phosphorus Alloys by Chemical Reduction in Aqueous Solution)*, A.S.T.M. Special Technical Publication No. 265, Philadelphia (1959)
66. GORBUNOVA, K. M. and NIKIFORROVA, A. A., *Physicochemical Principles of Nickel Plating*, Akademii Nauk SSSR, Moscow, 1960 (Translated by Israel Program for Scientific Translations, Jerusalem, 1963)
67. GUTZEIT, G., *Plating*, **46**, 1158, 1275, 1377 (1959); **47**, 63 (1960)
68. RANDIN, J. P. and HINTERMANN, H. E., *J. Electrochem. Soc.*, **117**, 160 (1970)
69. FELDSTEIN, N. and LANCSEK, T. S., *Trans. Inst. Metal Finishing*, **49**, 156 (1971)
70. JARRETT, G. D. R., *Industrial Finishing*, **18** No. 204, 32 (1966); **18** No. 218, 41 (1966)
71. BARTLETT, B. C., CANN, L. and HAYWARD, J. L., *Plating*, **56**, 168 (1969)
72. KHOPERIYA, T. N., *Zashchita Metallov*, **3**, 328 (1967) (English translation in Protection of Metals, **3**, 272 (1967)
73. HOLBROOK, K. A. and TWIST, P. J., *Plating*, **56**, 523 (1969)
74. GREENWOOD, J. D., Chapter 10 of *Heavy Deposition (Chemical and Electrodeposition of Copper, Nickel and Chromium for Engineering Purposes)*, Robert Draper Ltd., Teddington (1970)
75. MURSKI, K., *Metal Finishing*, **68** No. 12, 36 (1970)
76. BROWN, L. D. and JARRETT, G. D. R., *Trans. Inst. Metal Finishing*, **49**, 1 (1971)
77. GRAHAM, A. H., LINDSAY, R. W. and READ, H. J., *J. Electrochem. Soc.*, **112**, 401 (1965)
78. BALDWIN, C. and SUCH, T. E., *Trans. Inst. Metal Finishing*, **46**, 73 (1968)
79. PARKER. K. and SHAH. H., *J. Electrochem. Soc.*, **117**, 1091 (1970)
80. WIEGARD. H. and HEINKE. G., *Metalloberfläche*, **24**, 163 (1970)
81. RANDIN, J. P. and HINTERMANN. H. E., *Plating*, **54**, 523 (1967)
82. ANDREW, J. F. and HERON. J. T., *Trans. Inst. Metal Finishing*, **49**, 105 (1971)
83. LANG. K., *Galvanotechnik*, **55**, 728 (1964); **56**, 347 (1965); *Electroplating and Metal Finishing*, **19**, 86 (1966)
84. *Nickel-Berichte*, **25**, 75 (1967)
85. NIEDERPRUM, H. and KLEIN, H-G., *Metalloberfläche* **24**, 568 (1970); *Metal Finishing Journal*, **17**, 18 (1971)
86. LEVY, D. J., *Electrochem. Tech.*, **1**, 38 (1963)
87. DINI. J. W. and CORONADO. P. R., *Plating*, **54**, 385 (1967)

88. MULLER, G. and BAUDRAND, D. W., *Plating ABS Plastics*, 2nd edn, Robert Draper Ltd., Teddington (1971)
89. GOLDIE, W., *Metallic Coating of Plastics*, Electrochemical Publications Ltd., Hatch End (1969)
90. WEAVER, E. P., Paper in Proc. Rotofinish Symposium, Nottingham (1969)
91. NARCUS, H., *Metallizing of Plastics*, Reinhold, New York (1960)
92. SAUBESTRE, E. B., *Trans. Inst. Metal Finishing*, **47**, 228 (1969)
93. EBNETH, H., *Metalloberflache*, **24**, 2 (1970)
94. SCHWARZ, G. K., *Metalloberflache*, **24**, 475 (1970)
95. HASKO. F. and FATH, R., *Galvanotechnik*, **59**, 32 (1968)
96. WIEBUSCH, K., Paper in Proc. Rotofinish Symposium, Oxford (1966); Proc. 7th International Metal Finishing Conference. Hanover. 147 (1968)
97. HEPFER, I. V., U.S. Pat. No. 3515649 (2.6.70)
98. HEYMANN. K.. *Galvanotechnik*. **56**. 413 (1965)
99. SAUBESTRE. E. B.. DURNEY. L. J.. HAJDN, J. and BASTENBECK, E., *Plating*, **52**, 982 (1965)
100. LOGIE, L. R. and RANTELL, A., *Trans. Inst. Metal Finishing*, **46**, 91 (1968)
101. RANTELL, A., *Trans. Inst. Metal Finishing*, **47**, 197 (1969)
102. SAUBESTRE. E. B. and KHERA, R. P.. *Plating*. **58**. 464 (1971)
103. RANTELL. A.. *Product Finishing*, **23**, 37 (1970)
104. ZÄHN. E. and WIEBUSCH. K.. *Kunststoffe*, **56**, 773 (1966)
105. SLATER, J. R. and PATON. W.. Chapter 16 in *Design Engineering Product Finishing Handbook*. Ed. Philpott. B. A.. Product Journals Ltd.. West Wickham (1968)
106. REMPEL, D., EBNETH, H., CHRISTOPH, J. and HOYER, W., Paper in Proc. of Symposium on Plating in the Seventies, Cambridge (1970)
107. SIMPKINS, D., *Metal Finishing Journal*, **16**, 184 (1970)
108. SHIPLEY CO. INC., U.S. Pat. 3 011 920 (5.12.61) and U.K. Pat. 929 799 (26.6.63)
109. HEYMANN, K. and WOLDT, G., *Galvanotechnik*, **61**, 221 (1970)
110. WOLDT, G., Paper in Proc. of Symposium on Plating in the Seventies, Cambridge (1970)
111. MINE SAFETY APPLIANCE CO., U.S. Pat. 3 507 681 (21.4.70); SCHERING AG, U.K. Pat. 1 209 179 (21.10.70); KNAPSACK AG, U.K. Pat. 1 214 429 (2.12.70)
112. NARCUS. H.. *Plating*. **55**. 816 (1968)
113. SHADDOCK. A. W.. *Trans. Inst. Metal Finishing*. **47**. 217 (1969)
114. WIEBUSCH. K.. *Galvanotechnik*, **61**, 984 (1970)
115. POLLARD. E.. *Product Finishing*. **23** No. 5. 24 (1970)
116. LOWENHEIM. F. A.. *Metallic Coating of Plastics*, Noyes Data Corpn., Park Ridge (1970)
117. ATKINSON, E. B., BROOKS, P. R., LEWIS, D., SMITH, R. R. and WHITE, K. A., *Trans. Plastics Inst.*, **35**, 549 (1967)
118. SARD, R., *J. Electrochem. Soc.*, **117**, 864 (1970)
119. RANTELL. A.. *Trans. Inst. Metal Finishing*. **48**. 191 (1970)
120. *Standards and Guidelines for Electroplated Plastics*, American Society for Electroplated Plastics (1970)
121. *Recommended Practice for Method of Test for Peel Strength of Metal Plated Plastics*. A.S.T.M. B533–70
122. *Electroplated Coatings of Nickel Plus Chromium on Plastics Materials*. BS 4601:1970
123. *Testing of Electroplated Plastics (Thermal Cycling Test)*, DIN 53 496 (1970)
124. KATO, K., *Polymer Eng. and Sci.*, **7**, 39 (1967)
125. WIEBUSCH, K., *Galvanotechnik*, **61**, 704 (1970)
126. CARTER, V. E., *Trans. Inst. Metal Finishing*, **46**, 49 (1968); **48**, 58 (1970); **49**, 29 (1971)
127. WIGGLE, R., HOSPADARUK, V. and FITCHMUN, D. R., *J. Electrochem. Soc.*, **118**, 158 (1971)
128. ROOBOL, N. R., DELANO, T. and MEYER, B. F., *Plating*, **57**, 1122 (1970)
129. SUMNER, G. P. A., Project Report, Metallurgy Department, University of Aston in Birmingham (1971)
130. SMITH, R. R. and LEWIS, T. D., paper in Proc. Rotofinish Symposium, Oxford (1967)
131. FITCHMUN, D. R., NEWMAN, S. and WIGGLE. R., *J. App. Polymer Sci.*, **14**, 2441, 2457 (1970)
132. ELMORE, G. V. and DAVIS, K. C., *J. Electrochem. Soc.*, **116**, 1455 (1969)
133. SYKES, J. M. and HOAR. T. P., *J. Polymer Sci.*, **7**, 1385 (1969)
134. *Plastics Technology*, **12** No. 12, 7 (1966)
135. MILLER, W. G., *British Plastics*, **42** No. 9, 139 (1969)
136. KONO, M., *Trans. Inst. Metal Finishing*, **49**, 227 (1971)
137. PETROW, C. and BAEWA, W., *Metalloberflache*, **24**, 200 (1970)
138. AVISUN CORP., U.K. Pat. 1 216 748 (23.12.70)

139. MCGREGOR. A. and PERRINS. L. E.. *Plastics and Polymers*. **38**. 192 (1970)
140. INNES. W. P.. *et al.*, *Plating*. **56**, 51 (1969); **57**, 1006 (1970)
141. *Products Finishing*. **32** No. 11. 72 (1968)

BIBLIOGRAPHY

High-speed Nickel Plating

EISNER, S. and WISDOM, N. E., 'Electroplating Accompanied by Controlled Abrasion of the Plate', *Plating*, **58**, 993, 1099, 1183 (1971)
RUBINSTEIN, M., 'Modern High Speed Selective Plating', *Prod. Fin.*, **24** No. 10, 39 (1971)

Plating on Aluminium

INDIRA, K. S., SUBRAMANIAN, R. and SHENOI, B. A., *Met. Fin.*, **69** No. 4, 53 (1971)

Electroless Nickel

KLEIN, H.-G., NIEDERPRUM, H. and HORN, E. M., 'Investigation of Electroless Nickel/Boron Deposits on Metals, *Metalloberflache*, **25**, 305 (1971) and **26**, 7 (1972) (in German)
MALLORY, G. O.. 'The Electroless Nickel/Boron Plating Bath: Effects of Variables on Deposit Properties'. *Plating*, **58**, 319 (1971)
PARKER, K. and SHAH, H., 'Residual Stresses in Electroless Nickel Plating', *Plating*, **58**, 230 (1971)
SCHMECKENBECHER, A. F., 'Autocatalytic (Electroless) Nickel and Nickel–Iron Plating for Electronic Applications', *Plating*, **58**, 905 (1971)

Plating on Plastics

ARROWSMITH, D. J., 'Aspects of Adhesion'. *Prod. Fin.*, **24** No. 1, 40 (1971)
CARTER, V. E., 'Thermal Stability and Corrosion Resistance of Plated Polypropylene and Polysulphone', *Trans. Inst. Met. Fin.*, **50**, 28 (1972)
CHRISTOPH, J., REMPEL, D. and EBNETH, H., *Proc. Plating on Plastics and Metals Symposium*, Tech. Conf. Org., Hemel Hempstead (1971)
HEPFER, I. C., *et al.*. 'Electroplated Plastics: Environmental and Accelerated Corrosion Studies,' *Plating*, **58**, 333 (1971)
HEYMAN, K., RIEDEL, W. and WOLDT, G., 'Electroplating of Plastics—Theory and Practice, *Prod. Fin.*, **24** No. 10, 20 (1971)
MATSUNAGA, M. and HAQIUDA, Y., 'Mechanical Properties of Electroplated ABS Plastics', *Met. Fin.*, **69** No. 4, 36 (1971)
PERRINS, L. E. and PETTET, K., 'Mechanism for the Adhesion of Electroplated Copper to Polypropylene', *Plastics and Polymers*, **39**, 391 (1971)
PETROW, C. and BAEWA, W., 'Chemical Etching of Polypropylene Using a Radioactive Chromium Isotope', *Galvanotechnik*, **63** No. 1, 3 (1972) (in German)
RANTELL, A., 'Recent Developments in Plated Plastics', *Electroplating and Met. Fin.*, **24** No. 10, 5 (1971) and **24** No. 11, 5 (1971)
REINHARD, D. L., 'Performance of a New Plateable Thermoplastic (Polyphenylene Oxide), *Plating*, **58**, 1161 (1971)
SASSE, H. P., 'Microporous Chromium Plating of Plastics', *Oberflache*, **11**, 676 (1971) (in German)

Appendix 1

Combinations of Organic Compounds that Produce Semi-bright or Bright Nickel Plate

The following examples of combinations of organic compounds, which when present together in Watts nickel baths produce semi-bright or bright nickel electrodeposits, are taken from the voluminous patent literature. Some are also claimed to be satisfactory in baths based mainly on the chloride, sulphamate or fluoborate salts of nickel. The fact that particular examples are chosen from the patents quoted, many of which are still in force, does not imply that they are actual formulations that are utilised industrially. They are given to illustrate the types and concentrations of organic compounds which are used in modern commercial nickel plating baths.

SEMI-BRIGHT NICKEL

U.K. Pat. 622761 (1949)
    Coumarin   0·15 g/l (leveller)
    Sodium lauryl sulphate   0·1 g/l (wetting agent)

U.K. Pat. 880056 (1961)
    3-bromo coumarin   0·125 g/l (leveller)
    2-butyne-1,4-diol   0·1 g/l (brightener of the second class)

U.K. Pat. 1093490 (1967)
    Coumarin   0·1 g/l (leveller)
    Formaldehyde   0·1 g/l ⎫
    Diethylene glycol monopropargyl   ⎬ (brighteners of the second class)
       ether   0·02 g/l
    Chloral hydrate   0·05 g/l ⎭

BRIGHT NICKEL

U.K. Pat. 894190 (1962)
    Bi-dibenzene sulphonamide   3 g/l (brightener of the first class)
    β-cyano ethyl thio ether   0·002 g/l (brightener of the second class)

    Sodium tetradecyl sulphate   0·02 g/l (wetting agent)

U.K. Pat. 1023201 (1966)
    o-benzoic sulphonimide   1 g/l (brightener of the first class)
    3-dimethylamino-1-propyne   0·075 g/l (brightener of the second class)

    Dithiodimalic acid   0·01 g/l (depolariser)
    Sodium 2-ethyl hexyl sulphate   0·04 g/l (wetting agent)

U.S. Pat. 2409120 (1946)
    Naphthalene 1.5-disodium sulphonate   2 g/l (brightener of the first class)
    o-sulphobenzaldehyde   3 g/l } (brighteners of the second
    Allyl thiourea   0·08 g/l }   class)

U.S. Pat. 2647866 (1950)
    Benzene sulphonamide   2 g/l }
    o-benzoic sulphonimide   2 g/l } (brighteners of the first class)
    N-allyl quinolinium bromide   0·03 g/l (brightener of the second class)

U.S. Pat. 2712522 (1955)
    Naphthalene 1.3.6-trisodium sulphonate   4 g/l (brightener of the first class)
    2-butyne-1.4-diol   0·2 g/l (brightener of the second class)

U.S. Pat. 2782152 (1957)
    Bi-dibenzene sulphonimide   1 g/l (brightener of the first class)
    Triamino triphenyl methane   0·0005 g/l (brightener of the second class)

    Coumarin   0·1 g/l (brightener of the second class and leveller)

    Sodium lauryl sulphate   0·2 g/l (wetting agent)

U.S. Pat. 2782154 (1957)
    o-benzoic sulphonamide   1 g/l (brightener of the first class)
    Tetra ethylene pentamine   0·02 g/l (brightener of the second class)

    Coumarin   0·1 g/l (brightener of the second class and leveller)

    Sodium lauryl sulphate   0·06 g/l (wetting agent)

U.S. Pat. 2782155 (1957)
    Di-benzene sulphonamide   1 g/l (brightener of the first class)
    Polyethylene glycol (where $n = 20$, see Table 6.2)   0·01 g/l (brightener of the second class)

    Coumarin   0·1 g/l (brightener of the second class and leveller)

    Sodium lauryl sulphate   0·06 g/l (wetting agent)

U.S. Pat. 2800441 (1957)
    o-benzoic sulphonimide   1·5 g/l (brightener of the first class)
    Allyl sulphonic acid   1 g/l (brightener both of the first and second class)
    Propynoxy acetic acid   0·05 g/l (brightener of the second class)
    o-sulpho benzaldehyde   2 g/l (brightener of the second class)

U.S. Pat. 2841602 (1958)
    o-benzoic sulphonimide   1·5 g/l (brightener of the first class)
    Sodium allyl sulphonate   1·5 g/l (brightener both of the first and second class)
    2-butynoxy-1.4-diethane sodium sulphonate   0·1 g/l (brightener of the second class)

U.S. Pat. 3041256 (1962)
    Benzene sulphonamide   1 g/l (brightener of the first class)
    Sodium benzene sulphonate   1 g/l (brightener of the first class)
    Sodium 1.4-dihydroxy-2-butene-2-sulphonate   0·4 g/l (brightener both of the first and second class)
    1.4-di-(β-hydroxyethoxy)-2-butyne   0·14 g/l (brightener of the second class)

It must be noted that all the organic compounds listed in these examples as acting solely as brighteners of the first class are also stress-relievers. It will be seen that brighteners of the second class often function as levellers, and vice-versa.

Appendix 2

Analysis of Deposition Solutions—Selected Methods

NICKEL ELECTROPLATING SOLUTIONS

Estimation of nickel

Reagents required:

0·02M E.D.T.A. standard solution
Ammonium hydroxide (s.g. 0·880)
Murexide indicator (mix 0·2 g of dry indicator powder with 100g of sodium chloride).

Procedure. Dilute 10·0 ml of plating solution to 100 ml in a standard flask. Pipette a 5·0 ml sample of the diluted solution into a 250 ml Erlenmeyer flask, add 50 ml of distilled water and 5 ml of ammonium hydroxide. Add 0·5g of Murexide indicator and titrate with the standard E.D.T.A. solution until the colour changes from greenish-yellow to a bright magenta tint.

Using an 0·5 ml equivalent sample (i.e. as described above): volume (ml) of 0·02M E.D.T.A. solution $\times 2\cdot348$ = g/l nickel, and g/l nickel $\times 4\cdot478$ = g/l nickel sulphate ($NiSO_4 \cdot 6H_2O$).

Estimation of chloride

Reagents required:

Potassium chromate solution (5% w/v)
0·1N silver nitrate standard solution.

Procedure. Dilute 5·0 ml of plating solution to 150 ml in a 400 ml Phillips beaker. Add 3 ml of the solution of potassium chromate and titrate with 0·1N silver nitrate solution until a faint red tinge persists. The end point can be seen more clearly if the titration is carried out against a white background.

Volume (ml) of 0·1N silver nitrate solution $\times 2.38 =$ g/l nickel chloride ($NiCl_2.6H_2O$). Volume (ml) of 0·1N silver nitrate solution $\times 1.17 =$ g/l sodium chloride.

Estimation of boric acid

Reagents required:

Saturated potassium ferrocyanide solution (≈ 320 g/l)
Bromocresol purple indicator solution (0·1% w/v)
Phenolphthalein indicator solution
Mannitol or glycerol
0·1N sodium hydroxide standard solution
Hydrochloric acid (≈ 0.1N).

Procedure. Pipette 2·0 ml of plating solution into a 250 ml Erlenmeyer flask. Add 50 ml water. 10 ml saturated potassium ferrocyanide and 8–10 drops of bromocresol purple indicator solution. If the indicator turns green*, add 0·1N sodium hydroxide solution from a burette until a blue end point is reached. Add 5g of powdered mannitol or 5 ml of glycerol and 10–15 drops of phenolphthalein indicator. Shake thoroughly and titrate the solution with 0·1N sodium hydroxide solution until a violet-purple end point is reached.

Volume (ml) of 0·1N sodium hydroxide solution used in the titration after adding mannitol $\times 3.09 =$ g/l boric acid.

Estimation of ammonium sulphate

Reagents required:

0·1N sulphuric acid or hydrochloric acid standard solution
Sodium hydroxide solution (25% w/v)
Methyl red indicator (0·1% w/v in alcohol)
0·1N sodium hydroxide standard solution.

Procedure. Pipette 10·0 ml of plating solution into a 300–500 ml long-necked flask. Add a few pieces of broken porcelain and dilute to ≈ 150 ml with water. An ammonia distillation unit is required for this estimation and into the receiver is introduced 50 ml of 0·1N sulphuric or hydrochloric acid. Dilute this acid to ≈ 100 ml with distilled water and adjust the condenser

---

* If the bromocresol purple indicator turns blue instead of green at this stage, add dilute hydrochloric acid until the solution has a definite green colouration. Then, add 0·1N sodium hydroxide using a burette, until a blue end point is reached. The mannitol or glycerol can then be added and the titration conducted as described above.

so that the outlet dips below the surface of this acid in the receiver. Add ≈30 ml of 25% w/v sodium hydroxide solution to the flask and close immediately. Distil until the volume of the residue is ≈50 ml. On completion of the distillation, remove the trap and wash out the condenser into the receiver with distilled water. Add 3–4 drops of methyl red indicator to the distillate and titrate with 0·1N sodium hydroxide solution until the indicator changes colour.

Volume (ml) of 0·1N sodium hydroxide (titration reading) ×0·66 = g/l ammonium sulphate.

ELECTROLESS NICKEL PLATING SOLUTIONS

Estimation of nickel

Procedure is exactly the same as that described for electroplating solutions, except that a 5 ml sample of undiluted solution is used.

Volume (ml) of 0·02M E.D.T.A. solution ×0·235 = g/l nickel.

Estimation of hypophosphite

Many methods of analysis for hypophosphite have been recorded in the literature, but some of these give high results in the presence of a large excess of phosphite. An example of these is the rapid but inaccurate colorimetric technique based on the development of the molybdenum blue complex[1]. However, the gasometric method devised by Surash and Lansing[2] is satisfactory whatever the concentration of phosphite, since it is based on the catalytic oxidation of hypophosphite ions by palladium ions. The hypophosphite decomposes into hydrogen gas and phosphite anions, and it is the volume of gas evolved that is measured.

The following iodimetric method has been selected for detailed description, as it is only affected by the presence of phosphite to a very slight extent, and is therefore considered the most convenient to use for routine control.

Reagents required:

   0·1N iodine standard solution
   0·1N sodium thiosulphate standard solution
   Concentrated hydrochloric acid (s.g. 1·18)
   Starch indicator.

Procedure. Pipette a 2·0 ml sample of the electroless nickel plating solution into a 250 ml stoppered flask. Add 5 ml of concentrated hydrochloric acid and mix well (the solution must be acidic before the iodine solution is added). Pipette 25·0 ml of 0·1N iodine solution into the flask, stopper, mix well and allow to stand for 30 min. Titrate the excess iodine with 0·1N sodium thiosulphate solution, adding starch as indicator near the end of the titration.

(25 − titration value) × 2·65 = g/l sodium hypophosphite
(NaH$_2$PO$_2$.H$_2$O)

Estimation of phosphite

Reagents required:

Sodium bicarbonate solution (5% w/v)
0·1N iodine standard solution
0·1N sodium thiosulphate standard solution
Acetic acid solution (10% v/v)
Starch indicator.

Procedure. Pipette a 2·0 ml* sample of the electroless nickel plating solution into a 250 ml stoppered flask. Add 20 ml of distilled water and 20 ml of sodium bicarbonate solution. Pipette 50 ml of 0·1N iodine solution into the flask, replace the stopper and allow to stand for 20 min with occasional shaking. Acidify with 20 ml of 10% acetic acid solution and titrate the excess iodine with 0·1N sodium thiosulphate solution. adding starch solution as indicator near the end of the titration.

50 − titration value (ml) × 5·4 = g/l sodium phosphite
(Na$_2$HPO$_3$.5H$_2$O)

CHROMIUM ELECTROPLATING SOLUTIONS

Estimation of hexavalent chromium, Cr (VI)

Reagents required:

Ferrous ammonium sulphate ($\approx$0·1N)
0·1N potassium permanganate standard solution
Sulphuric acid (20% v/v).

Procedure. Transfer 10·0 ml of the chromium plating solution into a 250 ml graduated flask and make up with distilled water. Pipette 5 ml of the diluted plating solution into a 400 ml Erlenmeyer flask; add $\approx$100 ml distilled water and 20 ml of dilute sulphuric acid. Add a measured excess of ferrous ammonium sulphate solution (x ml). (Excess is present when the colour of the liquid is a bluish-green with no indication of a yellowish tint of unreduced dichromate; 50 ml is usually sufficient). Titrate the excess ferrous ammonium sulphate with 0·1N potassium permanganate to a steel grey end point (y ml).

---

* If the titration value is less than 2 ml, repeat using a 1·0 ml sample of electroless nickel solution but use the appropriate factor (10·8) for the subsequent calculation.

Standardisation of the ferrous ammonium sulphate solution. Pipette 25 ml of the ferrous ammonium sulphate solution into a 400 ml flask. add 100 ml of distilled water and 20 ml of dilute sulphuric acid. Titrate with 0·1N potassium permanganate solution to the first appearance of a pink colour (z ml).

Normality of ferrous ammonium sulphate solution $= z \times 0{\cdot}1\text{N}/25$

Calculation of hexavalent chromium content. 1 ml 0·1N ferrous ammonium sulphate $= 0{\cdot}00333$g chromic acid. Using an 0·2 ml sample:

$$\left(\frac{x \times z}{25} - y\right) \times 16{\cdot}67 = \text{g/l chromic acid. } CrO_3$$

Estimation of sulphate[*]

Reagents required:

Acetic acid solution (50% v/v)
Concentrated hydrochloric acid (s.g. 1·18)
Industrial methylated spirits or isopropyl alcohol
Barium chloride (10% w/v).

Procedure. After filtering an aliquot of chromium plating solution through a sintered glass filter or allowing it to stand so that solids settle out. transfer a 25 ml portion to an 800 ml squat beaker equipped with a boiling stick and a watch-glass cover. Place the beaker in a fume cupboard and add. with stirring. the following quantities of reagents in the order listed:

100 ml acetic acid solution (50% v/v)
50 ml concentrated hydrochloric acid (s.g. 1·18)
20 ml industrial methylated spirits or 30 ml isopropyl alcohol

Boil gently for 15 min. after which time there should be no indication of a brown or yellow tinge in the solution. Dilute to ≈ 500 ml with distilled water. raise to the boil. then add slowly from a burette. 30 ml of 10% w/v barium chloride solution. Simmer gently for 1–2 h and allow to stand. preferably overnight. Then heat to just below the boiling point and filter off the barium sulphate using a Whatman No. 42 paper or equivalent. 'Bobby out' the beaker and wash the precipitate and the beaker thoroughly with boiling distilled water. Transfer the paper and precipitate to a weighed silica crucible and dry carefully; finally burn off in a muffle furnace or over a burner until all traces of carbon are removed. Cool and weigh as barium sulphate. For 25 ml, weight increase (g) $\times 16{\cdot}8 = $ g/l of sulphate as H_2SO_4.

---

[*] A very rapid but not so accurate method is based on the measurement of the volume of precipitate centrifuged out of a sample to which barium chloride has been added without prior reduction of Cr (VI). Previous preparation of a calibration graph is necessary.

Estimation of trivalent chromium, Cr(III) (Titrimetric method)

Reagents required:

Ammonium hydroxide (30% v/v)
Sulphuric acid (20% v/v)
Ferrous ammonium sulphate, ≈ 0.1N
0.1N potassium permanganate standard solution
Silver nitrate, solid
Ammonium persulphate, solid.

Procedure. Transfer 100 ml of the plating solution into a 250 ml squat beaker, add ≈ 100 ml of distilled water and raise to the boil. Add, slowly, dilute ammonium hydroxide until just alkaline and leave in a water bath for about 10 min. Filter through a Whatman No. 541 filter paper, or equivalent, 'bobby out' the beaker and wash thoroughly with hot distilled water until the filtrate is free from chromium.

Place the funnel and precipitate over a 400 ml beaker and pour hot dilute sulphuric acid over the precipitate in order to dissolve it. Wash thoroughly with water and transfer the solution and washings into a 250 ml standard flask. Make up to the mark with distilled water and denote as solution I. Transfer 50 ml of solution to a 400 ml beaker; add ≈ 100 ml distilled water, 20 ml dilute sulphuric acid, about 0.1 g of silver nitrate and 2 g of ammonium persulphate. Heat to boiling point and maintain at that temperature for about 20 min until all the oxygen has been evolved. Cool to room temperature and then add a measured excess of ferrous ammonium sulphate solution (x ml). Titrate excess ferrous ammonium sulphate solution with 0.1N potassium permanganate solution (y ml).

Standardisation of ferrous ammonium sulphate solution and correction for hexavalent chromium retained in precipitate. Transfer 50 ml of solution I to a 400 ml beaker, add ≈ 100 ml of water and 25 ml of dilute sulphuric acid. Add x ml of the ferrous ammonium sulphate solution, then titrate the excess ferrous ammonium sulphate solution with 0.1N potassium permanganate solution to the first appearance of a pinkish tint (z ml).

Calculation of Trivalent Chromium. The difference between the two permanganate titrations (z and y ml) represents the volume of ferrous ammonium sulphate solution (as 0.1N) required to reduce the oxidised trivalent chromium in a 2 ml equivalent sample.

$(z-y)$ ml of 0.1N potassium permanganate $\times$ 0.867 = g/l trivalent chromium as Cr

Estimation of trivalent chromium (visible absorption method)

First it is necessary to plot a calibration curve at a particular wavelength; in this case either the absorption of a diluted solution is measured at 600 nm, or, an orange filter is used (depending on the instrument available) with water in the reference cell. A series of solutions having known concentrations of trivalent chromium can be prepared by adding calculated weights of oxalic acid to chromic acid solutions (see page 380 of *Analysis of Electroplating and Related Solutions* by Langford and Parker[3]).

To estimate the trivalent chromium concentration in a plating solution, dilute a sample by the appropriate factor, measure the absorption either at 600 nm or use an orange filter, and read off the concentration from the calibration curve.

Estimation of silicofluoride (titrimetric method)

Reagents required:

N sodium hydroxide solution
Silver nitrate solution (10% w/v)
Alizarin S indicator in water (0·1% w/v)
0·04N thorium nitrate standard solution
Dilute nitric acid—5 ml concentrated acid diluted to 250 ml
Buffer solution—dissolve 7·56 g of monochloroacetic acid in 100 ml of water and 40 ml of N sodium hydroxide solution. Add 2 ml of N silver nitrate solution, mix thoroughly and dilute to 200 ml.

Procedure. Transfer 5·0 ml of the plating solution to a 250 ml beaker and add N sodium hydroxide until the liquid is just alkaline to litmus. Then add 10% silver nitrate solution slowly, until the yellow chromate colour is no longer visible after allowing the precipitate to settle. Thorough stirring is essential during both additions. Filter through a No. 40 Whatman paper into a titration flask and wash the beaker and precipitate well with distilled water. Retain the filtrate and washings, which should be colourless. Add 10 drops of indicator, when a pink colouration should appear, and then add dilute nitric acid dropwise until the colour of the solution changes to yellow. Add 2·5 ml of the buffer solution and titrate with thorium nitrate solution to the pink end point.

Volume (ml) of thorium nitrate $\times 0.192$ = silicofluoride as g/l hydrofluosilicic acid (H_2SiF_6)

Estimation of silicofluoride (potentiometric method[4-6])

Equipment:

Fluoride selective ion electrode
Reference electrode—this is filled with saturated potassium chloride solution, which is saturated with respect to silver ion by adding silver nitrate solution, a drop at a time, until a milky silver chloride precipitate forms.

A combination fluoride electrode may be used so that a reference electrode is not required

A pH meter with expanded scale or Specific Ion Meter

Distilled water, which is essential, since fluoride is added to many mains water supplies.

Reagents required:

For silicofluoride determination in unknown:
Buffer solution prepared by dissolving 100g of sodium acetate and 5g of sodium citrate in 1 litre of solution

For preparation of calibration curve:
Buffer as above
Standard solutions of sodium silicofluoride
Chromic acid solutions of approximately the same strength as the plating solutions concerned.

Theory

The fluoride ion electrode is used for measurement of silicofluoride concentration in solution in the same way as a glass electrode is used for measurement of the hydrogen ion concentration of a solution, i.e. the measurement of pH. The fluoride electrode and the reference electrode are used in conjunction with an expanded scale pH meter or a specific ion meter and solutions of known silicofluoride concentration are used in the same way that buffered standards are used in conventional pH measurement. When using a pH meter, a graph is plotted of potential against concentration on semi-logarithmic paper, with the potential on the linear axis. The unknown can then be read directly from this curve. (If a specific ion meter is used, calibration must be as recommended in the instruction manual for that particular instrument).

Procedure. Prepare three or four standard solutions of sodium silicofluoride in chromic acid to cover the required concentration range, e.g. 0·5, 1·0, 2·0 and 4·0 g/l. For measurement, pipette 1·0 ml of each of these solutions into a standard 100 ml flask, add 25 ml of buffer solution and dilute to the mark. Measure the potential of each of these solutions and then plot the calibration curve. These standards should be retained and stored in polythene bottles for use as checks before any estimation of unkown concentrations is carried out.

For estimation in the plating solution, to a 1·0 ml sample add 25 ml of buffer and dilute to 100 ml with distilled water. Measure the potential of two standards and then the unknown. Read off the concentration from the calibration curve.

REFERENCES

1. ANTON, A., *Analytical Chemistry,* **37** No. 11, 1422 (1965)
2. SURASH, J. J. and LANSING, R. H., *Plating,* **50**, 221 (1963)

3. LANGFORD, K. E. and PARKER, J. E., *Analysis of Electroplating and Related Solutions*, 4th edn, Robert Draper Ltd., Teddington (1971)
4. FRANT, M. F., *Plating*, **54,** 702 (1967) and **58,** 686 (1971)
5. OEHME, F., ERTL, S. and DOLEZALOVA, L., *Oberflache (Surface)*, **10,** 597 (1969)
6. TAUBINGER, R. P., *Product Finishing*, **24** No. 3, 32 (1971)

Note. In addition to these references the authors have consulted other textbooks which contain analytical methods, in particular *Handbook on Electroplating*, 21st edn, W. Canning & Co. Ltd., Birmingham (1970).

Appendix 3

Properties of Chromium, Nickel and Copper

PROPERTIES OF CHROMIUM, NICKEL AND COPPER

Metal	Chromium*	Nickel	Copper
Symbol	Cr	Ni	Cu
Atomic number	24	28	29
Atomic weight	51·996	58·710	63·546
Valency states	2, 3, 4, 6	2	1, 2
Density, kg/m^3 at 20°C	7·14 × 10^3	8·90 × 10^3	8·96 × 10^3
Melting point, °C	1893	1455	1084
Mean specific heat, J/gK	0·461	0·452	0·386
Thermal conductivity, W/mK	69·08	87·92	393·98
Coefficient of thermal expansion, × 10^{-6}/°C (0–100°C)	6·5	13·3	17·0
Resistivity, μΩ cm at 20°C	12·90	6·84	1·67
Type of lattice	b.c.c.	f.c.c.	f.c.c.
Lattice parameter, Å	2·710	3·517	3·608
Closest interatomic distance, Å	2·498	2·491	2·556
Atomic radius, Å	1·250	1·244	1·276

* Chromium may occur other than in the b.c.c. form, and consequently this will affect some of the properties. Close packed hexagonal and manganese types of structure have been detected.

Note also that electroplated forms of the metals may have properties significantly different from those listed above.

Index

Abrasive particles, codeposition of, 81
ABS plastics *see plating onto ABS (acrylonitrile butadiene styrene)*
Accounting and costing, 66
Acetic acid–salt spray test, 241, 246
Acetylenic compounds, 97
Acid treatment, 102
Activated carbon treatment, 105, *143*, 216
Activation overpotential, 16, *18*, 19
Activation treatments, 48, 73, *101*
  for plating onto plastics, 290
Activity, *11*, 21, 58
Addition agents, 21, 22, 85, *111–112*
  acetylenic compounds, 97
  break-down products, 105
  butyne diol, *95*, 112, 226, 303
  coumarin, 29, 52, 94, 107, *109*, 113–115, 129, 142, 143, 216, 303, 304
  effect on stress, ductility and hardness of electrodeposits, 119
  effect on structure of electrodeposits, 52, 118
  melilotic acid, *110*, 142, 143
  naphthalene sulphonates, 60, 93, 119, 304
  saccharin, 29, 93, 98, *109*, 111–113, 115, 281, 304, 305
  thiourea, 29, 96, *108*, 112, 115, 119
Adhesion, 6, *101*, 155
  causes of poor adhesion, 155
  effect of stress on, 103
  electrodeposited coatings, 156
  qualitative tests for determination of, 157
  quantitative tests for determination of, 157–161
Adsorption, 109, 112
Adsorption–diffusion mechanism, 111
Agitation, 4, 21, 81, 99, 114, 232, *266*
Alkyl amine boranes, 289, 290
All-chloride nickel plating solution, 29, 52, 59, 77
Aluminium, plating onto, 267–277
  as an impurity in nickel plating solution, 127, 129
  nickel plating, 159, *267–276*, 286
Aluminium alloys, plating onto, 272, 275
Ammonium ions, hardening effect of, 138
Ammonium sulphate, 359
  estimation of, 307
Analysis
  of chromium plating solutions, 309–313
  of electroless nickel plating solutions, 308
  of nickel plating solutions, 306–308
Anode
  auxiliary 67

Anode *continued*
  bags, 62, 67, 144
    failure of, 68, 99
  basket design, *65*, 66
  baskets, titanium, 6, *66*
  diaphragms, 201
  films, 17
  inert, 3, 62, *67*
  internal, 67
Anode potential
  in acetic acid–salt spray solution, 214
  measurement of, 215
Anode processes, 61
Anodes for chromium plating
  reactions at, 206
  types of, 205
Anodes for nickel plating
  carbon-containing, 6, *62*
  carbonyl shot, 65
  corrodants, 4, *58*, 60
  depolarised, 6, *62*
  effect of additions on chemical reactivity and type of corrosion, 63
  electrolytic, 3, *62*, 65
  mode of dissolution, 62
  polarisation, 63
  primary, 6, *62*, *65*
  purity, 62
  sulphur containing, 63, 65
  types of nickel, 6
Anodic etching, 6, 76, 145, 278
Anodic oxidation in plating aluminium, 269
Anti-pit agents—*see wetting agents*
A.S.T.M. method of assessing corrosion, *248*, 251, 256
Atomic absorption spectrophotometer, 140
Atomic numbers, 315
Atomic radii, 315
Atomic weights, 315
Auto-catalytic process—*see electroless plating*

Barrel chromium plating, 73
Barrel nickel plating, 69–73
  barrel design, 70
  plating solutions, 72
  problems encountered, 72
  types of load, 70
Bend test, 147, 151
Bent cathode, 25
σ-Benzoic sulphonimide—*see saccharin*
Biaxial stress, 152
Black chrome plating, 199
  effect of sulphate ions on, 200
  solution formulation, 200

318 INDEX

Black nickel plating, 68
  solution formulation, 69
B.N.F.M.R.A. jet test equipment, 237
B.N.F.M.R.A. trivalent chromium plating solution, 201
Boric acid, in nickel plating solutions, 4, *58*, 60
  estimation of, 307
Bornhauser tetrachromate chromium bath, 198
Borohydrides, 286
Brass deposition, 270
  preparation for plating, 103
Brenner and Senderoff spiral contractometer, 163
Brighteners
  'break-down products of organic brighteners, 105, 142. 143, 216
  of the first class, 92, *93*, 219, 220, 226, 304, 305
  of the second class, *94–97*. 226. 303–305
Brightness, 98, 118
Bright nickel electrodeposits
  definition of, 92
  introduction of, 209
  properties of, 98–105
  sulphur content of, 100, 108, 209
Bright nickel plating, 1, 5, *92–121*
  organic additions to, 304–305
  properties of baths, 105–107
  types of base solutions, 107
Brinell hardness, 174
B.S.I. method of assessing corrosion, *250*, 255
Brittleness—*see ductility*
Bromide, 60
Bronze plating, 276
Buffers, 22, 58, 283
Bulge test, 151
Bunching hypothesis, 42
Burnt deposits, *21*, 58
Butyne diol, *95*, 112, 226, 303

Cadmium as a brightener in nickel plating bath, 5, 127
Calcium, in nickel plating baths, 138, 142
Calomel electrode, 13
Carbon content of nickel deposits, 110
Carbon treatment of solution, 105, *143*, 216
CASS test, *242*, 246, 255, 296
Cathode films, 21
  examination of, 17
  in chromium plating, 185, 186
Cathode potential, 29, *113*, 114
Cathodic reduction of organic compounds, 109
Cavitation, 102
Centre line average (C.L.A.), 31
Chemical etching, *269*, 277, 278
Chemically reduced nickel—*see electroless plating*
Chloride in nickel plating solutions, 4, *58*, 60, 129, 303

Chloride in nickel plating solutions *continued*
  estimation of, 306
Chlorine evolution at anodes, 61, 68
Chromability, 25, *101*, 107
Chromate films, 259
Chromatography, thin layer and paper, 143
Chromic acid, 7, *184*, 194
Chromium anhydride—*see chromic acid*, 184
Chromium contamination, 126, 129, *132*
  removal of, 141
Chromium electrodeposits,
  coefficient of friction of, 174
  etching techniques for metallographic purposes, 55
  grain size of, 55
  micro-discontinuities of, 221–231
  structure and surface topography, 55
  use as intermediate layer in multi-layer coating, 217
Chromium hydride, 48, 170
Chromium, micro-cracked—*see micro-cracked chromium*
Chromium, micro-porous—*see micro-porous chromium*
Chromium plating
  anodes employed in, 205
  bright plating range, 188
  covering power, 25
  early baths, 7
  effect of plating variables, 186–187
  effect of substrate, 197, 204, *208*
  efficiency, 8, 89, 186, *188*, 202
  history of, 7
  mechanism of deposition from chromic acid baths, 185
  nucleation, 47
  on aluminium, 277
Chromium plating solutions,
  bath formulation, 192
  black chrome, 200
  conventional decorative, 186
  crack-free, 191, 221
  hard chromium, 189
  hexavalent, 186
    estimation of, 309
  high temperature high ratio (HTHR), 193
  micro-cracked, 193, *222–226*
  micro-porous, 198, *226–231*
  self-regulating high-speed (SRHS), 190
  tetrachromate, 199
  trivalent, 7, *200*
Chromium trioxide—*see chromic acid*
Cobalt, as impurity, 138
Cobalt nickel alloy (by electrodeposition), 5, 138
Co-deposited hydrogen theory, 170
Coefficient of friction, chromium, 174
Coefficient of thermal expansion, 315
Colloidal suspensions, 221
Colorimetric analysis, 139
Columnar structure, 34, 35, 39, *49*

INDEX

Commercial chromium plating, 8
Commercial nickel plating, 5
Composite coatings, 81
  fibre reinforcement, 82
  on aluminium, 268
  production of diamond impregnated tools, 81
  properties, 82
Concentration overpotential, 16, 17, *20*, 30
Contaminants, 124
Contamination
  inorganic, 124–126
    by chromium, 126, 129, *132*
    by copper, *129*, 134, 136, 139
    by iron, 124, 126
    by lead, 129
    by zinc, 128, 129
  organic, 142
Contour high-speed plating machine, 266
Control
  of organic constituents of plating baths, 122
  of inorganic constituents of plating baths, 123
Copper contamination, *129*, 134–136, 139
Copper plating, 49
  cyanide bath, 232
  pyrophosphate bath, *232*, 274, 290, 291
  sulphate bath, 42, 210, *231*, 291
Copper, preparation for plating, 103
Copper undercoats, 231, *257–258*, 274
Corrodkote paste, formulation of, 243
Corrodkote test, 242, 246
Corrosion pits
  examination of as a method for corrosion assessment, 248, 251, 256
  in nickel plus chromium coatings, 209, *211*, 212, 215, 217, 218, 223, 224, 227–229, 253
Corrosion resistance of
  aluminium plated with nickel plus chromium, 276
  deformed nickel plus chromium coatings, 259–264
  multi-layer nickel plus chromium coatings, 254
  nickel plus chromium coatings, 100
  satin nickel plus decorative chromium coatings, 220
Corrosion testing
  accelerated tests, 240
  choice of test, 246
  comparison of test results on various coatings, 247
  degree of acceleration of tests, 246
  methods of evaluating degree of corrosion, 248–253
  outdoor tests, 244
  preparation of samples for testing, 242
  treatment before rating, 251
Coulometric thickness testing, 236

Coumarin, 29, 52, 94, 107, *109*, 113–115, 118, 129, 142, 143, 216, 303, 304
  effect on properties of, 54
Covering power, 25, 107
Crack density, 193
Crack-free chromium, *191*, 222, 260
  engineering applications of, 88
Current density, limiting, 17, *20*, 21, 49, 58, 266
Current distribution, 23
Current efficiency, 28
  of chromium plating baths, 186
Current source—square wave a.c., 59

Decomposition voltage, 22
Decorative chromium (conventional), 186
Deformation of nickel plus chromium coatings, 259
  corrosion resistance, 200
  cracks induced, 262
Deformed nickel structure, 154
Dendritic growth, 129
Densities, 315
Density of plating solutions, measurements of, 122
Depolarisers, 92, *97*, 304
Diamond dust, co-deposition with nickel, 81
Dichromate treatment (cathodic), 258
Diffuse reflection, 178
Diffusion, 20, 30
  coefficient, 21, 58
  control mechanism, 108
  films, 17
Direct plating on aluminium, 269
Discharge potential, 22
Dislocations, 51
Dislocation theory, 171
Double layer nickel, 5, 50, 100, *211*, 214, 216, 244, 254, 298
Double nickel salt plating solutions, 3
Double zincate dip process, 273
'Drag-out', 21, *122*, 126
Droplets, use in production of satin nickel, 221
Dubpernell test, *225*, 230
Ductility, 103, *147*
  effect of test procedure on value obtained, 153
  types of failure, 151, 152
  types of test employed, 148
  comparison of bend and tensile tests for evaluation, 151
Dull nickel—*see Watts nickel*
Du Nuoy ring detachment technique, 181
Duplex nickel—*see double layer nickel*

E.C. test, 243, 246
Eddy current method, 239

Efficiency
  anode, 15, 60, *106*
  cathode, 15, *28*, 58, *106*, 202
Electric double layer, 13
Electrode equilibria, 10
Electrode potentials, 11, 12
Electrodeposition on worn components, 2, 75
Electrodeposits, orientation of, 41, 54, 118–119
Electrodes of the 'second kind', 12
Electroforming, 2, 60, 85–87
Electroforms, 86
Electroless copper, 290, 295
Electroless nickel
  boron alloys, 285–287
    typical bath formulation, 285
  phosphorus alloys, 279–285
    deposit composition, 280
    mechanism, 280
    properties, 282
    typical bath formulations, 281
  pure nickel, 287
Electroless plating, 2, *279–287*, 290, 294
Electrolytic cleaning, 102
Electron diffraction, 39, *52*
Electron microscopy
  application for metal finishing purposes 36, 50
  limitations of, 36
  preparation of thin foils for transmission work, 38
  replica technique, *36*, 47
  scanning electron microscopy, *36*, 254, 272
Electron probe microanalysis, *39–41*, 254, 272
Electroplating cell, 14
Electropolishing, *54*, 145, 180
Emulsification, 102
Engineering applications of
  chromium electrodeposition, 87
  nickel electrodeposition, 1, 75
Epitaxial growth, *49*, 55, 155
Equal strain hypothesis, 150
Equipotential surfaces, 23
Erichsen test, 151.
Etching techniques, 34, 50, *55*

Fatigue strength, 83
  effect of chromium plating, 90
  effect of nickel plating, 83–84
Ferric ion, 132
Fibre axis, 54
'Filling in', 116
Filters and filter media, 145
Filtration of electroplating baths, 99, *144*
  reasons for, 144
'Flexible strip' techniques, 162
  method of automatic control, 168
Fluoaluminate in chromium baths, 188

Fluoborate
  in chromium baths, 188
  nickel bath, 29, *61*, 303
Fluoride, in chromium baths, 185, 186, *188*, 189, 193
  complex ions, 191
Fluorinated organic compounds, 205
Fluosilicate—*see silicofluoride*
Foil, examination of, 38

Gabe and West theory, 172
Gardam grid
  application, 32, *176*
  illustration, 175
Geometric levelling, 30
Glossmeter, 178
Grain sizes of electrodeposits, 48, 118
  illustrations of, 38, 39
Guild reflectometer, 98, *176*
  illustration of, 177
'Hard' chromium
  baths—formulations of, 89, 189
  definition of, 87
  hardness, 88, 190
  operating conditions of baths, 89
  optimum plating conditions, 189
  properties of, 88
  rate of deposition of, 89

'Hard' nickel, 59
Hardness, 104, *173*
  application of hard deposits, 174
  Brinell, 174
  causes of, 173
  Knoop, 173
  methods of measurement, 173
  Vickers, 174
Haring-Blum throwing-power box, 26
'Heavy' nickel plate, 2, 6, 60, *75*
  applications of, 75
  baths employed, 76
  definition of, 75
  preparation prior to deposition, 76
  properties of, 77
Helmholtz double layer, *13*, 18, 19, 77
High-chloride nickel baths, 59
  tolerance to impurity, 129
High-speed nickel plating, 266–267
  baths employed for, 266
History of
  chromium plating, 7
  nickel plating, 2
Hoar and Arrowsmith stress method, 164
  equation for calculation, 165
  illustration of, 165
Hull cell, *23*, 98, 139
  application of, 125
Hydraulic bulge test, 151
Hydrazine, 287
Hydrogen absorption, 104, 202

INDEX

Hydrogen embrittlement, *84*, 102, 105, 188, 263
  alleviation by heat treatment, 89–90
  sources of, 84, 89
Hydrogen ion concentration, 14
Hydrogen peroxide, 141, 143
Hydrogen pitting, 54, *99*, 106, 181
Hydroxyl ion concentration, 15
Hypophosphite, 279–281
  estimation of, 308

Immersion deposits
  on aluminium, 270–277
  on zinc alloy diecastings, 231
Impurities in chromium plating solutions, 196
Impurities in nickel plating solutions, 105 124
  effect on appearance, 125, *126*, 128
  effect on corrosion resistance, 139
  effect on mechanical properties, 129, 138
  effect on structure, 129
  effect on surface topography, 130
  methods of removal, 140, 143
  rate of removal, 135
  types of impurities, 124, 142
Inert anodes, 3, 62, 67
Inert particles
  in satin nickel, 219
  in special nickel—to produce micropores in chromium. 198, 226
Inorganic brighteners
  cadmium, 5, 127
  zinc, 5
Inorganic impurities
  estimation in plating solutions, 139
  types, 124
Interference microscope, 254
Interference patterns, 180–181
Intermediate layers, 215–219
  chromium, 217
  copper, 218
  gold, 218
  high sulphur-containing nickel, 215
  silver, 218
Internal stress, 60, 104, *161*, 189
  explanatory theories, 169
Iron contamination, 141
  effect on properties, *125–126*, *138*
  illustration of effect. 132
  removal, 141
Irreversible electrode reactions, 11

Jacquet peel test, 159
Jet plating, 266
Jet test method, 236
Jigging arrangements. 24
Jigs, as a source of contamination, 126

Kaolin, use in Corrodkote slurry, 243
Kesternich test, 240
Knoop hardness, 173
Kushner stressometer, 166
  equation for calculation of stress, 167
Kushner theory, 172

Lamellar structure, 34, 35, *49*
Lattice
  constants, 41
  misfit resulting in internal stress, 170
  parameters. 315
  types, 315
Lead
  anodes for chromium plating, 205
  anodes for nickel plating, 68, 267
  impurity in nickel plating baths, 129
Levelling, 1, *30*, *112–118*, 179, 303
  addition agents, 97
  addition agents, incorporated in deposit, 115
  calculation of, 31
  effect of plating variables on, 114
  geometrical, 31
  investigation by radiotracer technique, 115
  'perfect', 116
  'true', 30, *116*
  Watts and bright nickel, 117
Light profile microscope, 180
Liscombe process, 141, 144

Macro-crack pattern, *189*, 193, 222
Macro throwing power, *26*, 107. 116
  calculation of, 26
  factors which influence, 27
  of nickel solutions, 29
Magnesium, plating onto, 277
Magnetic measurement of thickness, 239
Magnetic particles, 145
Mandrel bend test, 147
  equation for calculations of ductility, 148
Mandrels, 2, *85*
Manganese, 129
Measurement of potentials, 215
Mechanical etching
  of aluminium, 268, 277
  of titanium, 278
Melilotic acid, *110*, 142, 143
Melting points, 315
Metal cleaning prior to plating, 101–103
Metallic contamination of nickel plating baths, 125–141
  effect on corrosion resistance. 139
  effect on structure and properties of nickel deposits, 129–139
  methods of removal, 127, *140*
  rate of removal from Watts bath, 135
  relationship between the concentration in solution and deposit, 136
Metallurgical microscopes. 33

Micro-cracked chromium, 193–198
  catalysts, 193
  corrosion behaviour of, 223–224
  crack density of, 225
  crack pattern—effect of substrate, 196
  illustration of, 37
  oil retention property of, 88
  single stage, 223
  stress, 203
  thickness, 222–225
  two layer system, 222
  variables affecting micro-cracking, 194
Micro-cracked nickel, 196, 198, *226*
  corrosion behaviour of, 226
  solution, 226
Micro-porous chromium, 198, *226*
  corrosion behaviour of, 227–230
  equipment required and method of deposition of, 230
  examination of, 230
  pore density, 227
  technique, 231
  thickness, 227
Micro-stresses, 169
Micro throwing power, 30, 116
'Missing', 107
Mobile corrosion tests, 244, 246
Molybdenum, plating onto, 278
Morphology, of corrosion pits, 253
Motorcar trim, 255
Moulding of ABS, 290

Naphthalene sulphonates, 60, 93, 119, 304
Nernst equation, *10*, 11, 17
Nickel
  consumption of, 1
  shortage of, 210
Nickel electrodeposits
  internal stress, 97
  mechanical properties of heavy nickel deposits, 80
  open-circuit potential of, 213, 214
  passivity, 25, 47, *101*, 107
  properties of bright nickel, 98–105
  structure and surface topography, 34–39, 49
  sulphur content, 108, 209
Nickel estimation
  in electroless plating solutions, 308
  in plating baths, 306
Nickel plating
  all chloride and high-chloride baths, 29, 52, 59, 79
  early baths, 2
  early patents, 3
  fluoborate bath, 29, *61*, 80
  hard Watts bath, 59
  history of, 2
  industrial practice, 216
  satin, 174, 219–221
  sulphamate baths, 53, 60, 79, 87, 138, 266, 280

Nickel plating *continued*
  sulphamate bath—high density, 79
  sulphate bath,*59, 68
  Wood's bath, 163
Nickel plus chromium coatings, 208–234
  bright nickel plus decorative chromium, 209, 210
  double layer or duplex nickel, 211
  micro-cracked chromium produced by 226
  micro-cracked chromium—single stage system, 223
  micro-cracked chromium—two layer system, 222
  micro-porous chromium, 226
  multilayer coatings of dissimilar metals, 217
  satin nickel, 219
  single layer nickel plus chromium, 219
  three layer nickel systems, 215
  uses of copper undercoats, 231
Nickel sulphate, 57
Non-conductors, plating onto, 288
N.S.U. Wankel rotary engine, 267
Nucleation and growth of electrodeposits, 41–48
  chromium, 47
  copper, 42
  nickel, 44

Ohmic overpotential, 16, *22*
Ollard adhesion test, 158
  modifications, 159
Operating sequences
  cleaning, 102
  plating onto aluminium, 275, 276
  plating onto plastics, 287
Optical microscopy, 33–35
Organic addition agents—*see addition agents*
Organic brighteners—*see brighteners*
Organic contamination
  methods of analysis of, 142
  methods of removal of, 143
  types of, 105, 142
Outdoor corrosion tests, 244
Overpotential, *15*, 16, 18, 22, 28, 48
  activation, 16, *18*, 19
  concentration, 16, 17, *20*, 30
  ohmic, 16, *22*
  resistance, 16, *22*
Overvoltage—*see overpotential*

Palladium chloride, 290, 291
Palladium, colloidal, 291
Passive nickel electrodeposits, 25, 47, *101*, 107
Patents
  bright nickel plating, 304
  semi-bright nickel plating, 303
Peel test, 159, 295
pH control and estimation, 58, 123

INDEX

Phosphite, estimation of, 309
Pitting, 54, 81, 99, 106, 142. 181
   effect of addition agents on, 181
   effect of agitation on, 181
   methods of evaluation, 181–182
   prevention in thick deposits, 81
Plating on ABS, 289–297
   activating, 290, 294
   effect of moulding conditions, 290
   etching process, 289, 291
   mechanism of pretreatment process. 291
   processing sequence. 287
   quantity processed, 288
   sensitising, 290, 294
   structure of etched ABS, 292
   transmission electron micrograph of ABS, 294
Plating on aluminium, 267–277
   applications, 267
   typical process sequences. 270, 275, 276
Plating on plastics, 287–299
   adhesion, 289, 294, 295–296, 298
   corrosion testing. 295
   crack-free chromium, unsuitability of. 222
   ductility of plated plastics, 297
   electroplated coatings applied, 291
   history, 288
   peel testing, 295
   polypropylene. 289, 297, 298
   polysulphone, 289, 298
   reasons for process, 288
   thermal cycling, 295
Plating on plastics other than ABS, 298–299
Plating out, 134, 140
Platinised titanium, 67
   failure of anodes made from, 67
   use in chromium plating baths, 206
Polarisation—*see overpotential*
Polarography, 139
Polishing, 208, 209, 219
Polypropylene, 289, 297, 298
   pretreatment process for plating, 298
Polysulphone. pretreatment process for plating, 299
Porosity, 100
   methods of determination, 100
Potassium, effect in nickel plating solutions, 138
Potassium permanganate, 143
   treatment of nickel baths with, 129
Potassium salts, 191
Potential measurements, for evaluation of double layer nickel systems, 212
Potentials of metals in aqueous solutions, 10
Primary current distribution. 23–27
Properties of bright nickel baths, 105–107
   cathode and anode efficiencies, 106
   operating range, 106
   simplicity of operation, 107
   stability, 105
   throwing power, 107

Properties of electrodeposited bright nickel, 98–105
Purification of plating solutions, 42, 45, 105, *140–145*
   methods, 140–145

Radioactive tracer techniques, 108, 115,
Reference electrodes, 12
Reflectivity, 99, 175, *176*
Replica technique, *36*, 47
Resistance overpotential, 16, *22*
Resistivities, 315
Reversible equilibria, 10, 11
'Robbers'. 24
Roughness, *99*, 145, 179
'Rule of mixtures', 150

Saccharin, 29, 93, 98, *109*, 111–113, 115, 281 304, 305
Salt spray test, neutral, 240
Sandwich coatings. 215–219
Satin nickel, 219–221
Saturated calomel electrode (S.C.E.), 13
Scanning electron microscopy, 36, 254
Secondary current distribution, 24, 27
Selenium, 193
Semi-bright nickel electrodeposits, 5
   appearance and surface topography, 54
   corrosion resistance, 244
   definition, 92
   levelling, 97
   structure, 39
   sulphur content, 211
Semi-bright nickel plating
   industrial practice, 216
   organic additions, 303
   ratio semi-bright to bright nickel, 211
   use in double layer nickel coatings, 211–215
Sensitising of non-conductors, 290
Shear adhesion tests. 157
Shot peening, 84, 90
Sign conventions, 12
Silica
   as impurity, 138
   to produce satin nickel, 219
Silicofluoride, 8, 73, *188*, 193
   estimation of, 312
'Skipping', 107, 142
Sodium, 138
Solution purification, 143
Solvent degreasing, 102
Specific heats, mean, 315
Spectrophotometer,
   use for analysis of plating solutions, 124
   use for estimation of organic impurity in plating baths, 142
Specular reflectivity, 178
Spiral bend test apparatus, 148
Spiral contractometer, 163

Spray control, 205
Stalagmometer, 181
Standard reversible electrode potential, 11.
Stannate process, 275–277
  process sequence, 276
Stannous chloride, 290, 291
Static outdoor corrosion tests, 244, 246
Steel, preparation for plating, 102
Steady state potential, 212
Strain gauge, 168
Stress in chromium deposits, 170, 172, 189, 191, 194, *202–205*
  effect of substrate, 202
  micro-cracked type, 222
Stress
  internal, 60, 97, 104, *161*, 162
  in deposits from nickel sulphamate bath, 61
  methods of determination, 162–169
  sources of, 161–162
  theories proposed to account for stress, 169–173
Stressometer (Kushner), 166–168
Stress relief, 85
Stress relievers, 92, 97, 162
Strike plate, 231
Strontium, 191
Structure and surface topography, 34–49
  methods of examination, 33
Sulphamate nickel bath, effect of current density on hardness, 78
Sulphate in chromium plating baths, 7, 185, *186*, 188, 189, 191, 193, 194
  estimation of, 310
Sulphide treatment, 141
Sulphur dioxide test, 240
  modification, 240
Surface analyser, 179
Surface profile, 179
Surface quality, 179
Surface tension, 181, 205

Tafel equation, *18*, 214
'Talysurf' surface roughness analyser, 179
Tantalum, plating onto, 279
Tapered-pin adhesion test, 158
Tensile adhesion tests, 157
Tensile ductility tests, 149, 151
Tensile internal stress—*see stress, internal*
Tetrachromate plating baths, 199
  bright deposits, 199
Thermal conductivities, 315
Thermal cycling tests, 157, 295
Thermoelectric method of thickness determination, 239
Thickness determination
  destructive methods, 235–238
  non-destructive methods, 235, 238–240
Thickness testing, 235–240
  average thickness methods, 235

Thickness testing *continued*
  back scatter of β particles method, 239
  classification of methods, 235
  coulometric method, 236
  destructive methods, 235–238
  eddy current method, 239
  jet test method, 236
  magnetic methods, 239
  microscope method, 235
  non-destructive methods, 235, 238–240
  thermoelectric method, 239
Thick nickel plate, 2
  applications, 75
Thiourea, 29, 96, *108*, 112, 115, 119
Three layer nickel system, 215
Throwing power
  macro, *26–29*
  calculation of, 26
  factors influencing, 27
  for chromium plating solutions, 186
  for nickel plating solutions, 29
  micro, 30
Time-lapse photography, 34, 45
Titanium
  industrial applications of, 65
  for anode baskets, 6, 65
  limitations of, in nickel plating solutions, 65
  plating onto, 278
  platinised, 67
p-Toluene sulphonamide, 109
Torsion tests, 150
Treeing, 21
Trivalent chromium, estimation of, 311
Trivalent chromium plating baths, 200
  non-aqueous solvents, 200–201
Tungsten, plating onto, 278
Twinning, 51, 154

Ultracentrifuge techniques, 161
Ultrasonic agitation
  of cleaning solutions, 102
  of copper plating baths, 232–233
  of electroplating baths, 266
Ultrasonic techniques, 161
Ultraviolet range spectrophotometer, 124, 142
Uranium, plating onto, 279

Valency states, 315
Vapour phase chromatography, 143
Vickers hardness, 174
Vogt process, 269–270
  typical process sequence, 270

Wankel rotary engine, 267
Watts electrodeposits
  corrosion resistance after deformation of, 15

INDEX

Watts electrodeposits *continued*
    effect of current density on structure of, 38
    effect of solution and operating conditions on properties of, 78
    structure of, 50
Watts nickel plating bath
    control of, 5
    electrochemistry of, 15
    formulation of, 4, *57*
    use for heavy nickel deposition, 76
Watts O.P., 4
Wear resistance of electrodeposits, 173–174
Weighting factors used in proposed A.S.T.M. method, 248
Wet blasting, 277
Wetting agents, 21, 81, 92, *98*, 221, 277, *303, 304*
Wood's nickel solution, 163
Wrought nickel, structure of, 51

X-ray techniques
    applications to metal finishing, 41, 44
    line broadening, 169
    measurement of stress, 169

Zinc alloy die castings
    as a source of contamination, 106, 129
    copper plating of, 210, 232
    corrosion resistance of, 256
    immersion coating, 231
    plating of vs plated ABS, 288
    preparation for plating, 103
Zinc alloy immersion deposits, *271–275*, 277
    appearance of, 273–274
    thickness of film on various alloys, 272
Zinc, as a brightener, 5
Zinc immersion deposits, 271
Zinc impurity. 106
    effect of, 129–131, 139
    removal of, 134